SCIENTIFIC KNOWLEDGE

BOSTON STUDIES IN THE PHILOSOPHY OF SCIENCE

EDITED BY ROBERT S. COHEN AND MARX W. WARTOFSKY

VOLUME 69

JAMES H. FETZER

New College of the University of South Florida

SCIENTIFIC KNOWLEDGE

Causation, Explanation, and Corroboration

D. REIDEL PUBLISHING COMPANY

DORDRECHT : HOLLAND / BOSTON : U.S.A.
LONDON : ENGLAND

Library of Congress Cataloging in Publication Data

Fetzer, James H., 1940—
Scientific knowledge.

(Boston studies in the philosophy of science ; v. 69)
Bibliography: p.
Includes indexes.
1. Science—Philosophy.   I. Title.   II. Series.
Q174.B67 vol. 69 [Q175]   501s   81-12062
ISBN 90-277-1335-9   [501]   AACR2
ISBN 90-277-1336-7 (pbk.)

Q
174
.B67
v.69
1981

Published by D. Reidel Publishing Company,
P.O. Box 17, 3300 AA Dordrecht, Holland.

Sold and distributed in the U.S.A. and Canada
by Kluwer Boston Inc.,
190 Old Derby Street, Hingham, MA 02043, U.S.A.

In all other countries, sold and distributed
by Kluwer Academic Publishers Group,
P.O. Box 322, 3300 AH Dordrecht, Holland.

D. Reidel Publishing Company is a member of the Kluwer Group.

All Rights Reserved
Copyright © 1981 by D. Reidel Publishing Company, Dordrecht, Holland
No part of the material protected by this copyright notice may be reproduced or
utilized in any form or by any means, electronic or mechanical,
including photocopying, recording or by any informational storage and
retrieval system, without written permission from the copyright owner

Printed in The Netherlands

EDITORIAL PREFACE

With this defense of intensional realism as a philosophical foundation for understanding scientific procedures and grounding scientific knowledge, James Fetzer provides a systematic alternative to much of recent work on scientific theory. To Fetzer, the current state of understanding the 'laws' of nature, or the 'law-like' statements of scientific theories, appears to be one of philosophical defeat; and he is determined to overcome that defeat. Based upon his incisive advocacy of the single-case propensity interpretation of probability, Fetzer develops a coherent structure within which the central problems of the philosophy of science find their solutions. Whether the reader accepts the author's contentions may, in the end, depend upon ancient choices in the interpretation of experience and explanation, but there can be little doubt of Fetzer's spirited competence in arguing for setting ontology before epistemology, and within the analysis of language. To us, Fetzer's ambition is appealing, fusing, as he says, the substantive commitment of the Popperian with the conscientious sensitivity of the Hempelian to the technical precision required for justified explication. To Fetzer, science is the objective pursuit of fallible general knowledge. This innocent characterization, which we suppose most scientists would welcome, receives a most careful elaboration in this book; it will demand equally careful critical consideration.

Center for the Philosophy and　　　　　　　　　　ROBERT S. COHEN
　　History of Science,　　　　　　　　　　　　　MARX W. WARTOFSKY
Boston University

October 1981

TABLE OF CONTENTS

EDITORIAL PREFACE — v

FOREWORD — xi

ACKNOWLEDGEMENTS — xv

PART I: CAUSATION

1. *The Knowledge Context* $\mathcal{K}zt$ — 3
   - Personal Possibilities — 4
   - Minimal Rationality — 9
   - Epistemic Possibilities — 16

2. *The Language Framework:* $\mathcal{L}$ *or* $\mathcal{L}^*$? — 23
   - Peirce's Theory of Signs — 24
   - The Theoretical Objective — 28
   - A Dispositional Ontology — 36

3. *Syntax, Semantics, and Ontology* — 46
   - Nomological Conditionals — 49
   - A Probabilistic Causal Calculus — 59
   - Alternative Interpretations — 67

PART II: EXPLANATION

4. *Statistical Explanation and Statistical Relevance* — 77
   - Reichenbach's Reference Classes — 78
   - Salmon's Statistical Relevance — 86
   - Hempel's Maximal Specificity — 94

5. *A Single Case Theory of Causal Explanation* — 104
   - "Long Run" Dispositional Concepts — 105
   - Alternative "Single Case" Concepts — 115
   - A Single-Case Theory of Explanation — 124

TABLE OF CONTENTS

6. *The Dispositional Construction of Theories* — 137
   Causal and Non-Causal Explanations — 139
   Theories and Theoretical Explanations — 147
   "Instrumentalism" and Theoretical Realism — 161

PART III: CORROBORATION

7. *The Justification of Induction* — 175
   The Traditional Problem of Induction — 177
   The "Paradoxes" of Confirmation — 187
   A Critique of Hume's Critique — 197

8. *Confirmation and Corroboration* — 202
   Bayesian Conceptions of Confirmation — 204
   Traditional Principles of Induction — 208
   Popperian Procedures of Corroboration — 222

9. *Acceptance and Rejection Rules* — 231
   "Orthodox" Hypothesis Testing — 234
   An Inductive Acceptance Rule — 244
   In Defense of this Conception — 255

10. *Rationality and Fallibility* — 264
    Scientific Rationality — 265
    Personal Probabilities — 274
    Scientific Fallibility — 287

REFERENCES — 297

INDEX OF NAMES — 310

INDEX OF SUBJECTS — 313

To KARL POPPER

FOREWORD

The aim of the philosophy of science, generally speaking, is to develop a theory about science which is consistent, complete, and illuminating: it should be consistent, since from inconsistent premises every conclusion follows, thereby establishing no distinction between true and false assertions about science; it should be complete, in the sense that what is wanted is a theory that provides answers to a variety of important questions that may be raised concerning scientific inquiry, such as what scientific explanations, scientific predictions, and scientific theories are, or, more broadly, what it is that differentiates a *scientific* inquiry from other modes of human inquiry; and it should be illuminating, since answers to these questions that are either trivial or false do not enhance our understanding of what science is all about. Such investigations are primarily, though not exclusively, semantical in their character; they are also partially, but not entirely, empirical in foundation.

The analysis of causation, explanation, and corroboration that will be offered here has at least three major theoretical objectives, namely:

first, to elaborate a dispositional ontology for the physical world, including an account of the logical structure of lawlike propositions, of subjunctive conditionals, and of causal connections, employing a possible world semantics which is based upon that ontology;

second, to determine the necessary and sufficient conditions for the adequacy of scientific explanations for either (a) the having-a-property-at-a-time by a particular individual thing (in which no change of properties occurs) or (b) the coming-to-have-a-property-at-a-time by a particular individual thing (in which some change of properties occurs); and,

third, to explicate the conception of scientific knowledge which is supported by this analysis, especially with respect to (i) the classical problem of induction advanced by Hume; (ii) the contemporary "paradoxes" of confirmation formulated by Hempel, by Goodman and by Kyburg; and (iii) the appropriate principles that should properly apply within the context of scientific inquiry in general (with special concern for the epistemic role of conjectures and assumptions in the development and evaluation of scientific hypotheses and theories).

These priorities reflect the author's belief that the philosophy of language (including problems of ontology) logically precedes the problems of knowledge, since an empirical investigation of hypotheses and theories presupposes the determination of their truth-conditions. This point, of course, will appear quite obvious to those who acknowledge that evidence acquires its significance for hypotheses as a function of their meaning; indeed, it would hardly merit explicit mention were it not now subject to a direct assault by Isaac Levi [1977], [1979], whose criticism of efforts to specify the meaning of probability hypotheses and statistical laws is as unwarranted as it is vigorous. For, unless Levi possesses "privileged access" to the answers to these questions of meaning and truth, he should pause to consider the prospects for *epistemology without ontology*, rather than assail those who attempt to fulfill the desiderata that arise in the pursuit of more promising and less perplexing conceptions.

Not the least of the benefits this inquiry should provide, moreover, is a clarification of the relations that obtain between the kind of knowledge upon which we depend in our daily lives ("ordinary knowledge" let us say) and the kind of knowledge which the sciences are supposed to furnish (as "scientific knowledge") with respect to their conditions, limitations, and similarities. Although the most important differences distinguishing *ordinary* from *scientific* knowledge concern certain formal conditions, such as logical consistency, deductive closure, and available evidence requirements, which they satisfy to varying degrees, at least some of the themes the author develops should be enumerated here; for example,

(1) that scientific language is best understood as an *extension* of ordinary language;

(2) that this extension generally comes about through the addition of *technical* language to an ordinary language core;

(3) that the difference in question does *not* coincide with the distinction between quantitative and qualitative language;

(4) that the difference in question is certainly *not* coincidental with the observational/theoretical language distinction;

(5) that the difference in question is better understood as one of *degree* (once again), rather than as one of kind;

(6) that this difference in degree largely concerns the extent to which ordinary and scientific discourse fulfill a *special* condition;

(7) that ordinary language exhibits significant differences between users which make the concept of an *idiolect* of central importance; and,

(8) that reliance upon technical language promotes the objective of *successful communication* among the members of the scientific community.

I shall not be surprised if some philosophers find certain aspects of this book very much open to question; in fact, I expect that the most serious reservations will be felt, first, by those who believe, not just that whatever can be said can be said clearly, but that whatever can be said clearly must be expressible, in principle, solely by means of truth-functional language alone; second, by those who believe that theoretical terms do not designate properties of the physical world, that theoretical sentences are neither true nor false, and that the purpose of science is to describe and to predict, but not to explain, the phenomena of our experience. The first class, of course, consists of those who are committed to an extensionalist thesis regarding the language of science, while the second consists of those who are similarly committed to an instrumentalist rather than a realist conception of scientific theories. My only request of those who fall into one or the other of these classes is that they consider my arguments before they reject them.

If this work were to be identified with an historical tradition, it would have to be that of Peirce, Popper, and Hacking, at least to the extent to which they represent the acceptance, in principle, of theoretical realism and epistemic fallibility, while rejecting, in principle, dependence upon the probabilistic framework and the conditionalization theorem of "The Bayesian Way". Others whose work has exerted a strong influence on the development of my views are Reichenbach, Hempel, and Salmon, from whom we all have much to learn, however much we may ultimately disagree. With respect to substantive commitments, this work bears testimony to the Popperian conception of science, while with respect to analytical methodology, it reflects the Hempelian conception of explication. Although my efforts have benefitted from this rich intellectual inheritance, I would hesitate to claim that any of those philosophers would necessarily agree with the positions I have presented here, including the extent to which I have pursued the systematic integration of ontology with epistemology.

Although I believe that the investigation which follows sufficiently differentiates between traditional principles of induction, probabilistic conceptions of confirmation, and Popperian procedures of corroboration, I am less confident that enough has been said to completely delineate those procedures which I recommend from those advocated by the orthodox statistical hypothesis-testing tradition, which is associated with Fisher, Neyman, Pearson, and Wald. For those who would like to consider this issue in somewhat greater detail, therefore, I would suggest at least reviewing several available discussions; in particular, (i) for a positive account of the statistical tradition, see Giere [1975], [1976], and [1977]; and, (ii) for a negative critique (with which I share substantial sympathies), see Gillies [1973], especially Chapter 11. In view of this shortcoming, I take consolation in the thought that, if this analysis is nevertheless a qualified success, it will thereby have reduced the number of rational epistemic alternatives from very many to relatively few, a matter which others must judge for themselves.

Sarasota, Florida J.H.F.

ACKNOWLEDGEMENTS

Parts I and II represent the integration and refinement of positions I have previously advanced, while Part III reflects the extension of my views with respect to the difficult problems of confirmation and of corroboration. The measure to which I am indebted to Carl G. Hempel, Wesley C. Salmon, and Karl R. Popper will be apparent to all who read this work. I am also grateful to Ronald N. Giere, especially for convincing me of the logical distinction between the 'single case' and the 'long run' propensity constructions, after I had convinced him of their ontological difference (as discussed, for example, in Fetzer [1971], Giere [1973a], and Kyburg [1974]). There are no footnotes, since any references important enough to cite were also valuable enough to incorporate directly into the text, including, in particular, several acknowledgements for specific suggestions which others have proposed on my behalf.

Directly or indirectly, the present version of this analysis of the theoretical character of scientific knowledge has been influenced by the comments and criticism of numerous friends and colleagues, among whom I must mention James Cargile, Susan Haack, and Paul Humphreys of the University of Virginia, William "Ted" Morris and Charles E. M. Dunlop of the University of Cincinnati, Ilkka Niiniluoto and Raimo Tuomela of the University of Helsinki, and Teddy Seidenfeld of the University of Pittsburgh. I am also grateful to my friend and collaborator, Donald E. Nute of the University of Georgia, for his valuable criticism and technical advice. Without the financial support of the National Science Foundation, this investigation could not have been undertaken; without the emotional support of my family, it would never have been completed. If this work contributes toward the resolution of the problems with which it is concerned, therefore, they will have made it possible.

For the benefit of intellectual archaeologists, many of the most important ontological and explanatory conceptions presented in this monograph have their roots in several earlier papers, in particular: Fetzer [1971], Fetzer [1974d], Fetzer [1976], Fetzer [1976a], Fetzer [1977], Fetzer [1977a], Fetzer and Nute [1979], and Fetzer and Nute [1980]. I am grateful to the D. Reidel Publishing Company for permission to draw upon this material.

ACKNOWLEDGEMENTS

The bibliography presented with this manuscript cites only works actually referred to in the text, except for other articles which I happen to have written. That this should occur, of course, is not altogether a matter of accident, since it serves to suggest the possibility that one who has labored so long and so hard upon problems such as these *ought* to have something worthwhile to say about them by now. Whether or not that expectation is rationally justifiable, it is one with which I myself wholeheartedly agree: if you do not like what you find here, then you have no one to blame but me.

J. H. F.

PART I

CAUSATION

1. THE KNOWLEDGE CONTEXT $\mathcal{K}zt$

Perhaps the central concept of any theory of knowledge is that of a *knowledge context* $\mathcal{K}zt$, understood as the set of beliefs (or propositions) $\mathcal{B}$ which a particular person z happens to know at a specific time t, where that same person z at different times $t^1, t^2, \ldots$, and different persons $z^1, z^2, \ldots$, at that same time t may perfectly properly "happen to know" different sets of propositions (or beliefs). The most important feature distinguishing one theory of knowledge from another, therefore, concerns the conditions under which beliefs are properly accepted into or rejected from such a knowledge context $\mathcal{K}zt$, where the members of these belief sets are supposed to satisfy certain additional relations among themselves as well. For convenience of reference, we may refer to these as "external" and as "internal" requirements, respectively. The exact significance of the conditions thereby imposed, however, tends to depend upon the general character of the conception under consideration, i.e., whether the notion of knowledge in question should be construed as *descriptive* of actual persons or as *normative* of ideal persons, since they may be intended as specifying requirements that should be fulfilled, rather than as a description of those that actually are satisfied, by ordinary human beings.

Whether descriptive or normative accounts are the object of inquiry, the concept of a knowledge context $\mathcal{K}zt$ presupposes the existence of some corresponding *language framework* $\mathfrak{L}zt$ for the individual z at the time t. Without denying the theoretical importance of Ryle's distinction between "knowing how" and "knowing that", which we shall subsequently reconsider, there are at least two fundamental reasons for assuming such a relation:

first, the set of beliefs $b_1, b_2, \ldots$, i.e., $\mathcal{B}zt$, that individual z accepts at time t may be defined, relative to the language framework $\mathfrak{L}zt$, by adopting the principle that z accepts b_i at t if and only if z assumes that some sentence S_i in the language $\mathfrak{L}zt$ is true, where S_i is an eternal sentence asserting that b_i is the case, i.e., by a set of sentences, Szt;

second, the relations of entailment (or of implication) that obtain between sentences S_i and S_j within $\mathfrak{L}zt$ then may likewise be unequivocally specified, where "propositions" are themselves interpreted as equivalence classes of sentences relative to different language frameworks, since S_i may entail S_k within framework $\mathcal{L}^1$, not-S_k within framework $\mathcal{L}^2$, and so on. Moreover, insofar as these language frameworks $\mathcal{L}^n$ (distinguished by reliance upon specific grammatical principles and vocabulary postulates) vary from individual to individual and from time to time, the concept in question is that of the language framework for a particular person z at a specific time t, i.e., z's *idiolect* at t, rather than that of a community of language users, all of whom rely upon the very same language framework. The existence of relations of entailment (or of implication) within such a knowledge context, therefore, presupposes the existence of a language framework of this kind: if there were no language framework $\mathcal{L}zt$, there could be no knowledge context $\mathcal{K}zt$. The significance of this conception, however, still varies with the particular interpretation of the requirements thereby imposed, specifically, upon whether a descriptive analysis or a normative analysis is intended, as the following considerations are supposed to explain.

PERSONAL POSSIBILITIES

The classical conception of possibility, of course, interprets that notion as a relation between a language framework $\mathcal{L}$ and a proposition 'p' where 'p' is possible, relative to $\mathcal{L}$, if and only if 'p' is not logically inconsistent (syntactically or semantically), relative to $\mathcal{L}$. The concept thus defined, of course, corresponds to the modal category of *logical possibility* within those schemes of classification in which modal categories for *physical possibility* (or for *historical possibility*) might also occur; thus, for example, assume that N is a set of lawlike sentences in $\mathcal{L}$ which includes every such proposition that is true, then,

(i) for any sentence S in $\mathcal{L}$, S describes a *logically possible* state-of-affairs (or "world") if and only if it is not the case that $\mathcal{L}$ entails (or implies) not-S; and,

(ii) for any sentence S in $\mathcal{L}$ and any such set N, S describes a *physically possible* state-of-affairs (or "world") if and only if it is not the case that $\mathcal{L}$-and-N entails (or implies) not-S;

where these conceptions are completely objective, in the sense that they univocally determine logical and physical possibilities, respectively, if the language framework $\mathcal{L}$ and the set of true lawlike sentences N are thus specified.

It is intriguing to consider the possibilities that might arise from the adoption of (what Hacking [1967] refers to as) a "more realistic" conception, which takes into account the varied and limited mental abilities of ordinary human beings; in particular, Hacking suggests that an element e of the *personal language* $\mathfrak{L}$ of the individual z at t *personally implies* an element h, if $e \& -h$ or $-h \& e$ or e or $-h$ is not personally possible for z at t. Precisely what elements of the personal language $\mathfrak{L}$ may or may not be entailed (or implied) for z at t, therefore, not only depends upon the specific elements e and h in question but also varies with z's deductive powers; thus, Hacking adopts "the very harsh view",

that a man can know how to use *modus ponens*, can know that the rule is valid, can know p, and can know $p \supset q$, and yet not know q, simply because he has not thought of putting them together. We should call this an examiner's view of knowledge. (Hacking [1967], p. 319)

A conception of this kind, of course, is not only plausible but even indispensable for a descriptive conception of knowledge, even though there can be no univocal determination of logical (or of physical or whatever) possibilities for the individual z at t without an inventory of the specific contents of z's "state of mind" at t, a necessity which never arises within the context of a normative conception.

On the basis of this foundation, therefore, it is relatively simple to explicate the concepts of *personal possibility, personal probability*, and *personal necessity* for an individual z at time t, in relation to z's personal language $\mathfrak{L}zt$ and some set of beliefs accepted by z at t, which we may refer to as "facts" f (following Hacking), provided that we bear in mind that these accepted sentences, as sincere beliefs, are not invariably true; thus, in particular,

(a) for any sentence S in $\mathfrak{L}zt$, S describes a *personal possibility* for z at t, given facts f, if and only if z accepts f at t and it is not the case that z at t believes that not-S is true;

(b) for any sentence S in $\mathfrak{L}zt$, S describes a *personal probability* for z at t, given facts f, if and only if z accepts f at t and it is the case that z at t believes that S is probable; and,

(c) for any sentence S in $\mathfrak{L}zt$, S describes a *personal necessity* for z at t, given facts f, if and only if z accepts f at t and it is the case that z at t believes that S is true.

Since the logical relations that obtain between the members of the set of beliefs which z accepts at t are determined by z's intellectual abilities at t, it should come as no surprise that, even when z accepts the fact f as true at t, it might nevertheless happen that not-f, for example, is a *personal necessity* for z at t, namely: when z believes that f is true at t and also believes that not-f is true at t, i.e., when it happens that z holds logically inconsistent beliefs.

Although the defense of this conception might appear to present some uninviting difficulties, there are those, such as Wolgast [1977], who are spirited in their endorsement of its implications. Thus, while conceding that it is "a paradox of knowledge" that, "we claim to know things whose consequences we deny knowing; yet this denial indicates our uncertainty, and from this it follows that we do not know what we claim" (Wolgast [1977], p. 23), she maintains that, under various conditions, a question may mean different things; in particular, the question, "But do you know that q?", raised when someone claims to know that p (and 'p' logically entails 'q'), may merely mean, "Do you have independent evidence for 'q', i.e., which differs from that you have for 'p'?" For, in these circumstances, Wolgast contends, no inconsistency arises and this "paradox" simply disappears. Wolgast's analysis, however, should demarcate more clearly between "knowing that p" and "claiming to know that p", "believing that p" and "claiming to believe that p", and so on, since an individual z might *claim to know* that p for any of an enormous variety of reasons (such as to deceive, to mislead, to entertain, to amuse, to reassure, to soothe, and so forth) *without* knowing that p (and conversely). The issue, after all, concerns knowledge and what *it* entails, rather than knowledge claims and what *they* may entail.

The underlying phenomenon, of course, has not passed by previously undetected; for sentences of the form, 'z knows that p', in general, do not entail or imply others of the form, 'z knows that q', even when 'p' itself does entail or imply 'q', on a realistic conception (since that may be something which z either does not realize or else fails to infer). Such contexts are referred to as *intensional*, since the truth or falsity of sentences constructed by the application of the phrase or expression, 'z knows that', to another sentence, 'S', is not completely determined by the truth value of 'S' alone. [The converse, however, is ordinarily supposed to be the case and, indeed, Wolgast herself endorses the principle that, if 'z knows that p', then 'p', i.e., the view that *truth* is a necessary condition for having knowledge (of the ordinary variety, at least).] Analogous considerations attend the deductive relations which may obtain between 'p' and other propositions 'p' itself may entail or imply; for even if 'z knows that S', when 'S' and 'S^*' happen to be logically equivalent within the language framework $\mathfrak{L}zt$, it does not follow that, 'z knows that S^*', for reasons similar to those already described. Notice, in particular, that if 'S' and '$S*$' *are* logically equivalent, it may happen that 'z knows that S' is true, while 'z knows that $S*$' is not, even though z could not have *evidence* for '$S*$' which differs from that z has for 'S'. [Compare the discussion of Hempel's paradox in Chapter 7.]

An avenue of escape from implications such as these, of course, may be thought to lie in the direction of generalization from an *individual z* to a *community of individuals C*, in the optimistic expectation that numbers might make a difference. A suggestion of this kind, for example, is advanced by Teller [1972], who recommends the following basic conception:

It is possible that p if and only if:

(i) it is not the case that p is known to be false by any member of community C;
(ii) nor is there a member $z*$ of community C such that, if $z*$ were to know all of the propositions known to community C, then he could, on the strength of his knowledge of these propositions as basis, data, or evidence, come to know that p is false.

In order to refine this definition, Teller introduces the notion of the set of propositions known to community C, which we may represent as the conjunction of all of the "facts" f known to the members z of community C, such that the (maximal) class of consequences which could be derived by some member $z*$ of community C, were $z*$ to be cognizant of the set of propositions in question, $\{F\}$, is, let us say, the *F-consequence class for community C*. Substituting the sentence letter "S" for the propositional variable "p", therefore, we may propose a set of definitions for the community C which parallels those previously proposed for an individual z along these lines, namely:

(a') for any sentence S entertained by any member z of community C, S describes a *community possibility* for C at t, given facts $\{F\}$, if and only if it is not the case that not-S is a member of the F-consequence class for community C at t;

(b') for any sentence S entertained by any member z of community C, S desribes a *community probability* for C at t, given facts $\{F\}$, if and only if it is the case that S is probable is a member of the F-consequence class for community C at t; and,

(c') for any sentence S entertained by any member z of community C, S describes a *community necessity* for C at t, given facts $\{F\}$, if and only if it is the case that S is a member of the F-consequence class for community C at t.

The issue thus arises of ascertaining whether such a proposal brings about some improvement in the realistic conception of knowledge by introducing a community of individuals C in lieu of an individual z, where at least three aspects of this exchange warrant additional discussion, specifically:

(1) Reliance upon the role fulfilled by the individual member z^* of C in ascertaining the F-consequences class for the community C, of course, ultimately dictates that all community possibilities, probabilities, and necessities, are hypothetically relativized to some particular member of C, namely: the "smartest" member of C, who will be z^* (or the class of "smartest" members of C, if there happens to be more than one). Notice that the "smartest" members of C are not necessarily those who happen to possess the most theoretical knowledge as, say, the Aristotles, Newtons, or Einsteins of their time, but rather those who happen to be possessed of the greatest deductive powers as, say, the Euclids, the Newtons, and the Churchs of *their* times. This condition is necessary to ensure that the F-consequence class includes all the knowledge available to C at t.

(2) Representing the set of propositions known to the members of C by a conjunction of all of the "facts" known to its individual members, moreover, presumes that the set of propositions in question, $\{F\}$, will always turn out to be a logically consistent set, since the F-consequence class for the community C is otherwise either inconsistent or ill-defined. While it might appear temping to resolve this predicament by means of an appeal to *truth* as a condition necessary for knowledge, such a defense is at least hazardous and perhaps self-defeating for a descriptive analysis: since the time of Copernicus, Kepler, and especially Galileo, it has been conspicuous that the members of ordinary, academic, or scientific communities do not invariably agree upon what does and what does not qualify as knowledge in their time, which tends to undermine a realistic conception.

(3) Formalizing the requirements for community possibilities, probabilities, and necessities by means of the F-consequence class for community C, furthermore, implicitly presupposes that the members of community C rely upon essentially the same language framework, namely: that there exists a core of language $\mathfrak{L}$ such that, for every member z of community C at t, the language framework $\mathfrak{L}zt$ for z at t includes $\mathfrak{L}$. As a depiction of the sentences employed by each such member, however, such a condition may or may not be non-trivially satisfied; indeed, it would be untrue if there were families of overlapping idiolects *without* a significant common core. Moreover, although it might be enticing to insist upon the indispensability of propositions in the role assigned to sentences in the formulation of this requirement, that would be descriptively unjustifiable.

It may be worth remarking that the motivation behind Teller's analysis originated with difficulties encountered in contending with specific examples concerning the location of a sunken shipwreck and anticipating the outcome of a lottery. The first of these problems (which Hacking [1967a] initially proposed) involves a mate and salvage crew who sought for the shipwreck at a mistaken location through an error in calculation. Hacking and Teller judged that what *appeared* to be possible (on the basis of a mistaken calculation) later turned out *not* to have been possible (by discovering and correcting this error). Their response to this problem, therefore, was to relativize the concept of possibility to the community at large, supplemented by (some version of) a practicable investigations requirement, in order to obtain a univocal solution. Since both assumed that the basis, data, or evidence remains the same throughout the period in question, there would appear to be little recourse other than to make the move to some individual z^*, whose intellectual capabilities are able to detect and correct this error.

Although this maneuver may seem plausible, it would appear to be at least equally reasonable to conclude instead that two *different* knowledge contexts were involved, insofar as the first set of beliefs included the presumption that the (first) calculations were correct, while the second set of beliefs included the presumption that those original calculations were *not* correct. Consequently, at least this one belief differed from case to case; and, presuming that these calculations themselves (or such beliefs about their accuracy) were *included* among the evidence (just as the premises and derived lines of unsound deductions serve as "evidence" for their conclusions), what was initially supposed to be possible was based upon one body of evidence e, whereas what was later determined to be possible was based upon another body of evidence e^*. Surely that is exactly what we should expect; for, there is no good reason to suppose that what is possible relative to one knowledge context $\mathcal{K}zt$ must also be possible relative to another knowledge context $\mathcal{K}^*zt$. Their apparent "problem", therefore, is no problem at all and thus requires no "solution".

MINIMAL RATIONALITY

The fundamental difficulty, of course, appears to be that Hacking, Teller, and Wolgast are attempting to provide a descriptive and psychological conception of the character of knowledge where only a normative and analytic conception will do. From this point of view, the accounts they recommend strike one as essentially misconceived. It is therefore intriguing to discover that Hacking ([1967], p. 324) has raised doubts concerning the very distinction at issue:

> Let me, ..., question whether there are *any* normative theories. I think there are only descriptive models of reasonable behavior. If any were normative, I do not see how they could be approximately valid. But I believe there are not any. There are models of reasonable behavior, and all models only approximate the truth.

It should be borne in mind, of course, that, in Hacking's language, all *models* involve both *idealization* and *approximation*. When this is taken into account, however, together with the evaluative component necessary for the specification and identification of instances of (what qualifies as) "reasonable behavior" (or as "rational belief"), it becomes a matter of debate whether Hacking's denial should be interpreted as significant: if the actions and beliefs of a special class of agents and thinkers is adopted as the paradigm for a *descriptive* inquiry where those actions or beliefs are subjected to an idealized and approximate description, there would appear to be little point to denying its *normative* dimensions — especially when those individuals are selected on the basis of the extent to which they tend to display the desirable properties in question (for example, the rare geniuses, but not the common half-wits, of their time).

What does not appear debatable, by contrast, are the *desiderata* that underlie these "descriptive" investigations, insofar as the function performed by the "smartest" member z^* of community C is to implement (or to enforce) the highest standards of deductive reasoning actually available at that time, while the objectives secured through identification of the F-consequence class for community C include both logical consistency and maximizing knowledge. Perhaps it would be worthwhile, therefore, to consider the prospects for an explicitly *normative* and *idealized* analysis of knowledge, in which conditions such as these serve as requirements which *should be satisfied* for the set of beliefs $\mathscr{B}zt$ accepted by any person z at time t to properly qualify as a *knowledge context* $\mathscr{K}zt$; otherwise, the ongoing pursuit of a descriptive conception is likely to result in an analysis where "knowledge" itself remains an unexplicated primitive and no systematic solutions are provided for problems such as these: If an individual z *knows* that a certain sentence S is true, does z therefore *know* all of the propositions that are valid deductive consequences of S? If z *knows* a body of evidence e that affords inductive support for a specific generalization S, does z therefore *know* S? If z *knows* S at one time t^1, can z subsequently *know* not-S at another time t^2? If different persons, z^1 and z^2, both possess the same body of evidence e, is it therefore the case that z^1 and z^2 both *know* S, where S is inductively supported by that evidence with a strength equal to r? Can an individual z *know* both S and not-S at one and the same time t? Is it impossible for different persons to possess the same body of evidence e and yet ultimately disagree about what they do and do not *know* on the basis of that evidence? The answers to these questions, we may assume, should be both explicit and systematic.

It would not be unwarranted, therefore, to focus upon the concept of knowledge with the expectation of exhibiting the requirements that should be satisfied, in principle, for an individual z to be in a knowledge context $\mathscr{K}zt$ at time t. As we have already implicitly discovered, the extent to which these conditions are actually satisfied by ordinary human beings varies from person to person and from time to time: babies, infants, and children, for example, like those who suffer from brain damage and mental retardation, should be expected to fulfill these conditions rather seldom and partially indeed; those from impoverished environments, who are illiterate and culturally deprived, may fare better but not very well; persons with ordinary backgrounds and average educations, relatively well; others, who have benefitted from nourishing environments and superior educations, may satisfy them in general and to a high degree. Nevertheless, although the varying extents to which various individuals meet these requirements serves as empirical evidence supporting such an explication, the account in question is intended to clarify and improve upon ordinary usage, without attempting to recapitulate common practice in every detail. An investigation of this kind, therefore, considers the possibilities for getting beneath the surface of linguistic usage in order to develop and refine a set of concepts that display the theoretical significance of such notions as "evidence", "inference" and "rationality", thereby affording a basis for exploring and elucidating potential connections of a syntactical, semantical, and pragmatical kind by means of which knowledge *per se* may be more adequately understood. Thus, in its characteristic respects, these investigations are perfectly analogous to the development within science itself of a *technical language* more useful and precise than the language of daily life, for it is precisely this language which we are trying to rectify and improve that it might better serve the purpose we pursue in the development of a normative conception of scientific knowledge.

On the basis of our previous inquiry, we may entertain the preliminary conception of the *knowledge context* $\mathscr{K}$ for an individual z at a time t as the set of beliefs $b_1, b_2, ...$, i.e., $\mathscr{B}zt$, (a) which the individual z accepts at time t (understood as it has previously been defined), and (b) which satisfies as well *at least* the following conditions of rational belief (where Szt is a set of sentences in $\mathfrak{L}zt$ representing the set of beliefs $\mathscr{B}zt$), namely:

(CR-1) the set of sentences Szt is deductively closed, i.e., any logical consequence S_k of any member S_i or any conjunction of members of Szt is also a member of Szt; and,

(CR-2) the set of sentences Szt is logically consistent, i.e., it is not the case that, for any sentence S_i, both S_i and its negation not-S_i are logical consequences of Szt.

A set of belief-sentences Szt for an individual z at a time t which satisfies (CR-1) and (CR-2) may be referred to as a *(minimally) rational set of beliefs*. An individual z will be in a knowledge context $\mathscr{K}zt$ at time t, therefore, *only if* the set of belief-sentences representing the set of beliefs $\mathscr{B}zt$ accepted by z at t forms a (minimally) rational set of beliefs. [Unless otherwise indicated, the knowledge context $\mathscr{K}$ for an individual z at time t will be assumed to consist of an infinite set of sentences Szt.]

Notice, in particular, that (CR-1) and (CR-2) both concern the logical relations that may obtain among the members of an accepted set Szt of sentences with a language $\mathfrak{L}zt$, i.e., these conditions are both "internal" requirements, which invites our attention to the consequence that, given a specification of the grammatical principles and vocabulary postulates, $\mathfrak{L}$, upon which z relies at t, i.e., $\mathfrak{L}zt$, the deductive relations between the sentences $S^1, S^2, \ldots$, that are well-formed formulae of $\mathfrak{L}zt$ are preserved within any set of beliefs z accepts at t. Consequently, the need to take account of the intensional character of contexts of knowledge or of belief is substantially diminished; specifically, if z is in a knowledge context $\mathscr{K}zt$ at t, then if z accepts S^1 at t, when S^1 entails or implies S^2, then z accepts S^2 as well, *necessarily*, by virtue of these two conditions. Wolgast's "paradox of knowledge", according to which some z might accept S^1 at t but not accept S^2 at t, when S^1 entails or implies S^2 (as something which z either does not realize or fails to infer) cannot occur under this normative conception. This consideration suggests the desirability of distinguishing between *personal* possibilities, and so on, which are not thus restrained, and (let us say) *impersonal* possibilities, and so on, which are [relative to an individual z at time t, given the language framework $\mathfrak{L}zt$, some set of "facts" f (not necessarily true), and assuming satisfaction of (CR-1) and (CR-2)], as follows:

(a″) for any sentence S in $\mathfrak{L}zt$, S describes an *impersonal possibility* for z at t, given facts f, if and only if z accepts f at t and it is not the case that f entails or implies not-S;

(b″) for any sentence S in $\mathfrak{L}zt$, S describes an *impersonal probability* for z at t, given facts f, if and only if z accepts f at t and it is the case that f entails or implies S is probable; and,

(c″) for any sentence S in $\mathfrak{L}zt$, S describes an *impersonal necessity* for z at t, given facts f, if and only if z accepts f at t and it is the case that f entails or implies S.

These impersonal possibilities, probabilities, and necessities, of course, represent normative and idealized conceptions, since some person who actually satisfied the conditions imposed by (CR-1) and (CR-2) would be incomparably "smarter" than any other person who has ever been alive; indeed, the closest approximation to their complete fulfilment that comes to mind would appear to be the most recent IBM, CAC and UNIVAC computers, which, in pure deductive power, exceed even the ability of Alonzo Church. Nevertheless, although the "impersonal" achievements of these electronic marvels represent remarkable attainments, their limited memories and non-instantaneous performance falls short of their perfect satisfaction. Insofar as these two conditions only concern the relationships that obtain between the members of a set of accepted beliefs (or propositions), however, they do not represent the most important features distinguishing a theory of knowledge from its epistemic alternatives, namely: conditions under which beliefs, b_1, b_2, ..., are themselves properly accepted into and rejected from such a knowledge context $\mathcal{K}zt$, i.e., the "external" requirements for rational beliefs (which Lehrer [1974] tends to ignore).

These conditions for acceptance and rejection themselves, moreover, likewise reflect distinct aspects of the epistemic situation, in particular, (a) the non-deductive "rules of inference" that properly apply to the acquisition and evaluation of those individual beliefs b_1, b_2, ..., by the person z at time t; and, (b) the circumstances of application of these "rules of inference" to the set of beliefs $\mathcal{B}zt$ thereby acquired, in order to ascertain whether any other beliefs warrant acceptance by z at t (where, as before, Szt is a set of sentences in $\mathfrak{L}zt$ representing the set of accepted beliefs $\mathcal{B}zt$); specifically,

(CR-3) if sentential (or experiential) evidence available to z at t warrants the acceptance of a proposition b_k at t, relative to the "rules of inference" z should properly apply in framework $\mathfrak{L}zt$ at t, then z must accept a sentence S_k of $\mathfrak{L}zt$ into $\mathcal{K}zt$ at t, where S_k asserts that b_k is the case; and, finally,

(CR-4) the acceptance of any (singular or general) proposition b_k by z at t must also be warranted on the basis of all of the relevant evidence available to z at t, relative to the "rules of inference" z should properly apply in framework $\mathfrak{L}zt$ at t; otherwise, b_k may not be accepted into $\mathcal{K}zt$, after all.

A set of belief-sentences Szt for an individual z at a time t which satisfies (CR-1), (CR-2), (CR-3), and (CR-4), therefore, may be referred to as a *(maximally) rational set of beliefs*. Accordingly, *the individual z is in the knowledge context $\mathcal{K}zt$ at a time t if and only if the set of belief-sentences Szt representing the set of beliefs $\mathcal{B}$ accepted by z at t forms a (maximally) rational set of beliefs*. These requirements, therefore, are individually necessary and jointly sufficient for an idealized person z to be in a *knowledge context $\mathcal{K}zt$ at a time t*.

At least three aspects of these conditions merit immediate comment. The first, of course, is that (CR-3) and (CR-4) rely upon (unspecified) "rules of inference", which obviously fulfill a crucial role within the explication of knowledge under consideration above. This, of course, is the principal explanation for according this conception the status of a "preliminary explication", since its ultimate adequacy will depend upon the detailed characteristics of the "rules of inference" that shall be proposed in the analysis that follows. The second is that a belief, b_k, which z does not presently accept at t, must be accepted into $\mathcal{K}zt$ at t when *some* of the evidence available to z at t warrants its acceptance at t, so long as it remains warranted by *all* of the evidence available to z at t. Consequently, it is not strictly necessary for z to explicitly inventory *every* belief-sentence S_i belonging to the set of beliefs that z accepts at t for each such proposition b_k; however, it *is* strictly necessary that, for b_k to be acceptable for z at t, (CR-4) has to be true. The third is that none of these conditions, individually or in conjunction, implicitly or explicitly, entail or imply that the set of beliefs Szt which z accepts at t qualifies as a knowledge context $\mathcal{K}zt$ for z at t *only if* each of the members of that set of beliefs Szt accepted by z at t is *true*. This deliberate omission reflects the author's conviction that the explication of a conception of scientific knowledge which illuminates the *history* of science requires its rejection, even though a conception which clarifies its *objective* requires its retention.

No doubt this final point deserves further elaboration. There are a number of reasons for hesitating to incorporate a truth condition for the explication of scientific knowledge, including, for example, these:

(i) Virtually every paradigm of scientific knowledge, from the past unto the present, has found or will find itself the victim of scientific progress, including, of course, the astronomical and cosmological speculations of the pre-Socratic philosophers, the elaborate epicyclic theory of Plato and Ptolemy, the crystalline spheres of Aristotle, the complex heliocentric account of Copernicus, and the astoundingly elegant theory of classical Newtonian mechanics — all of which have long been superseded by more adequate relativistic analyses which we presently accept as true. An explication of scientific knowledge according to which only currently accepted theories qualify as "scientific knowledge" does not appear to be especially promising, since we have no guarantee that the theories we now accept are destined for a different fate than their predecessors endured.

(ii) We ordinarily assume that the condition of truth possesses some explanatory significance with respect to epistemic situations in which an hypothesis whose acceptance is evidentially warranted in one situation is nevertheless not evidentially warranted in another. The cases exhibited by Gettier [1963], for example, intended as a critique of the "standard conception" of knowledge, i.e., as warranted true belief, may reasonably be construed as necessitating a truth condition, not just for the belief that is accepted, but for all of those other beliefs on which its acceptance itself depends as well. Since we have no means for discriminating a warranted truth from a warranted falsehood, however, while all such examples are amenable to analysis as illustrating that there are some circumstances in which the satisfaction of (CR-4) does not invariably accompany satisfaction of (CR-3), this argument does not seem especially compelling.

(iii) Since it is always a mistake to confuse truth and confirmation, however, perhaps the absence of such truth conditions from the present explication might best be explained as follows: the conception of knowledge as having to be true, necessarily, affords an idealized standard, relative to which our accepted beliefs must be subjected to appraisal. For if our beliefs are true, then their consequences cannot be false; and if our beliefs are inconsistent, then they cannot possibly all be true. The "evidence" at our disposal at any specific time, of course, is inevitably fragmentary and incomplete (and by no means is it guaranteed to be true). Thus, let us embrace Carnap's [1949] distinction between 'perfect' and 'imperfect' knowledge, where *perfect knowledge* cannot possibly be weakened or refuted by any future experience; then we must admit — at least with respect to all empirical knowledge — that *imperfect knowledge* is the best we can have. Yet imperfect knowledge is not sufficient to determine with absolute assurance the truth or the falsity of any belief about ourselves or the world around us. Apparently, the risk of error is the price of "scientific knowledge"

In order to relieve this tension without creating needless diversion from the essential enterprise of investigating the conditions of rational belief, let us introduce a *fifth* condition for the explication of the concept of a *(perfect) knowledge context* $\mathcal{K}zt$ for the individual z at a time t (assuming, once again, that Szt is a set of sentences in the language framework $\mathfrak{L}zt$ representing the set of beliefs that z accepts at t), i.e.,

(CR-5)   the set of sentences Szt is true, necessarily, i.e., any sentence S_i included in Szt and any consequence S_k of Szt must be true.

A set of belief-sentences Szt for an individual z at a time t which satisfies (CR-1), (CR-2), (CR-3), (CR-4), and (CR-5), therefore, will be referred to as a *(perfect) knowledge context* $\mathcal{K}zt$ for individual z at time t, whereas one which satisfies all expect for (CR-5) may be referred to as an *(imperfect) knowledge context* $\mathcal{K}zt$ for individual z at time t. Thus, those who desire to retain the truth condition will find that all which follows, *mutatis mutandis*, also applies to that conception of scientific knowledge; indeed, as we shall subsequently discover, this interpretation provides an indispensable ingredient for understanding theoretical realism, in general.

EPISTEMIC POSSIBILITIES

Our explication of the conception of the knowledge context $\mathcal{K}zt$ for z at t, of course, affords an opportunity to introduce definitions for both individual and community-relative notions of possibility, probability, and necessity which are completely objective in the sense that, assuming some specification of the set of "rules of inference" which properly apply in the language framework $\mathfrak{L}zt$ together with some description of the evidence available to z at t by means of a set of evidence-sentences $\{E\}$ in $\mathfrak{L}zt$ as well, these *epistemic possibilities* (let us say) are thereby univocally determined for an individual z in $\mathcal{K}zt$ at t; specifically,

(a*)   for any sentence S in $\mathfrak{L}zt$, S describes an *epistemic possibility* for z at t, relative to evidence $\{E\}$ and $\mathcal{K}zt$, if and only if z accepts $\{E\}$ at t and it is not the case that $\{E\}$-and-$\mathcal{K}zt$ entails or implies not-S;

(b*)   for any sentence S in $\mathfrak{L}zt$, S describes an *epistemic probability* for z at t, relative to evidence $\{E\}$ and $\mathcal{K}zt$, if and only if z accepts $\{E\}$ at t and it is the case that $\{E\}$-and-$\mathcal{K}zt$ entails or implies S is probable;

(c*)   for any sentence S in $\mathfrak{L}zt$, S describes an *epistemic necessity* for z at t, relative to evidence $\{E\}$ and $\mathcal{K}zt$, if and only if z accepts $\{E\}$ at t and it is the case that $\{E\}$-and-$\mathcal{K}zt$ entails or implies S.

Since the same conclusions hold for *any* such z in $\mathscr{K}zt$ at t, relative to the "rules of inference" that properly apply and the set of evidence $\{E\}$ available at t, the generalization of these conditions for some *community Z* (all of whose members employ the same "rules of inference" while accepting the same evidence $\{E\}$ available at t) is a trivial undertaking; therefore, let us simply stipulate that the definitions obtained by substituting such a community Z for such an individual z in (a*), (b*), and (c*) specify the corresponding concepts of *scientific possibility, scientific probability*, and *scientific necessity*, without stating the corresponding formulations. The conception of a *scientific community Z* represented here, moreover, is one in which every member of Z employs the same "rules of inference" relative to a common language framework and consequently would derive *all and only* the very same inferential consequences under the very same evidential circumstances, i.e., it is an "impersonal" conception. It might be worthwhile, therefore, to compare this approach with an alternative interpretation, i.e., a somewhat "more personal" conception.

Perhaps the most intriguing example of an analysis of the "personal" variety has been advanced by Polanyi [1964], who has articulated the role of each member of a scientific community in upholding its standards:

... a system of scientific facts and standards can be said to exist only to the extent to which each scientist trusts all the others, to uphold his own special sector of the system in respect of his research, his teaching, and his administrative actions. Though each may dissent (as I am myself dissenting) from some of the accepted standards of science, such heterodoxies must remain fragmentary if science is to survive as a *coherent system of superior knowledge, upheld by people mutually recognizing each other as scientists, and acknowledged by modern society as its guide.* (Polanyi [1964], p. 375, original italics)

While Polanyi has undoubtedly delineated rather clearly the necessity for scientific integrity among the members of the scientific community, it is important to consider whether the standards of science are *those that are relied upon by the members of the scientific community*, or are the members of the scientific community *those who rely upon the standards of science*? Let us therefore appraise the prospects for ascertaining the appropriate standards of scientific inquiry by consulting the scientific community.

Our first problem would be to identify the members of that community in order to subject the manner in which they conduct an inquiry to observation and scrutiny. We might, for example, review the membership lists of professional societies, such as the American Physical Society, or the American Chemical Society, or the Association of Engineers and Scientists. But what about the Society of American Astrologers, the American Pyschological Association, or the American Association for the Advancement of Science? If *Galileo* cast horoscopes for the Grand Duke of Tuscany, how can astrologers be excluded? And if some psychologists accept the work of Adler, Jung, and Freud, should *they* be included? Some members of the AAAS, after all, are not scientists but philosophers: surely they would *have* to be discounted. What about the members of the Society for Library Science? the Mortuary Science Association? the Christian Science Society? Moreover, should we accept what the members themselves *say* about science and its methods or focus instead upon what they actually do? Are all of their activities *scientific* activities, or only some? If not all, which are? What if they are Rosicrucians, philanderers, or Democrats? How are we to distinguish *good* science from bad? Clearly, the prospects for such a determination of scientific standards are not exceptionally promising.

Suppose, nevertheless, that, in spite of enormous obstacles, it were possible to establish a list of persons belonging to the presumptive scientific community, to interview each at length and to observe their conduct as they work. Suppose, moreover, that they often *agree* upon their conception of scientific inquiry, maintaining, for example, that science invariably begins with Observation, proceeds to Classification, Generalization, Derivation, and Experimentation in an orderly, logical, sequence or progression. If, say, 89%, 95%, or 99% of the members of this ersatz community asserted this position during interviews, defended these views in their classrooms, and apparently practiced such procedures within the confines of their laboratories, would the questions under consideration thereby have been settled? Obviously not. Fortunately for the progress of science, even the endorsement of inadequate conceptions does not seem to exert an especially damaging influence upon the actual conduct of inquiry; but it is impossible to calculate the benefits that might accrue from the general acceptance of more appropriate standards. Even if *all* those subjects completely agreed upon every detail of such a conception, moreover, that would still not ensure its adequacy; for surely there are no built-in guarantees that one or another view of things must certainly be correct if only enough of us believe it.

It should come as no surprise, therefore, that the issues involved here are virtually identical to those we have considered in evaluating Hacking's suggestion that there might not be *any* normative theories at all. Let us henceforth admit the necessity for a normative approach in matters such as these and consider in detail the *role* to be fulfilled by "rules of inference" within the context of the present explication. As it happens, these "rules of inference" appear to be of at least two very different kinds, yet both are *non-demonstrative* in relation to language framework 𝔏zt; specifically,

(i) non-demonstrative rules for the acceptance and rejection of a proposition b_k on the basis of *evidence-sentences* $\{E\}$ within 𝔏zt; and,

(ii) non-demonstrative rules for the acceptance and rejection of a proposition b_k on the basis of *experiential findings*, relative to 𝔏zt.

Our "rules of inference" of type (i) thus regulate *inductive* inferences, such as those from a finite set of singular evidence-sentences $\{E\}$ to a logically unrestricted generalization S_k, within the language framework 𝔏zt; while "rules of inference" of type (ii) regulate *perceptual* inferences, which describe the results of (more or less) focused observations and/or simple measurements by means of singular sentences S_k, within 𝔏zt.

All "rules of inference" of these two types are "non-demonstrative", in the sense that, even when the "evidence" to which they are applied is *true*, a proposition b_k whose acceptance (or rejection) is then warranted may, nevertheless, be *false* (or true), respectively [where the antecedent of this condition is vacuously satisfied by perceptual inferences], i.e., unlike *deductive* inferences, they are *not* truth-preserving. The absence of this property, of course, reflects the fact that these non-deductive inferences *are* ampliative, i.e., their proper application, in principle, warrants the acceptance of propositions, such as b_k, which possess *more content* than does the evidence, such as $\{E\}$, upon which they are based, where this content itself may be represented by some class of deductive consequences. Moreover, these non-deductive inferences are also unlike their deductive counterparts insofar as, when a proposition b_k *follows from* a set of evidence-sentences $\{E\}$, that relation will remain even if additional members are added to that set; but propositions which happen to be non-deductively warranted may or may not remain so warranted when the evidence upon which they are based increases. Thus, we may say that deductive inferences are evidentially additive, while the others are not.

Altogether, therefore, we have three types of "rules of inference", namely, deductive, inductive, and perceptual. The latter two should be distinguished from the former, insofar as inductive and perceptual rules are non-demonstrative, ampliative, and non-additive, while all deductive rules are demonstrative, non-ampliative, and additive. Perceptual rules, moreover, may be further differentiated from all inductive and deductive rules, insofar as they alone do not require evidence in the form of some set of *evidence-sentences* $\{E\}$ for their proper application; consequently, their ampliative dimension arises directly from the use of language for the purpose of formulating and describing experiential findings relative to some language framework $\mathcal{L}$. Their application, whether successful or not, therefore, thus reflects upon the ability of some language user z to properly employ some language framework $\mathcal{L}$ under various experiential circumstances at different times t, i.e., their successful use reflects the *linguistic competence* of z at t, relative to $\mathcal{L}zt$, under appropriate conditions, as we shall subsequently ascertain.

With this clarification of the conception of "rules of inference" at hand, therefore, we should review our conditions of rational belief, (CR-3) and (CR-4), in order to enhance our appreciation of their exact significance; thus, for example, (CR-3) imposes the following condition:

(CR-3)   if sentential (or experiential) evidence available to z at t warrants the acceptance of a proposition b_k at t, relative to the inductive, deductive, and perceptual rules that z should properly apply in language $\mathcal{L}zt$, then z must accept a sentence S_k of $\mathcal{L}zt$ into $\mathcal{K}zt$ at t, where S_k asserts that b_k is the case;

while the same substitutions exhibit the requirement enforced by (CR-4):

(CR-4)   the acceptance of any (singular or general) proposition b_k by z at t must also be warranted on the basis of all of the relevant evidence available to z at t, relative to the inductive, deductive, and perceptual rules that z should properly apply in language $\mathcal{L}zt$; otherwise, b_k may not be accepted into $\mathcal{K}zt$, after all.

Consideration of the significance of these conditions in application to perceptual rules of inference, especially, may prove to be instructive. (CR-3), of course, requires that, if *some* of the results of observations and measurements available to z at t warrant the acceptance of a proposition b_k by z at t, relative to $\mathcal{L}zt$, then z must accept some sentence S_k of $\mathcal{L}zt$ into $\mathcal{K}zt$ at t, where S_k provides a description of those experiential findings, unless *all* of the observations and measurements that are available to z at t would not warrant the acceptance of S_k into $\mathcal{K}zt$ at t. These normative and idealized conditions thus establish (what we perhaps may refer to as) *a rational responsibility* for an individual z at time t with respect to the formulation and description of experiential findings. But they also reflect the "imperfect" character of the knowledge context $\mathcal{K}zt$ to the extent to which these requirements are not actually fulfilled by the members of any ordinary, academic, or scientific community C at t.

The introduction of *inductive* rules of inference in the development of the knowledge context $\mathcal{K}zt$, moreover, effects a subtle but nevertheless important equivocation with respect to "probability" as it appears within the context of definitions of personal, community, impersonal, epistemic, and scientific possibilities, probabilities, and necessities; for in the absence of these ampliative and non-demonstrative rules of inference, it would be logically possible that the only sentences that are *probable* are those that describe the results or outcomes of singular and multiple sets of experiments or observations, i.e., strictly speaking, it might happen that the only "probabilities" that occur within such contexts are "probabilities of events" within an object language $\mathfrak{L}zt$, rather than "probabilities of hypotheses" within its meta-language, say, $\mathfrak{M}$-$\mathfrak{L}zt$. For example, the definition of *impersonal probability* explicitly asserts as follows:

(b″) for any sentence S in $\mathfrak{L}zt$, S describes an *impersonal probability* for z at t, given facts f, if and only if z accepts f at t and it is the case that f entails or implies S is probable.

Let us assume that the facts f are exclusively object language sentences or propositions which attribute properties to the physical world and that the outcome or event described by S in $\mathfrak{L}zt$ is *probable* only if it is neither impossible nor necessary, relative to f and $\mathfrak{L}zt$. Then when f includes among its members such propositions as that the probability for obtaining an even-number as the outcome of a single toss with a fair die is ½ and that the next toss is one made with a fair die, then f entails or implies, "The result of the next toss is even-numbered", is probable. Nevertheless, as we shall subsequently ascertain, these "probabilities" are *not* ones that could only obtain with "inductive rules of inference".

Another technical matter deserving brief preliminary consideration, no doubt, is that inductive rules of inference *are* quite frequently supposed to be *probabilistic*. The most important representatives of this influential point of view, moreover, are the so-called "Bayesians", all of whom attribute a central role to Bayes' theorem as a fundamental rule of inductive inference within a probabilistic framework. However, while the consistency of a set of beliefs, such as f, accepted by z at t might properly be supposed to require conformity to the calculus of probability as a necessary condition *with respect to the probabilities of events*, it is by no means obvious that the principles of acceptance and of rejection themselves must be similarly constrained *with respect to measures of support for hypotheses* as a necessary condition for their own adequacy; for the consistency of a set of beliefs, such as f, after all, concerns the contents of the accepted sentences of some *object language* $\mathfrak{L}zt$, while the principles of acceptance and rejection belong to the *meta-language* $\mathfrak{M}$-$\mathfrak{L}zt$ of inductive procedure. Thus, soothsayers, astrologers, and crystal-ball gazers, for example, rely upon non-probabilistic modes of knowledge acquisition, but the results of their inquiries (concerning games of chance or otherwise) are not *therefore* inconsistent. Indeed, if the arguments that follow are well-taken, the adoption of a probabilistic framework for our rules of induction represents what may well qualify as *the great blunder of 20th century theories of knowledge.*

During the course of this inquiry, the general outline of (what may be referred to as) *a theory of epistemic resources* will emerge; that is, an analysis of the various kinds of resources upon which we all may draw in attempting to ascertain the truth about ourselves and the world around us, including, especially, the following:

(1) the language framework $\mathfrak{L}$ which each person z accepts at t;
(2) the deductive rules of inference upon which z relies at t;
(3) the inductive rules of inference upon which z relies at t;
(4) the experiential findings available to z at t, relative to $\mathfrak{L}$; and,
(5) the powers of imagination and conjecture which z can exert at t.

Indeed, although it will come as some surprise to those who suppose that science proceeds through a process of Observation, Classification, Generalization, Derivation, and Experimentation, the most important factor in the development and growth of scientific knowledge is our capacity to exercise the powers of *imagination and conjecture*, without which we cannot create, change, and improve our language framework $\mathfrak{L}$, upon whose concepts and structure we ultimately depend for the description and interpretation of our experiential findings and our scientific theories. The progress of science, as Popper [1965], [1968], [1972], so eloquently proclaims, *is* essentially a complex process of "conjectures and refutations", whose details and defense we shall attempt to discover in the investigations that now follow.

2. THE LANGUAGE FRAMEWORK: $\mathfrak{L}$ OR $\mathfrak{L}^*$?

The discovery that language can be used to talk about language, it may be said, qualifies as the most important innovation in the history of philosophy; certainly, at the very least, it serves to delineate the difference between the *analytical* approach reflected already by the Platonic dialogues, which history thus records as Socrates' enduring contribution to philosophy, the *speculative* approach of the Milesian "nature" philosophers and even the *persuasive* approach of the Sophistic argumentative tradition. Its explicit invocation as a principle of analysis, of course, has been a practice which has flourished over the past 100 years or so, especially as a result of the formal methods facilitated by the development of symbolic logic as a device for the exploration of the structure of language, an approach whose ultimate potential, no doubt, has yet to be fully realized but whose possibilities of application were spectacularly illustrated by Whitehead and Russell [1910], [1912], [1913]. Indeed, some of the most noted works of contemporary philosophy, including Tarski's semantical conception of truth and Carnap's logical interpretation of probability, are investigations that would have been inconceivable without the resources provided by careful differentiation between object language and meta-language.

Within the context of the present inquiry, therefore, it is extremely important to understand the object language/meta-language distinctions that are applicable to our analysis of the relations between the language framework $\mathfrak{L}zt$ and the knowledge context $\mathcal{K}zt$, especially with respect to the different classes of "rules of inference" which have been specified, namely:

(i) *deductive* rules of inference are meta-linguistic in their explicit form and regulate the "internal" relations imposed by conditions (CR-1) and (CR-2) upon the members b_i of any accepted set of propositions Szt within the language $\mathfrak{L}zt$ belonging to the knowledge context $\mathcal{K}zt$, for all such sets;

(ii) *inductive* rules of inference are also meta-linguistic in their explicit form and partially implement the "external" requirements (CR-3) and (CR-4) by identifying additional propositions b_k for inclusion within Szt, relative to $\mathfrak{L}zt$ and a knowledge context $\mathcal{K}zt$, given evidence-sentences $\{E\}$;

(iii) *perceptual* rules of inference, however, are *not* meta-linguistic in formalizing relations that obtain *between* various sentences within $\mathfrak{L}zt$, but instead further implement the "external" requirements (CR-3) and (CR-4) by ascertaining additional propositions b_k which should be included in Szt, relative to $\mathfrak{L}zt$ and knowledge context $\mathscr{K}zt$, given z's experiential findings.

As an informal depiction of these crucial differences, therefore, let us say that deductive rules concern *consistency* within the object language via Szt, inductive rules promote *completeness* with respect to evidence sentences $\{E\}$, and perceptual rules infuse *empirical content* in the form of propositions b_k relative to z's experiential findings, i.e., their respective benefits are: consistency, completeness, and content, in the senses which we have defined.

To the extent that scientific knowledge properly qualifies as empirical knowledge, therefore, perceptual rules of inference fulfill an indispensable role in *effecting a connection between language and the physical world apart from which the concepts and the theories of science would all lack empirical content*. The truth of this contention, let us note, does not depend upon an observational/theoretical language distinction, such as Carnap maintained as late as [1966], although it *does* demand some notion of experiential findings; and it does not entail (what Hempel [1952], pp. 23—24, has labeled as) "the narrower thesis of empiricism", to wit:

Any term in the vocabulary of empirical science is definable by means of observation terms; i.e., it is possible to carry out a rational reconstruction of the language of science in such a way that all primitives are observation terms and all other terms are defined by means of them;

even though perceptual rules of inference *are* directly applicable to (at least some of) the undefined, i.e., primitive, terms of the language $\mathfrak{L}zt$. In order to explain the differences and the similarities between what are ultimately distinct accounts of the relationship between language and the physical world, moreover, perhaps we should review several basic elements of Peirce's analysis of *semiotic*, i.e., the general theory of signs.

PEIRCE'S THEORY OF SIGNS

According to Peirce (Buchler [1955], esp. pp. 97—115 and pp. 274—289), a sign is something that stands for something, in some respect or other, for somebody; thus, the crucial relation of "standing for" which underlies this construction is envisioned by Peirce as being three-placed or triadic. Peirce supposes that signs stand for something *when* they stand in a certain relation to that something for which they stand and *create in the mind of a sign-user* other — equivalent or more developed — signs, which stand in the same relation to that for which they stand as do the original signs creating them. Peirce elaborates numerous distinctions concerning sign systems, but we will consider only those essential for our purposes; in particular, Peirce introduces three trichotomies of signs, each of which appears to reflect possible sub-divisions of the three branches of semiotic when that theory is divided on the basis of the relations between signs and *the respects in which they stand for something*, signs and *the relations they bear to what they stand for*, and signs and *the kind of signs which can stand for them*, respectively, which Peirce designates as pure grammar, logic proper, and pure rhetoric.

The three sub-divisions of these three branches of semiotic in general, therefore, could well exhibit the intersection of Peirce's major categories of Firstness, Secondness, and Thirdness, with the categories reflecting the fundamental relations obtaining for all signs, which we have just reviewed. This suggestion, of course, has the character of a conjecture, for the major categories themselves are notoriously difficult to render clear and precise. A possibly helpful but perhaps inadequate interpretation may be to identify Firstness with *presentness*, i.e., with the immediately accessible qualities of things, Secondness with *causation* (or physical lawfulness), i.e., with the effects or influence things and events causally exert upon other things and events, and Thirdness with *conventions* (or habitual associations), i.e., connections between things that arise as a result of acquiring varied habits. The outcome of such an appraisal can be schematically summarized as follows:

Divisions of Semiotic

	Pure Grammar	Logic Proper	Pure Rhetoric
Firstness	qualisigns	icons	rhemes
Secondness	sinisigns	indices	dicent signs
Thirdness	legisigns	symbols	arguments

Thus, *qualisigns* are signs which possess properties in common with what they stand for; *sinisigns* are signs which are effects brought about by what they stand for; and, *legisigns* are signs which are habitually (or conventionally) associated with what they stand for. Similarly, *icons* are signs which stand for things that may or may not, but, in any case, do not necessarily, exist; *indices* are signs which stand for things upon which the signs themselves are dependent for their own existence and which therefore exist; and, *symbols* are signs which stand for things by virtue of conventional or habitual associations but that nevertheless exist necessarily, that is, "in the possibly imaginary universe to which the Symbol refers" (Buchler [1955], p. 103). Finally, *rhemes* are signs whose own representatives are signs of qualitative possibility; *dicent signs* are signs whose own representatives are signs of actual existence; and, *arguments* are signs whose own representatives are signs of general necessity (understood, of course, as Peirce thus intends).

Consideration of the categories of icon, index, and symbol, moreover, strongly suggests that ordinary languages, such as English, German, French, and so on, are neither iconic nor indexial in character, but primarily, if not exclusively, symbolic: pictures, portraits, statues, and so forth must be worth thousands of words because they are iconic, while words — even when we string them together — are not; heat, smoke, and light all signify fires, with respect to which they are both effects and symptoms, while words themselves — as opposed to the occasions of their utterance — are, intrinsically, neither effects nor symptoms of such things. Peirce's most important notion with respect to our inquiry, however, is his analysis of the role played by *habits* in imparting meaning to symbolic systems, such as ordinary language:

> The deliberately formed, self-analyzing habit — self-analyzing because formed by the aid of analysis of the exercises that nourished it — is the living definition, the veritable and final logical interpretant. Consequently, the most perfect account of a concept that words can convey will consist in a description of the habit which that concept is calculated to produce. (Buchler [1955], p. 286)

Here, in typically pellucid Peircean prose, he recommends that words as signs within symbolic language systems are ultimately best understood in terms of the *linguistic habits* those words represent; in particular, Peirce suggests that habits themselves are to be described by specifying the conditions and the motivation of the actions and behavior they may signify or require.

Peirce's concepts, of course, are very rich, which means that they are very fertile and perhaps occasionally ambiguous. Consider, for example, the familiar case of stop-signs in place at properly-marked and well-maintained street intersections. These signs are symbols standing for applying brakes, slowing down, and coming to a complete stop for vehicle drivers. The signs themselves are physical things which can create in the mind of a sign-user another (equivalent or more developed) sign, namely: a belief that a stop-sign is present, which may trigger a set of reactions and behavior patterns that implement the instructions a sign of this kind conventionally conveys. Notice, however, that several assumptions are involved in such an analysis; in particular, that the sign-user is familiar with such signs, is both physically and mentally able to detect and recognize their presence and capable of and motivated to drive safely and obey the law. Fleeing felons are more concerned to escape the law than to obey it; consequently, they may run the intersection, although they see the sign, to evade police pursuit. A nearsighted little old lady, by contrast, wants to obey the law, no doubt, but may be unable to do so, simply because she is incapable of detecting such a sign's presence. To the extent to which the acquisition and the exercise of a sign involves the acquisition and the exercise of a (possibly very complex) set of habits, therefore, it seems reasonable to conclude that success in the construction and employment of a sign system depends upon a presumptive class of *competent sign-users*, who are able to acquire and to exercise these habits.

A complete analysis of the language framework $\mathfrak{L}zt$, therefore, should take into account the crucial role of habits in effecting a relationship between signs, their users, and the world, without which our concepts and our theories (expressed in $\mathfrak{L}zt$) would lack empirical content. Notice also that such connections typically involve all three of Peirce's major categories, for the presence of symbols themselves is indicative of *Firstness*, causal interactions between symbols and their users is a manifestation of *Secondness*, and the habitual association of symbols with what they stand for represents the essential character of *Thirdness*. It therefore appears reasonable to assume that a language $\mathfrak{L}zt$ consists not only of the grammatical principles and the vocabulary postulates accepted by z at t, but also includes the linguistic practices employed by z at t, where these habits of language forge the linkage between the language $\mathfrak{L}zt$ and the physical world. Indeed, Austin [1950] has suggested the possibility of accounting for *truth* on the basis of a distinction between *two different kinds* of linguistic conventions (or habitual associations), namely:

(a) *descriptive conventions* correlating words (or sentences) with the various *types* of situations, *kinds* of things, *sorts* of events, and so forth, that may be encountered in the physical world; and,

(b) *demonstrative conventions* correlating the use of words (or sentences) with historical situations actually encountered in the physical world.

Thus, Austin suggests that a sentence S in $\mathfrak{L}zt$ is properly qualified as *true in* $\mathfrak{L}zt$ when the historical situation with which S is correlated by means of the demonstrative conventions of $\mathfrak{L}zt$ is of a type with which S is correlated by the descriptive conventions of $\mathfrak{L}zt$. In order to pursue this conception in greater detail, however, at least three additional assumptions are required.

The distinction between *pragmatics* (as a study of the relations between signs, what they stand for, and sign-users), *semantics* (as a study of the relations between signs and what they stand for), and *syntax* (as a study of the relations between signs and other signs), is doubtless so familiar to us all as to require no further elaboration. As Carnap [1939] has explained, each of these modes of inquiry, respectively, involves a higher degree of abstraction, successively, insofar as the available evidence concerning the logical structure of an ordinary language is necessarily pragmatical. Consequently, the investigation of the structure of languages also has a normative aspect, which, in the case at hand, should be embellished in the following respects:

first, we shall assume that the *deductive* rules of inference that apply within the knowledge context $\mathcal{K}zt$ are identical with or equivalent to the set of *transformation rules* of the language $\mathfrak{L}zt$, which therefore encompass both "rules of inference" and "rules of replacement" (in their ordinary senses);

second, we shall assume that the *inductive* rules of inference, as such, strictly speaking, are not elements of the language framework $\mathfrak{L}zt$, after all, but require *independent specification and theoretical justification* (where it is assumed that these rules, like their deductive counterparts, constitute a set of rules whose members do not change from language to language); and,

third, we shall assume that the *perceptual* rules of inference that apply within the knowledge context $\mathcal{K}zt$ are identical with or equivalent to the set of *conventions* (or *habitual associations*) which effect the connection between a language $\mathfrak{L}^n$ and the physical world, without denying that there may be many different distinct patterns of conventional association which are or would be successful in bringing about such a connection (because of which there could be innumerable non-equivalent language frameworks $\mathfrak{L}^1$, $\mathfrak{L}^2$, and so on).

THE THEORETICAL OBJECTIVE

Without a specification of the *object language* $\mathfrak{L}zt$ in which it is embedded, of course, our characterization of the knowledge context $\mathcal{K}zt$ would be substantially incomplete, since each belief-sentence S_i is a well-formed formula of that language. Choosing one language framework rather than some other, moreover, is a matter of considerable theoretical importance, as the following alternatives should demonstrate: Let $\mathfrak{L}$ be the standard first-order predicate calculus (without identity) employing the negation and disjunction signs as syntactical primitives; let $\mathfrak{L}^*$ be $\mathfrak{L}$ enriched by the addition of the subjunctive conditional and the causal conditional as syntactical primitives as well. Then the differences between $\mathfrak{L}$ and $\mathfrak{L}^*$ may be described (in part) as follows:

(i) the expressive-completeness properties of $\mathcal{L}$ and $\mathcal{L}^*$ differ greatly with respect to natural and scientific languages, since $\mathcal{L}^*$, but not $\mathcal{L}$, provides for the formalization of causal and of subjunctive conditionals (as well as of material conditionals);

(ii) this difference facilitates the (partial) explicit definition of dispositional predicates within the framework $\mathcal{L}^*$, while the same predicates receive only (partial) implicit definitions by means of reduction sentences as meaning postulates within the framework $\mathcal{L}$; and,

(iii) $\mathcal{L}^*$, but not $\mathcal{L}$, provides a sufficient foundation for distinguishing lawlike from accidental generalizations on the basis of their syntactical and semantical properties, while the employment of $\mathcal{L}$ for differentiating between sentences of these kinds is essentially dependent upon extra-linguistic (circumstantial and intentional) considerations.

All of these features are distinctive manifestations of what appears to be the fundamental difference between $\mathcal{L}$ and $\mathcal{L}^*$, namely: $\mathcal{L}$ is an *extensional* language, while $\mathcal{L}^*$ is not. Since extensional languages have a very simple structure, which has been thoroughly studied and is well-understood, there are excellent grounds for preferring their employment, other factors being equally balanced (between, say, $\mathcal{L}$ and $\mathcal{L}^*$). The choice between one language framework and another, however, may only be made on the basis of specified *pragmatical requirements*, in particular, by the (practical or theoretical) objectives which that language is intended to fulfill.

A fundamental problem of the theory of science is to develop an adequate explication of the concept of a physical law. Theoreticians such as Goodman [1965] and Hempel [1965] especially suggest that a sentence S is a *physical law* if and only if (a) S is lawlike and (b) S is true, where S is supposed to be lawlike only when (i) S is completely general and (ii) S is capable of providing support for counterfactual and subjunctive conditionals. Attempts to formalize lawlike sentences by employing the resources of wholly extensional languages such as the first-order predicate calculus, however, provide insufficient syntactical and semantical criteria for distinguishing between genuinely lawlike and merely accidental true generalizations, leading other theoreticians – including Braithwaite [1953] and Ayer [1956] – to the conclusion that this difference has to be viewed as pragmatical, for example, as a question of context or as a matter of attitude. There appear to be ample grounds for supposing, instead, that extensional frameworks may afford an inadequate foundation, in principle, for the logical formalization of any lawlike sentence. The arguments which follow, therefore, are intended to establish the inadequacy of such languages within theories of science.

In order to exhibit the deficiencies of extensional language for this purpose, the distinguishing features of extensional operators, such as the horseshoe (or material conditional), require explicit recognition, namely:

(a) that any *molecular* sentence constructed through the application of extensional operators is truth-functional (in the sense that its truth-value is exhaustively determined by the truth-values of its component sentences); and,

(b) that the truth-values of *atomic* sentences within such an extensional language are completely fixed in each instance by the history of the actual world (when they are not expressed in self-contradictory predicates).

The translation of 'if ___ then ...' sentences as material conditionals, therefore, fulfills this conception provided the truth-values of these sentences are determinate for *every combination* of truth-values for their constituent sentences, whose own truth-values do not vary with the occasions of their use, i.e., they are eternal, rather than occasion, sentences. Quine (in [1951], p. 11) has observed, "There is reason to believe that none but truth-functional modes of statement composition are needed in any discourse, mathematical or otherwise; but this is a difficult question". Indeed, as we shall ascertain, the obstacles encountered in formalizing lawlike sentences present at least three reasons for rejecting that belief altogether.

On an extensional conception, lawlike sentences — universal or statistical — characterize some relation that may obtain between the members of two classes, i.e., a *reference* class and an *attribute* class. Their descriptions must be completely general at least insofar as they may not be limited to any finite number of members on the basis of syntactical or semantical considerations alone; the other definitional properties of these classes, however, are logically arbitrary, i.e., they are pragmatically contrived. A universal generalization thus merely asserts that every member of the *reference class R* at time t will also be members of an *attribute class A* at t; while a statistical generalization similarly asserts that a certain proportion p more or less of the members of R at t will also be members of A at t. The appropriate syntax for a universal lawlike sentence (of the simplest kind), employing the resources of the predicate calculus, therefore, is,

(1) $\quad (x)(t)(Rxt \supset Axt)$;

which, of course, will be true if and only if every thing nameable, i.e., to which a name could be assigned, participating in the history of the physical world either does not possess the property R or does possess the property A.

Correspondingly, the appropriate syntax for a statistical lawlike sentence by employing predicate calculus and probability notation together will be,

(2) $P(Axt/Rxt) = r$;

that is, the probability that x is A at t, given that x is R at t, is equal to r, which, in its import, is equivalent to adopting a finite frequency interpretation (as do Russell [1948] and Sklar [1970]), if the world's history is finite and a limiting frequency interpretation (as do Reichenbach [1949] and von Mises [1964]), if it is not.

The criterion for distinguishing between merely accidental and *bona fide* lawlike generalizations of either kind, moreover, is supposed to be that the latter do, while the former do not, provide support for subjunctive and for counterfactual conditionals. The difficulty, however, is that there are no apparent syntactical or semantical criteria for separating the sentences that afford such support from those that do not. For example, the universal generalization, "All students who attend universities in tropical climates are less than 7' tall", which could be formalized as follows,

(3) $(x)(t)[(Sxt \cdot Uxt) \supset Lxt]$;

is supposed to be distinguishable from the universal generalization, "All pure pure water boils at a temperature of 212°F at sea level atmospheric pressure", which might be formalized analogously as follows,

(4) $(x)(t)[(Wxt \cdot Pxt) \supset Bxt]$;

on the grounds that (4) supports the counterfactual, "If the glass of water I just drank had been heated to 212°F at sea level atmospheric pressure instead, it would have boiled", while (3) does not support the counterfactual, "If Kareem Abdul Jabbar had attended a university in a tropical climate instead of U.C.L.A., he would have been less than 7' tall". As two instances of the same truth-functional forms, however, these assertions are syntactically identical and semantically indistinguishable; and, as true historical generalizations about the world's history, it would be surprising, indeed, if they were distinguishable on non-pragmatical extensional grounds.

The *first* significant reason for rejecting extensional formalizations of lawlike sentences, therefore, is that there are no syntactical or semantical criteria for distinguishing those sentences that do, from those that do not support counterfactual and subjunctive conditionals; for to explain the support for counterfactuals provided by sentences such as (4), but not by sentences such as (3), on the grounds that (4) is, while (3) is not, a lawlike generalization, is surely to beg the question. Quine has suggested the possibility that,

> what could be accomplished by a subjunctive conditional or other non-truth functional modes of statement composition can commonly be accomplished just as well by talking *about* the statements in question, thus using an implication relation or some other strong relation of statements instead of the strong mode of statement composition. Instead of saying:
> If Perth were 400 miles from Omaha then Perth would be in America
> one might say:
> 'Perth is 400 miles from Omaha' implies 'Perth is in America',
> in some appropriate sense of implication. (Quine [1951], p. 29)

But Quine is promising more than he can deliver; for with equal justification one might maintain that,
If Perth were 400 miles from Omaha then Omaha would be in Australia,
or, analogously, one might say:
'Perth is 400 miles from Omaha' implies 'Omaha is in Australia',
in that same appropriate sense. Quine's way out, therefore, supplies scant comfort for the extensional account: transferring problems from an object- to a meta-language does not automatically provide for their solution. The result would seem to be the choice between a pragmatical distinction and no distinction at all.

The *second* significant reason for rejecting extensional formulations of lawlike sentences is that the syntactical and semantical criteria which *are* provided clearly conflate principles (apparently) suitable for *explanation* with principles suitable for *retrodiction*. For *if* the sentence, "Whenever a bullet is fired into a piece of pine, it makes a hole", were *both* lawlike *and* adequately formalized in extensional language as,

$$(5) \quad (x)(y)(t)[(Bxt \cdot Pyt \cdot Fxyt) \supset Hxyt^*];$$

where t^* is simultaneous with or subsequent to t, then surely that law would be *equally adequately* formalized by its extensionally equivalent counterpart,

$$(6) \quad (x)(y)(t)[(Pyt \cdot -Hxyt^*) \supset -(Bxt \cdot Fxyt)];$$

which asserts, "Whenever a piece of pine has no hole in it at t^*, it is not the case that a bullet was fired into it at t". Yet while sentence (5) may initially *seem* to be suitable for explaining why a particular piece of pine displayed a hole at a certain time t^*, sentence (6) could hardly be invoked to *explain why* a bullet was not fired into some whole piece of pine at some time t. Not the least of the drawbacks to (6) as an adequate formulation of the law in question is that it portends the explanation of temporally prior states by reference to temporally subsequent states; and while (6) may serve a perfectly useful and unobjectionable role as a principle of retrodiction, i.e., for the purpose of inferring *what* has happened rather than *why* it has happened; it appears to be both useless and objectionable as a principle of explanation. Since (5) and (6) possess precisely the same syntactical and semantical significance, however, surely (5) is lawlike *if and only if* (6) is lawlike; so if it is an important desideratum that lawlike sentences but not accidental generalizations *are to be invoked to explain their instances*, then how can (5) justifiably be held to be lawlike to any greater or lesser a degree than (6)? If (6) is not a lawlike sentence, therefore, neither is (5) – on pain of contradiction.

The *third* significant reason for rejecting extensional formulations of lawlike sentences is that their application for the purpose of formalizing definitions for dispositional predicates produces inadequate and problematic outcomes. The "standard conception" of definition (as it might be referred to) within the logical positivist and logical empiricist philosophical traditions, at least, assumes the existence of three basic kinds of terms (or concepts), namely: observational, dispositional, and theoretical in kind, respectively, understood (more or less) as follows: an *observational* term designates an observable property of an observable entity; a *dispositional* term designates an unobservable property of an observable entity; while, a *theoretical* term designates an unobservable property of an unobservable entity – where "theoretical language" in its broad sense includes *both* dispositional *and* theoretical terms in these senses (Rudner [1966], pp. 21–23). Thus, in pursuit of the "narrower thesis of empiricism" previously specified, it became important to demonstrate the definability, in principle, of dispositional and theoretical terms by means of observational terms and extensional operators, exclusively, which led to unexpected difficulties. Indeed, the developments that motivated Carnap to propose *reduction sentences* as a method for the partial specification of the meaning of dispositional predicates are now exceedingly widely known, e.g., that the explicit definition of 'solubility' by reference to the test event of being submerged in water and the response event of dissolving, that is:

(7) x is soluble at t $=_{df}$ if x is submerged in water at t,
then x dissolves at t^*;

logically entails that any object – such as a green leaf, a yellow cow, or a red pencil – is soluble at any time it is not actually being subjected to the relevant test. As Pap ([1963], p. 560) has observed, however,

If Carnap succeeds in deriving from (the concept of solubility) the paradoxical consequence that any substance is soluble in any liquid in which it is never immersed, this is due to his formalizing the definiens, which is ordinarily meant as a *causal* implication, as a material conditional.

Thus, the lawlike character of these properties is not preserved by Carnap's extensional definitions; indeed, it seems quite evident that reduction sentences themselves provide no basis for overcoming the underlying difficulty; for, though dispositional attributions are thereby restricted to just those cases that actually satisfy test conditions of the relevant kind, these sentences remain purely extensional and therefore afford no support for counterfactual or for subjunctive affirmations.

Considerations such as these, I believe, reinforce the prospect that extensional languages themselves are not strong enough to adequately represent the lawlike properties of the physical world. Indeed, the fundamental difficulty might be explained intuitively as follows: extensional language, in principle, is *historical* and *descriptive*, but lawlike properties are *ontological* and *structural* in character. For it is the ontological structure of the physical world, so to speak, that generates the world's descriptive history *under the influence of a precipitating set of initial conditions*. As Popper ([1965], p. 433) has suggestively remarked,

A statement may be said to be naturally or physically necessary if, and only if, it is deducible from a statement function which is satisfied in all worlds that differ from our world, if at all, only with respect to initial conditions.

The basic inadequacy with the extensional approach, in other words, is that this world could have exhibited many different histories under varying sets of initial conditions; but what *would happen* if those conditions *were different* could never be expressed by extensional languages alone. The class of true generalizations of any one such history, as a result, inevitably encompasses a mixture of lawlike and accidental generalizations; for such an approach provides no basis for separating those statements true of all such worlds from those that are true of only one.

The reasonable alternative, of course, would appear to be the development of such an explication within an *intensional* language framework. The work of Lewis [1973], Stalnaker [1968], and perhaps Pollock [1974], [1976], for example, might be thought to hold the key to resolving these outstanding problems, an impression which is substantially strengthened by the realization that the principal difficulties we have discovered confronted by such extensional analyses involve ascertaining which of such generalizations are, and which are not, capable of supporting counterfactual and subjunctive conditionals. The application of their research to the problem before us, however, might not be quite so straightforward as it may initially appear. As Stern [1981] has observed, Lewis makes use of the concept of physical law in characterizing the notion of physical possibility in [1973], but the analysis he elaborates seems to be viciously circular. Moreover, although they have advanced a number of logical calculi intended to capture the logical properties of subjunctive and counterfactual discourse, these systems appear to be unsuited for the purpose of reflecting the use of these conditionals *within scientific language*; thus, Lewis's $\mathscr{VC}$, Stalnaker's $\mathscr{C}2$, and Pollock's $\mathscr{SS}$, for example, all contain as theorems the principle that,

(8)   if '$p \cdot q$' is true, then 'if p were, then q would be' is true;

which may or may not be unobjectionable within the context of an analysis of ordinary language but would hopelessly obscure the difference between lawlike and accidental generalizations in scientific discourse and could not possibly be satisfied in an indeterministic universe. For reasons such as these, therefore, it would appear to be overly optimistic to assume that their results could be directly transferable, in principle, to the resolution of those difficulties which confront the theory of science.

As an alternative approach, therefore, the author, independently [1974d], [1976a] and in collaboration [1979], [1980], has undertaken the development of an account of lawlikeness within an intensional language framework which is intended to clarify and illuminate characteristic features of scientific discourse. The present enterprise is intended to provide a clarification, elaboration, and refinement of the efforts which have gone before but also as an extension of this approach into the domain of "scientific knowledge" by attempting to evaluate the testability of the conceptions which tie together the analysis of objects, properties, events, and dispositions from the perspective of the difficult problems of induction, confirmation, and corroboration. The ultimate adequacy of the ontological foundations upon which the conceptions of causal connections, nomological conditionals and lawlike sentences are constructed, after all, does not reside only in its capacity to shed light upon the logical structure of explanations, predictions, and theories, but also upon the extent to which these analyses can clarify and illuminate the theoretical character of scientific knowledge. The objective in what follows, therefore, is not systematic metaphysics, but instead the systematic integration of the philosophy of science with both the philosophy of language and the theory of knowledge.

A DISPOSITIONAL ONTOLOGY

Although various theoreticians have acknowledged an intimate connection between the concepts of lawlikeness, dispositions, and subjunctive and counterfactual conditionals, the explication of lawlikeness that is to be presented here makes the relations involved fully explicit. Among the most important aspects of this intensional explication are, *first*, an ontological rather than an epistemological conception of dispositionality; *second*, the recognition of dispositional properties of probabilistic (or, statistical) strength as well as of universal strength; and, *third*, a distinction between dispositions as permanent, and dispositions as transient, properties of things. The most important among them, of course, is the concept of a disposition itself, which should be formulated as follows:

(D1) A predicate is *dispositional* within a language $\mathfrak{L}^*$ if and only if the property it designates in $\mathfrak{L}^*$ (i) is a tendency (of universal or statistical strength) to bring about specific outcome responses when subjected to appropriate singular tests, where this tendency (ii) is an actual physical state of some thing, individually, or of some arrangement of things, collectively (should that predicate happen to be satisfied by any thing at all).

The shape and color of a billiard ball, the malleability and conductivity of a metal bar, and the strength and charge of an electric current, are *all* dispositions in this sense; consequently, this conception is intended to represent the kind of property dispositions are as features of the physical world independently of any consideration for the ease with which their presence or absence may be ascertained on the basis of experiential findings. Not all tendencies are dispositional, moreover, as means, modes, and medians frequently reflect. A number of aspects of this conception warrant immediate elaboration, namely:

(a) It is important to differentiate between dispositional properties, *per se*, and the possession of a dispositional property by any single thing; for the definition of dispositionality recommended here precludes the (all too tempting) presumption that properties must necessarily be instantiated, which appears to be equivalent to the classical Aristotelian presupposition for categorical terms. For the possession of such a property by some thing or arrangement of things is (merely) an historically contingent happening, which should properly be classified not as a disposition but as an event.

(b) A specific collection of things, of course, may possess tendencies to bring about various outcomes with varying strength when arranged in relevantly different ways, as a loose ignition wire convincingly displays. It is therefore important to recognize that, although the specific properties of the individual members of a collection of things do not change simply by virtue of the way they are arranged, collections of things may be collectively disposed to bring about different results, when arranged in different ways, where "arrangements" are ordered sets (or sequences) of things.

(c) Since dispositional properties specify the outcome responses that would be brought about (either invariably or probably) by the occurrence of singular relevant tests, a dispositional predicate itself may be informally defined as a set of ordered triples, each of which consists of a test trial description T^i, an outcome response description O^j, and a numerical strength specification r^k, i.e., $\{\langle T^1, O^1, r^1\rangle, \langle T^2, O^2, r^2\rangle, \ldots\}$, where the number of members of these sets is determined by the variety of different trial tests and and response outcomes that are ontological constituents of each specific disposition — a possibly infinite set.

Thus, it is important to recognize that dispositions may be either of two distinctive kinds. For the sentences, "This die and tossing device is fair", and, "That metal bar is magnetic", are both dispositional by virtue of attributing to some thing or arrangement of things tendencies to display appropriate response behavior under relevant test conditions, yet they differ insofar as one property is probabilistic, while the other is universal. Failure to attract small iron filings within the vicinity of the metal bar would constitute conclusive evidence that that thing was not magnetic (provided, of course, interfering conditions were not present), but the failure to show an ace as the outcome of a single toss with a die and tossing device would certainly not suffice to establish that that die and tossing device is not fair (even though, once again, the test was a test of the relevant kind). What this means, therefore, is that dispositional properties are amenable to varying degrees of strength, where, for example, being magnetic should be viewed as a tendency of *universal* strength to produce a certain result on each and every singular trial, while being fair, by contrast, is a tendency of *statistical* strength to display one or another result on each and every singular trial.

Some of these considerations have interesting implications for understanding certain elements of language, such as, in particular, sentential functions. Consider, for example, atomic sentences, which may be assumed to consist of the concatenation of a predicate constant with an individual constant, e.g., a definite description or a proper name. Then from a dispositional perspective, it is apparently theoretically important to draw a distinction between *predicate constants* (of whatever degree), such as, 'M', and the *sentential functions* that may be constructed from them, say, 'Mx'. For while the predicate constant 'M' designates some specific disposition, say, magnetism, the sentential function, 'Mx', stands for, 'x *is* magnetic'; but since the instantiation of any disposition by any thing at all is an historically contingent happening (from an ontological point of view), *the sentential function 'Mx' exhibits the form of an event attribution* — where the concatenation operation itself is a linguistic representation of an instantiation relation that is historically contingent — rather than (merely) designating this specific disposition.

The importance of the distinction thus displayed is further reinforced when consideration is given to Quine's [1960] theoretical separation of *eternal sentences* and *occasion sentences*, where a sentence is eternal if and only if its truth value remains the same from speaker to speaker and for all times. An occasion sentence of the form, 'Mx', say, may be converted into an eternal sentence, say, 'Mxt^1', by means of its concatenation with a temporal constant, such as 't^1', where such phrases and expressions as, "this die and tossing device", or, "that metal bar", are contextually-dependent functional equivalents for definite descriptions and/or proper names, i.e., they can be employed in their place as individual constants, under proper circumstances. Since occasion sentences are sentences that are true on some occasions, but false on others, while events (of various kinds) may be presumed to occur on some occasions, but not on others, it seems plausible to adopt the principle that *an eternal sentence describes an event if and only if that sentence itself is the eternal form of some occasion sentence*, i.e., occasion sentences are the basic elements of language for the description of events. Thus, the sentence, "John F. Kennedy is President of the United States on 21 November, 1963" (in the tenseless sense of "is") describes an event, since it is an eternal sentence form of the occasion sentence, "John F. Kennedy is President of the United States".

A sentence of atomic form, say 'Mxt^1' for example, will be *true within a language* $\mathfrak{L}^*zt$, therefore, if and only if some particular thing denoted by the individual constant 'x' (or its functional equivalent) has the property designated by the predicate constant 'M' (at the appropriate time 't^1'), on the basis of the descriptive and demonstrative conventions, i.e., the linguistic practices, $\mathfrak{L}^*$, upon which z relies at t. In order to articulate the dispositional conception of an *object*, however, it is necessary to introduce one further definition, namely:

(D2) A property χ is a *permanent property* of every member of some reference class K (under a description 'K' within a language $\mathfrak{L}*$) if and only if (i) there is no process or procedure — whether natural or contrived — by means of which a member of K (as described by 'K' in $\mathfrak{L}$) could lose the property χ without also losing membership in K; and (ii) the possession of χ by a member of K is not logically entailed by the reference class description 'K' (in $\mathfrak{L}*$).

The permanent properties of members of the reference class described by this description, "things whose elements have the atomic number 15", thus include being soluble in turpentine, in vegetable oils and in ether; having a garliclike odor, producing skin-burns on contact and igniting at 30°C; but they do *not* include being employed for military purposes, being sold under restricted conditions, or being referred to by the predicate, "white phosphorous", i.e., those transient properties things may gain or lose independently of their membership in this reference class.

These distinctions thus afford a plausible foundation for theoretically differentiating between different kinds of *"natural kinds"* as follows:

(a) So-called mass nouns, such as "water" and "red", may be characterized as designating natural *property* kinds, in the sense that, when "water" is taken as designating a (pure) liquid whose molecules have the structure H_2O and "red" is taken to designate light whose wavelength falls between 6100 and 7500 Å, the members of both reference classes possess many different permanent dispositional properties, such as having a freezing point of 32°F and boiling point of 212°F at sea level atmospheric pressure in the case of water or such as exhibiting distinct interference and diffraction patterns under particular experimental conditions in the case of red.

(b) Ordinary general nouns, such as "planet" and "amoeba", by comparison, may be envisioned as designating natural *object* kinds, in the sense that, while these predicates likewise specify reference classes whose members possess (what might be referred to as) integrated sets of dispositional properties (such as reproductively multiplying by fission for example or exerting a gravitational attraction that is directly proportional to its mass), included among them is the tendency to take *impermeable (or, cohesive) external forms*, on account of which the members of these classes occur as numerically distinguishable things. The difference between natural things of these distinct varieties itself thus appears to be dispositional in character.

The point of classifying amoebas and water as *natural* kinds, presumably, is that they represent integrated sets or arrangements of dispositions which happen to have been instantiated during the course of this world's "natural" history, i.e., as features of its physical and biological evolution, independently of contrivance by man. The "naturalness" of natural kinds should not be made too much of here, however; for Stutz Bearcats or permanent waves, although "artificial" object and property kinds when viewed from this perspective, are ontologically on a par with "natural" kinds to the extent to which permanent dispositional properties attend membership in any of these classes. The important theoretical distinction, therefore, is really the difference between the transient properties (such as being frozen) and the permanent properties (such as a freezing point of $32°F$) of things (such as pure water), rather than the distinction between "natural" and "artificial" kinds. For the members of classes of both kinds possess permanent and transient dispositional properties, a difference (it should be stressed) which depends upon a (presupposed) reference class description.

Given these considerations, the dispositional conceptions of (particular) kinds of things and of things of (particular) kinds may be defined as follows:

(1) (particular) *kinds of things* are specific arrangements of (permanent and transient) dispositions, independently of whether or not these distinctive sets of properties happen to be instantiated during the course of the world's history; and,

(2) *things of* (particular) *kinds*, therefore, are instantiations of some specific arrangement of (permanent or transient) dispositions that happen to occur during the course of the world's history, independently of whether the arrangements they instantiate are object *or* property kinds.

Existential hypotheses, such as, "There are amoebas", or, "Something is red", are true, therefore, if and only if those object and property kinds happen to be instantiated *at least once* during the course of the history of this world; indeed, from this point of view, object kinds and property kinds are theoretically on an equal footing, for the truth conditions for property hypotheses, such as, "Something is omnipotent", appear no less — and no more — obvious than those for object hypotheses, such as, "There are unicorns".

It is significant to notice that the dispositional conceptions of kinds of things and of things of kinds do not logically entail that instantiations of specific arrangements of dispositional properties must necessarily happen to be instances of *object kinds* as numerically distinguishable things. They therefore leave open the logical possibility that properties may be manifest *in the form of things that are not objects* (a possibility that appears to be of more than hypothetical interest relative to developments in quantum theory, which suggest that radiant energy may be a phenomenon of precisely such a kind). Even independently of consideration for contemporary physics, however, it may be a benefit of this ontology that it does not beg the question with regard to this specific issue; for surely the existence of photons as instantiations of arrangements of dispositions which are *not* numerically distinguishable things (insofar as any number with the same properties may occupy the same place at the same time) would appear to be a physical possibility that certainly must require *empirical* investigation.

The dispositional conception of things of (particular) kinds, by comparison, does logically entail that two different things are objects of the same kind when (and only when) they are instantiations of the same object kind arrangements, which therefore presupposes having a theoretical specification of some reference class description for an unambiguous determination. Ice cubes and water ponds are things of the same kind as members of the reference class *water*, yet might not be things of the same kind as members of such reference classes as *frozen water, rectangular figures*, and *duck sanctuaries*. Indeed, since individual things may lose or gain properties during the course of the world's history without necessarily losing their identity as those individual things (as a book may become worn and its pages torn or a professor might gain weight and his hair turn gray), the continued existence of specific objects as nevertheless the same objects requires theoretical identification of those objects as members of an underlying reference class as follows:

(3) *individual objects* are continuous sequences of instantiations of particular arrangements of dispositions during the course of the world's history, where any object ceases to exist as an object of a particular kind whenever it no longer instantiates the corresponding (reference class) description.

Indeed, were individual objects invariably identified as instances of the totality of properties they happen to instantiate at any one particular time, no object could survive a change in any property at all.

As we have observed, occasion sentences are the basic elements of language for the description of events. In order to individuate between events, therefore, the relevant standard to employ appears to be definite descriptions for events by means of occasion sentence conjunctions; for certainly event names (such as "Lincoln's assassination", and, "*Titanic*'s sinking") and event definite descriptions (such as, "the death of the 16th President", and, "the greatest peacetime naval disaster") may fulfill their intended roles (of individuating these specific events) on the basis of the principle that *event names and event definite descriptions name or describe a single such event if and only if every occasion sentence true of one of those occasions is also true of the other*. It is therefore ironic to discover that the principle of identity for events has been with us right along but has been misinterpreted as the principle of identity for objects; for the most widely used principle of identity, namely:

(A) $(x)(y)[(x=y) \equiv (F)(Fx \equiv Fy)]$;

as applied to events, asserts that two different event names or event definite descriptions 'x' and 'y' name or describe the same event if and only if every property of one of those events is also a property of the other, i.e., every occasion sentence true of one of those occasions must also be true of the other. [The same point, of course, may be made in strictly first-order language, i.e., without quantifying over predicate variables, by relying on rules of identity such as those advanced by Copi [1973], p. 138.]

It is important to observe that singular events are similar to individual objects in the sense that, were they invariably identified as instances of the totality of properties that happen to be instantiated at any one particular time, no event could survive a change in any property at all. Since principle (A) establishes an identification of precisely this kind, however, a distinction should be drawn between *atomic events* and *molecular events* by envisioning molecular events as sequences of atomic events, where the basic principle of continuity for molecular events is provided by a reference class description [usually consisting of an event kind description, e.g., an assassination (of some person) or a sinking (of some ship), together with an identification of the individual object or collection of things which instantiate that event kind, e.g., (the person) Lincoln or (the ship) *Titanic*]. The continued existence of specific events as the same events in spite of some change in the arrangements of dispositions they instantiate at different times, therefore, requires theoretical identification of those events as members of an underlying reference class once again as follows:

(4) *singular events* are continuous sequences of instantiations of particular arrangements of dispositions during the course of the world's history, where any event ceases to exist as an event of a particular kind whenever it no longer instantiates the corresponding (reference class) description.

From this point of view, therefore, the singular event referred to as *the sinking of the Titanic* is a molecular event consisting of a continuous sequence of instantiations of the event kind *sinkings* (of ships) by the individual object *Titanic*, which includes instantiations of the event kinds *collisions of ships with icebergs, insufficient and poorly-manned lifeboats*, and *loss of over 1500 lives*, in a sequence terminating with an instantiation of the event kind *ship at rest on the bottom of the sea*.

THE LANGUAGE FRAMEWORK: ℒ OR ℒ*?   43

Since an individual object is a continuous sequence of instantiations of certain arrangements of dispositions during the course of the world's history — where any object ceases to exist as a thing of a particular kind whenever it no longer instantiates the corresponding (reference class) description — the object(s) named or described by different proper names or definite descriptions are *numerically identical*, i.e., a single thing, if and only if:

(B) $(x)(y)[(x = y) \equiv (F)(t)(Fxt \equiv Fyt)]$;

that is, "For all x and all y, x is identical to y if and only if for all properties F and all times t, x is an F at t if and only if y is an F at t", i.e., the names or descriptions 'x' and 'y' name or describe the same object if and only if *the objects 'x' and 'y' name or describe instantiate the same ordered sets of dispositions in the same sequence of historical events*. The object named "Lincoln", is the same object described by the description, "the 16th President", therefore, if and only if every disposition instantiated by the object named "Lincoln" (such as *human being of male sex, store clerk, log splitter*, and *speech maker*) is also instantiated by the object described by "the 16th President" and, indeed, in the same historical sequence (beginning with *birth in a log cabin* in Kentucky in 1809 and ending with *death from a gunshot wound* in Washington, D.C., 1865), and conversely. The function that is filled by temporal variables in formulating principles of identity for objects, therefore, is theoretically indispensable, since it serves to locate continuous sequences of atomic events as features of singular events in the history of the world. [These considerations thus suggest that rules of identity such as Copi [1973] has proposed (which do not incorporate any temporal referents) cannot possibly be adequate for the identity of objects, but they are easy to repair; otherwise, things with markedly similar life-cycles, such as electrons or ball-bearings, might turn out to be "identical" when they actually are not.]

One result of this ontology thus appears to be that *proper names* properly apply to *individual objects* as numerically distinguishable things. On perfectly general grounds, I take it, the satisfaction of a definite description is both necessary and sufficient for the introduction of such a name within a language framework ℒ*; for, if such a name denoted *less* than one individual object the result would be a violation of the law of excluded middle (since otherwise '$(Ex)(Et)Fxt \lor (Ex)(Et) - Fxt$' is true, necessarily, by existential generalization, contrary to the hypothesis of null denotation), and if any such name denoted *more* than one individual object the result would be a violation of the law of non-contradiction (since otherwise '$Fxt \cdot -Fxt$' is true, necessarily, by the hypothesis of multiple denotation). Thus, the (historical) *exis-*

tence condition is ontologically significant in covertly suggesting that Kripke's [1971], [1972] conception of proper names as *rigid designators* (which denote the same unique object in every possible world) is philosophically sound if and only if at least one law is of statistical strength, since otherwise names that denote objects with identical dispositions and identical histories prior to time t cannot possibly denote objects which differ in their histories and dispositions subsequent to t — an issue we pursue. The (historical) *uniqueness condition* is similarly significant in demonstrating the ontological consequences attending the selection of some particular definite description for a proper name introduction; for since every individual object is a unique instance of every property that it instantiates at any time, *the kind of object which is thereby being named requires explicit specification*. Thus, the existential condition should specify an underlying reference class, K, of which that object is supposed to be a *uniquely different* member, Ψ, as, say,

(C) $(Ex)(Et)\{(Kxt \cdot \Psi xt) \cdot (y)(t)[(Kyt \cdot \Psi yt) \supset (x = y)]\}$;

which would indicate that class by implication. Otherwise, since any individual object ceases to exist as an object of that kind when it no longer instantiates the corresponding reference class description, it would be theoretically impossible to ascertain whether or not the object named by any particular name continues to exist as an instance of an atom, a molecule, a certain shape or a certain size, and so on. The specific descriptions that fulfill these conditions, of course, may vary from idiolect to idiolect, i.e., from $\mathfrak{L}^i$ to $\mathfrak{L}^j$.

Another consequence, however, appears to be of major theoretical importance; for these preliminary considerations provide the basis for an explicit characterization of the intensional conception of lawlike sentences, as the following definition reflects:

(D3) A sentence S is *lawlike* within a language $\mathfrak{L}^*$ if and only if (i) S is completely general, i.e., S is not limited to any finite number of instances on the basis of syntactical or semantical features of $\mathfrak{L}^*$; and (ii) S is essentially dispositional, i.e., S attributes a permanent dispositional property χ to every member of a reference class K (under an appropriate description 'K') in $\mathfrak{L}^*$.

The class of accidental generalizations in relation to this interpretation, therefore, not only encompasses those sentences that, while possibly true, are not *completely* general, such as "All the coins in my pocket are silver" and "All the members of the Greenbury School Board for 1964 are bald", but also those that, while completely general, are not *essentially* dispositional, such as "All students who attend universities in tropical climates are less than 7' tall" and "All moas die before the age of fifty" (an example which is discussed by Popper [1965], pp. 427–428), as Chapter 6 explains. Lawlike sentences are thus understood as logically contingent but unrestrictedly general subjunctive conditionals attributing dispositional properties to members of classes under appropriate descriptions. In order to appreciate the potential benefits of such a conception, somewhat more formal methods of investigation are required, which we shall now consider.

3. SYNTAX, SEMANTICS, AND ONTOLOGY

Statistical laws are typically envisioned as attributing a particular limiting frequency to the occurrence of an event of a specific kind within a particular sequence, i.e., a sequence consisting of the ordered members of a certain reference class, where the ordering relation, at least in the case of physical events, is supposed to be the temporal relations that obtain between these events. The result of thus construing statistical laws as "relative frequencies" within temporally extended physical sequences is to reinforce the view that lawful generalizatons are essentially amenable to being determined by numerical count: to ascertain whether or not a particular regularity is universal or statistical in character, all one has to do is to determine the ratio between the number of members n belonging to the reference class, R, and the number of members m belonging to the attribute class, A. If every member of that reference class is also a member of that attribute class, the regularity is universal; if not, then the relative frequency indicates the statistical distribution of that attribute within that reference class.

What is needed, therefore, is to recognize that this way of looking at the world is fundamentally mistaken by virtue of failing to provide an adequate basis for distinguishing between (mere) correlations, i.e., accidental generalizations, and lawful regularities, i.e., genuine laws. From the perspective of a dispositional ontology, a conceptual shift is in order, since now statistical laws as well as universal laws are essentially of universal form; for a statistical law under the dispositional interpretation no longer simply affirms that a certain percentage of the reference class also belongs to the attribute class. What it asserts instead is that *every member of the reference class possesses a certain dispositional property*, which in the case of statistical laws should be envisioned as a *statistical* disposition, and in the case of universal laws as a *universal* disposition. The difference, therefore, is not a matter of the proportion of members of a reference class which possess the appropriate attribute, but rather of how strong the dispositional tendency is that is possessed by every member of that reference class.

Not the least of the reasons for adopting such a dispositional ontology, therefore, is that it establishes the foundation for a unified account of the logical structure of lawlike sentences from the point of view of dispositional properties. This analysis conforms to the characterization endorsed both by Hempel [1965] and by Goodman [1965] to the extent to which it assumes that a sentence S is a physical law if, and only if, (a) S is lawlike and (b) S is true, where S is supposed to be lawlike only when (i) S is completely general and (ii) S is capable of providing support for counterfactual and subjunctive conditionals. It goes beyond this account, however, in maintaining that only *essentially dispositional* sentences attributing permanent dispositional properties to every member of appropriately described reference classes are theoretically able to fulfill this conception, i.e., as logically contingent and unrestrictedly general sentences, which *are* capable of providing support for conditionals of these kinds. To determine whether or not a given sentence S is lawlike, therefore, consideration must be directed not only to its syntactical form but also to its semantical kind, as the following should explain.

Since a property χ is a permanent disposition of a member x of a reference class K (under a description 'K' in a language $\mathfrak{L}^*$) at t if and only if, although being a member of χ is not logically entailed by the description 'K' of that class in $\mathfrak{L}^*$, x would not be a member of K at t if x were not a member of χ at t, the *basic form* of a lawlike sentence under this dispositional interpretation is that of a *subjunctive conditional*, which may be represented by the non-extensional "fork" operator as follows:

(A) $(x)(t)(Kxt \mathrel{\ni\mkern-6mu-} \chi xt)$;

which asserts, ⌜For all x and all t, if x were K at t, then x would be χ at t⌝; for example, "For all x and all t, if x were gold at t, then x would be malleable at t", and, "For all x and all t, if x were polonium^{218} at t, then x would have a half-life of 3.05 minutes at t". Since the truth-value of a subjunctive conditional is not determined solely by the truth-values of its atomic constituents when *either* its consequent is true *or* its antecedent is false, moreover, the fork is *not* a truth-functional operator. Even if, say, "This is gold" and "This is sold in a jewelry store", for example, are both true, the truth-value of the subjunctive compound, "If this were gold, then this would be sold in a jewelry store", is not thereby determined. The fork is an essentially stronger operator than the horseshoe, since the truth of a corresponding material conditional is a necessary condition for the truth of a subjunctive. Nevertheless, the subjunctive conditional is not the strongest intensional connective this explication requires.

A (partial) explicit definition for a dispositional predicate χ may be obtained within the framework of an intensional language $\mathfrak{L}^*$ by introducing one further sentential operator whose logical properties are stronger than those of the subjunctive by virtue of embracing a primitive *brings about* relation as well, i.e., a *causal conditional*. A claim that the occurrence of an event of kind T^i causes the occurrence of an event of kind O^i – that dropping a fragile glass onto concrete, for example, causes its breaking – means not only that such an object *would* break if it *were* dropped, but also that dropping it would *bring about* its breaking. Since there are dispositions of statistical as well as of universal strength, the causal conditional is *probabilistic*, i.e., it is applicable with degrees of strength n whose values may range through varying statistical strengths n from zero to one, i.e., '$\ni_n$', to universal strength u, i.e., '$\ni_u$' (*not* to be confused with probabilities of one), where the appropriate numerical value is determined by the strength of the disposition that is involved. A dispositional predicate 'χ' may therefore be partially defined within $\mathfrak{L}^*$ by employing the non-extensional "n-fork" operator as follows:

(B) x is χ at $t =_{df} (T^1 xt \ni_m O^1 xt^*) \cdot (T^2 xt \ni_n O^2 xt^*) \cdot \ldots;$

where 'T^1', 'T^2', and so on describe relevant test conditions, 'O^1', 'O^2', and so forth describe appropriate outcome responses, and 'm', 'n', ..., assume the corresponding value of strength specification. When a melting point of 1063°C is a disposition of universal strength, for example, that predicate may be (in part) explicitly defined by specifying appropriate tests and outcome responses such as: "heating x to 950°C at t would invariably bring about x's becoming soft and pliable at t^*; heating x to 1013°C at t would invariably bring about x's becoming elastic and ductile at t^*; heating x to 1063°C at t would invariably bring about x's melting at t^*"; and so on.

Although the basic form of a lawlike sentence *is* that of a subjunctive conditional, therefore, the causal character of lawlike sentences only emerges in full force when the dispositional predicate 'χ' in sentences of form (A) is replaced by one of its defining conjuncts, thereby exhibiting the non-extensional sentence form,

(C) $(x)(t)[Kxt \ni (T^i xt \ni_n O^i xt^*)]$;

which asserts, ⌜For all x and all t, if x were K at t, then the strength of the dispositional tendency for T^i-ing x at t to bring about O^i-ing x at t^* is n⌝; or, less formally, ⌜For all x and all t, if x were K at t, then T^i-ing x at t would either *invariably* or *probably* bring about x's O^i-ing at t^*⌝. Among the sentences which reflect the causal significance of the lawlike sentence, "The melting point of gold is 1063°C", therefore, is one that asserts,

SYNTAX, SEMANTICS, AND ONTOLOGY

(1) For all x and all t, if x were gold at t, then heating x to $1063°C$ at t would invariably bring about its melting at t^*;

and among those sentences for the lawlike sentence, "The half-life of polonium^{218} is 3.05 minutes", is one that asserts,

(2) For all x and all t, if x were polonium^{218} at t, then a time trial of three minutes duration at t would very probably bring about the loss of nearly half the mass of x by t + three minutes.

Although lawlike sentences are therefore expressible either as subjunctive conditionals or as causal conditionals, their explanatory significance, for each case, depends upon the precise character of the phenomenon to be explained, as we shall subsequently ascertain.

NOMOLOGICAL CONDITIONALS

Let us differentiate terminologically between lawlike sentences (whether subjunctive or causal conditional in form) and instantiations of sentences of this kind by referring to the latter as "nomological conditionals" (of either subjunctive or causal conditional form). We may further distinguish the two kinds of nomological conditionals as "simple" and as "causal", respectively. It obviously follows that lawlike sentences are completely general nomological conditionals, i.e., all nomological conditionals are instantiations of lawlike sentences. It also follows (less obviously, perhaps) that nomic conditionals of neither kind are logically true, since the lawlike sentences which they instantiate attribute permanent properties to the members of reference classes, where the possession of those properties is never implied by the descriptions of those classes. Let us assume that 'at' represents the concatenation of an ambiguous name with an ambiguous time, i.e., names and times ranging over all nameable things and specifiable times; that 'ct' represents the concatenation of an individual constant with a determinate time; and, that 'bt' represents either 'at' or 'ct' indifferently. Then some of the basic forms of (what we may refer to as) *scientific conditionals* can be specified in general, namely:

(1)   *lawlike sentences:*
       (a) 'simple' forms: $(x)(t)(Kxt \ni \chi xt)$;
       (b) 'causal' forms: $(x)(t)[Kxt \ni (T^i xt \ni_n O^i xt^*)]$;
                                  $(x)(t)[(Kxt \cdot T^i xt) \ni_n O^i xt^*]$;
                                  $(x)(t)[T^i xt \ni (Kxt \ni_n O^i xt^*)]$;

(2) *nomological conditionals:*
(a) 'simple' forms: $Kbt \ni\!\!\!- \chi bt$;
(b) 'causal' forms: $Kbt \ni\!\!\!- (T^ibt \ni\!\!\!-_n O^ibt^*)$;
$(Kbt \cdot T^ibt) \ni\!\!\!-_n O^ibt^*$;
$T^ibt \ni\!\!\!- (Kbt \ni\!\!\!-_n O^ibt^*)$.

The crucial feature of these intensional formulations with respect to rules of inference, moreover, appears to be that a sentence of kind (2) logically entails a corresponding sentence of kind (1) *regardless of whether that sentential schema is instantiated by an individual constant or by an ambiguous name* [so long as the antecedent of that conditional as a whole does not entail the instantiation of its consequent (or, of its consequent's negation) for any specific individual case, i.e., it must be logically contingent for every such case], as the following considerations are intended to explain.

Let us assume that *a predicate 'F' is nomically relevant to an outcome predicate 'χ' or 'O^i', relative to reference class description 'K' or '$K \cdot T^i$', in $\mathfrak{L}^*$*, if and only if (a) neither 'χ' nor 'O^i' nor their negations is entailed by 'K' or by '$K \cdot T^i$' in $\mathfrak{L}^*$, respectively; and (b) *either* (i) if 'K' entails '$G \cdot F$' in $\mathfrak{L}^*$, then χ is a permanent property of every member of K, but not of every member of $G \cdot -F$, or conversely; and if 'K' does not entail 'F' or '$-F$' in $\mathfrak{L}^*$, then χ is a permanent property of every member of $K \cdot F$, but not of every member of $K \cdot -F$, or conversely; *or* (ii) if '$K \cdot T^i$' entails '$G \cdot F$' in $\mathfrak{L}^*$, the strength of the tendency for a single trial of kind $G \cdot F$ to bring about an outcome of kind O^i differs from that for a single trial of kind $G \cdot -F$; and if '$K \cdot T^i$' does not entail 'F' or '$-F$' in $\mathfrak{L}^*$, the strength of the tendency for a single trial of kind $K \cdot T^i \cdot F$ to bring about such an outcome differs from that for a single trial of kind $K \cdot T^i \cdot -F$. Then the following condition is satisfied by any scientific conditional which happens to be *true*:

> *The Requirement of Maximal Specificity:* If a nomically relevant predicate is added to the reference class description of a scientific conditional S which is true in $\mathfrak{L}^*$, then the resulting sentence S^* is such that either S^* is no longer true in $\mathfrak{L}^*$ (by virtue of the fact that its antecedent is now self-contradictory) or S^* is logically equivalent to S in $\mathfrak{L}^*$ (by virtue of the fact that that predicate was already entailed by the antecedent of S).

An alternative but equivalent formulation of this requirement (which some might find intuitively more appealing), moreover, may be advanced as follows: if 'p' is a true scientific conditional, 'K' is the reference class description of 'p' and 'F' is any predicate which is nomically relevant to the truth of 'p', then either '$(x)(t)(Kxt \supset Fxt)$' or '$(x)(t)(Kxt \supset -Fxt)$' is a logical truth; that is, for every nomically relevant predicate 'F', relative to 'p', either it or its negation must be entailed by the reference class description of 'p' if 'p' is true. In either formulation, however, this requirement is a generalization of Hempel's ([1968], p. 131) conception of *a maximally specific predicate related to a specific outcome predicate*, combined with an alternative analysis of the appropriate relevance condition, i.e., nomic rather than statistical relevance, as the chapter which follows conveys.

This requirement has considerable importance in scientific contexts, moreover, insofar as (i) it explains the significance of *ceteris paribus* clauses when some nomically relevant conditions are not specified; (ii) it facilitates the definition of the notion of a *severe test* of an hypothesis or theory (by varying the properties that must be irrelevant if that hypothesis or theory is true); and, (iii) it contributes to the clarification of the concept of *initial conditions* with respect to singular trials of statistical dispositions, in particular. Notice especially that, for scientific conditionals of 'causal' form, a predicate 'F' is *causally relevant* to an outcome predicate 'O^i', relative to a reference class description '$K \cdot T^i$' in $\mathfrak{L}^*$, for example, when

$$(x)(t)[(Kxt \cdot T^ixt \cdot Fxt) \ni_m O^ixt^*] \cdot (x)(t)[(Kxt \cdot T^ixt \cdot -Fxt) \ni_n O^ixt^*],$$

that is, so long as the strength of the tendency for $Kxt \cdot T^ixt \cdot Fxt$ to bring about O^ixt^* differs from that of $Kxt \cdot T^ixt \cdot -Fxt$ for that same outcome, i.e., provided $m \neq n$. No distinction is drawn here between (what is sometimes referred to as) "positive" causal relevance and "negative" causal relevance, notions which depend upon criteria requiring, in effect, that satisfaction of predicate 'F' either *increases* or *decreases* the strength of the tendency for $Kxt \cdot T^ixt$ to bring about O^ixt^*, respectively. In these circumstances, however, unless the requirement of maximal specificity is already satisfied by the original reference class description, '$Kxt \cdot T^ixt$', there will be no determinate strength for the tendency of $Kxt \cdot T^ixt$ to bring about O^ixt^* to compare with the tendency of $Kxt \cdot T^ixt \cdot Fxt$ to bring about $O^ixt^* -$ unless this comparison is assumed to be with $Kxt \cdot T^ixt \cdot -Fxt$! Although several other authors such as Suppes [1970] and Cartwright [1979] assign a crucial role to "positive" causal relevance, the considerations that follow, including, especially, those advanced in Chapters 4 and 5, indicate that "causal relevance" as such is the more fundamental notion.

The difference in emphasis between Suppes [1970] and Cartwright [1979] and the present investigation with respect to "positive" causal relevance, moreover, may arise (at least, in part) from divergent theoretical objectives. Suppes has characterized his program in relation to Hume's conception of 'causal relations' (Suppes [1970], p. 10) as follows:

> The claim being made here is that in restricting himself to the concept of constant conjunction, Hume was not fair to the use of causal notions in ordinary language and experience. Roughly speaking, the modification of Hume's analysis I propose is to say that one event is the cause of another if the appearance of the first event is followed with a high probability by the appearance of the second, and there is no third event that we can use to factor out the probability relationship between the first and second events.

The conception advocated here, by comparison, eliminates the requirement of *high* probability, for which, in general, there appears to be no theoretical warrant − except, perhaps, in relation to probabilistic (or "statistical") predictions of singular events within the knowledge context $\mathcal{K}zt$, as explained in Chapters 9 and 10. To paraphrase Suppes himself, therefore, the alternative conception that is one consequence of this investigation could be similarly formulated as follows:

> Roughly speaking, the modification of Hume's analysis recommended here is to say that one event is the (probabilistic) cause of another if the occurrence of the first event would be followed with probability by the occurrence of the second, where there are no other events which could be used to factor out this probabilistic relationship between the first and second events.

Abandoning such high probability requirements within the context of an explicitly intensional conception of causal relations satisfying maximal specificity conditions should therefore provide a more promising explication (cf. Salmon [1980]). [Hume's position is appraised in Chapter 7.]

Since the conjunction of predicates which are not nomically relevant will obviously not affect the truth of a scientific conditional, a scientific conditional that is true will remain true, regardless of the addition of any number of *nomically irrelevant* predicates to its reference class description (as, for example, the strength of the disposition for salt to dissolve in water or for men to become pregnant does not change when the salt is hexed or the men take birth-control pills). Analogously, although any specific individual c at any particular time t will possess unique characteristics unlike those of anything else at any other time, a nomological conditional will be true of c at t only if those additional characteristics are either logically redundant or nomically irrelevant (an object made of copper, for example, remains a conductor of electricity whether or not it happens to be the 1976-D penny in my pocket or the last coin inserted in a parking meter in New York City on 17 March, 1977). Since the truth-value of a nomological conditional may change if an extensionally equivalent description, 'K^*', is substituted for some original reference class description, 'K' (where every member of K is a member of K^* and conversely), while co-referring denoting expressions 'c' and 'c^*' are interchangeable without altering the value of a true nomological conditional, *nomological conditionals and lawlike sentences may be characterized as descriptively opaque, but referentially transparent.* [For related discussion, see Smokler [1979].]

The remarkable strength of the maximal specificity requirement is reflected by the striking triviality of appropriate principles of inference for quantified sentences regulating inferences from nomological conditionals to lawlike sentences and conversely, which may be illustrated by the following 'simple' examples:

EXISTENTIAL INSTANTIATION (EI): Where 'bt' represents the concatenation of an individual constant and a determinate time or the concatenation of an ambiguous name and an ambiguous time, from a previously obtained existentially quantified nomological conditional of the form,

$$(Ex)(Et)(Kxt \ni\!\!\!- \chi xt)$$

a singular sentence of the following form may be inferred,

$$Kbt \ni\!\!\!- \chi bt,$$

provided that ... bt ... results from ... xt ... by replacing each occurrence of xt with bt (making no other changes); and,

UNIVERSAL GENERALIZATION (UG): Where 'bt' represents either the concatenation of an individual constant and a determinate time or the concatenation of an ambiguous name and an ambiguous time, from a previously obtained singular nomological conditional of the form,

$$Kbt \ni\!\!\!- \chi bt,$$

a universally quantified sentence of the following form may be inferred,

$$(x)(t)(Kxt \ni\!\!\!- \chi xt),$$

provided that ... xt ... results from ... bt ... by replacing each occurrence of bt with xt (making no other changes).

The corresponding principles of inference for *Existential Generalization* (EG) and for *Universal Instantiation* (UI), of course, are obvious enough to require no exemplification; but notice, in particular, that these patterns of argument presuppose the existence of no more than (at least) one object during the course of the world's actual history.

The unrestricted generalization of nomological conditionals to obtain corresponding lawlike sentences, of course, does not hold for *every* dispositional attribution — for example, 'χbt' does not entail '$(x)(t)\chi xt$' — but *only* for nomological conditionals. The special provision concerning non-entailment for any specific instance of that consequent (or, its negation), moreover, appears to be important, since otherwise that principle would license such inferences as, for example, from '$(Kct \cdot Tct \cdot -Obt^*) \ni_n Oct^*$', to '$(x)(t)[(Kxt \cdot Txt \cdot -Obt^*) \ni_n Oxt^*]$' by generalization on the constants 'ct', to '$(Kbt^* \cdot Tbt^* \cdot -Obt^*) \ni_n Obt^*$' by universal instantiation (for $t = t^*$), an unacceptable result. Perhaps it should be observed as well that dispositional predicates are not being analyzed by means of simple subjunctives on this account; indeed, dispositional predicates as such appear intact in the *consequent* of subjunctive conditionals. Instead, an entire new family of connectives is being introduced for the purpose of representing causal conditionals of different strengths. Thus, it is important not to confuse the *fork* operator, '$\ni$', as a (trivially) innovative symbol for representing subjunctive conditionals, with the *n-fork* operator, '$\ni_n$', as a (non-trivially) innovative symbol for representing the (probabilistic) causal conditional employed to analyze dispositional predicates themselves.

The dispositional rationale for possible worlds emerges from the recognition that, relative to a specific reference class description, it may not be physically impossible for every member of that class to possess some transient disposition, but it is physically impossible for any member of such a class to not possess any of its permanent dispositional properties. As a result, from this point of view, *a world is physically possible (in relation to the actual world* **W**) *if and only if the permanent dispositional properties of things do not change from world to world*, i.e., they remain permanent properties of all and only things of the same kind as they are in the actual world **W**. Physical necessities are consequently described by sentences that cannot be made false (independently of definitions), but which are nevertheless not true on syntactical and definitional grounds alone. Thus, we may stipulate that *possible worlds* are described by sequences of sentences, where a world$_i$ is the same world as a world$_j$ if and only if every sentence true in world$_i$ is also true in world$_j$ and conversely. The concepts of logical, physical, and historical pos-

SYNTAX, SEMANTICS, AND ONTOLOGY 55

sibility, therefore, are themselves specifiable in relation to a language framework $\mathfrak{L}^*$ and the actual world **W**, on the assumptions that N is a set of lawlike sentences in $\mathfrak{L}^*$ containing every such sentence true in **W** and H is a set of historical descriptions (of atomic events occurring prior to t) in $\mathfrak{L}^*$ containing each such sentence true in **W**, relative to t, as follows:

(i) for any sentence S in $\mathfrak{L}^*$, *S describes a logically possible [necessary, impossible] state of affairs or 'world' (relative to $\mathfrak{L}^*$)* if and only if it is not the case that $\mathfrak{L}^*$ entails (or implies) not-S [it is the case that $\mathfrak{L}^*$ entails (or implies) S, it is the case that $\mathfrak{L}^*$ entails (or implies) not-S];

(ii) for any sentence S in $\mathfrak{L}^*$ and any such set N, *S describes a physically possible [necessary, impossible] state of affairs or 'world' (relative to $\mathfrak{L}^*$ and N)* if and only if it is not the case that $\mathfrak{L}^*$-and-N entails (or implies) not-S [it is the case that $\mathfrak{L}^*$-and-N but not $\mathfrak{L}^*$ alone entails (or implies) S, it is the case that $\mathfrak{L}^*$-and-N but not $\mathfrak{L}^*$ alone entails (or implies) not-S]; and,

(iii) for any sentence S in $\mathfrak{L}^*$ and any such sets N and H, *S describes an historically possible [necessary, impossible] state of affairs or 'world' (relative to $\mathfrak{L}^*$, N and H)* if and only if it is not the case that $\mathfrak{L}^*$-and-N-and-H entails (or implies) not-S [it is the case that $\mathfrak{L}^*$-and-N-and-H but not $\mathfrak{L}^*$-and-N alone entails (or implies) S, it is the case that $\mathfrak{L}^*$-and-N-and-H but not $\mathfrak{L}^*$-and-N alone entails (or implies) not-S];

where none of these sentences is true or false on definitional grounds alone. So, if N is the set of all true lawlike sentences and each member of N happens to be a general law of *universal strength*, i.e., a completely general sentence attributing a permanent dispositional property of universal strength to every member of a reference class, then the future history of any world is deducible from any such set H for any such time t, i.e., there will be only *one* historically possible world, given those conditions, relative to **W**; otherwise, at least one statistical law is true and different worlds will be possible at t, relative to **W**.

From a semantical point of view, of course, the major problem is that of providing an adequate set of truth conditions for nomological conditionals of 'simple' and of 'causal' forms, which poses a formidable task. An attempt to satisfy this objective (at least in part) has been advanced by Kyburg [1974], [1978] who focuses upon *causal conditionals of probabilistic (or statistical) character*, i.e., sentences of the form, '$(Kxt \cdot T^i xt) \ni_n O^i xt^*$', in relation to the schemata under consideration here. Kyburg suggests that a sentence of the form, '$(Kat^1 \cdot T^i at^1) \ni_n O^i at^1 *$', is true of the world **W** just in case, for almost every "lawful future world" in which there is an infinite sequence of repetitions of the test conditions, $Kxt \cdot T^i xt$, the limiting frequency with which the outcome, $O^i xt^*$, accompanies those trials "over the long run" is n.

These truth conditions may be made more precise, since for Kyburg a "lawful future world" is one in which (a) every non-statistical (or non-probabilistic) nomological conditional true of **W** in $\mathfrak{L}^*$, and (b) every historical statement true of **W** in $\mathfrak{L}^*$ (relative to the present t), is true. Thus, let $W_1, W_2, \ldots$ be a sequence of lawful future worlds such that, for each world W_j included in this sequence, there is an infinite sequence of individual constant/temporal constant pairs, $\langle c_j^1 t_j^1 \rangle, \langle c_j^2 t_j^2 \rangle, \ldots$ such that all sentences of the form, '$Kc_j^m t_j^m \cdot T^i c_j^m t_j^m$', are true of W_j. Then '$(Kat^1 \cdot T^i at^1) \ni_n O^i at^1 *$' is true of **W** if and only if "almost every world" W_j in the sequence of lawful future worlds satisfies the following condition:

(I) $\lim_{k \to \infty}$ [the number of members of $\{c_j^m t_j^m : m \leq k$ and $O^i c_j^m t_j^m *$ is true of $W_j\}]/k = n$.

The notion of "almost every world" incorporated into these truth conditions may likewise be made precise; for '$(Kat^1 \cdot T^i at^1) \ni_n O^i at^1 *$' is also true of **W** just in case the following (logically equivalent) condition is fulfilled:

(II) $\lim_{k \to \infty}$ [the number of members of $\{W_m : m \leq k$ and W_m satisfies (I)$\}]/k = 1$.

An interpretation of this kind, of course, is enormously appealing to those who adhere to a Humean "constant conjunction" analysis of universal causation or to a Reichenbachian "limiting frequency" analysis of statistical causation, since, in particular, although Kyburg's semantics is not wholly extensional in embracing possible worlds, it nevertheless remains remarkably faithful to the spirit of Reichenbach and of Hume by preserving extensionality across possible worlds. The difference between the (intensional) dispositional conception and the (extensional) frequency definition of "probability", therefore, appears to be both straightforward and easy to explain, namely: the frequency definition is extensional relative to the actual world **W**, while the dispositional conception is extensional relative to a well-defined set of possible worlds, i.e., the infinite set of lawful future worlds $\{W_1, W_2, \ldots\}$.

Precisely because Kyburg specifies a set of necessary and sufficient truth-conditions which is exhausted by the limiting frequency properties of infinite sequences of infinite sequences of trials, however, his semantical analysis is theoretically suspicious; for it is not at all obvious that this interpretation succeeds in capturing the conception of dispositional properties as single case causal tendencies. It is immensely tempting to suppose that the critical flaw in Kyburg's analysis is that the truth-conditions for sentences describing one specific trial, such as, '$(Kct \cdot T^1ct) \ni_n O^1ct*$', may be completely determined by sequences of trials in which that specific trial, i.e., $Kct \cdot T^1ct$, does not occur at all. Nevertheless, since we have assumed that nomological conditionals – instantiated either by individual constants or by ambiguous names – logically entail corresponding lawlike sentences (and conversely), such a criticism appears to be theoretically unsound. It does suggest, however, that Kyburg's set of possible worlds itself may be excessively restrictive; for as a description of a unique event in some world's history, surely the truth-conditions for such a sentence *only require that the nomically relevant properties of that specific individual trial be preserved from world to world*, i.e., just those properties and objects whose presence is necessary and sufficient for the truth of its unique trial description. Since such a trial description is maximally specific, necessarily, if that nomological conditional is true, *the additional characteristics of all logically possible worlds – whether otherwise physically or historically possible or not – are theoretically irrelevant to the truth of that specific nomological conditional*. Consequently, the only restriction that needs to be imposed upon a sequence of logically possible worlds, $W_1, W_2, \ldots$, for the purpose of specifying the truth-conditions of nomological conditionals is that they be '*p-worlds*' in the following sense, namely: where 'p' is the antecedent of a nomological conditional within a language $\mathfrak{L}^*$, a logically possible world W_j is a p-world (relative to the actual world W) if and only if 'p' is true in W_j and a specific set of dispositional properties possessed by objects (or events) in W mentioned in 'p' is also possessed by those same objects (or events) in W_j. Precisely which dispositional properties have to be preserved from world W to world W_j, however, depends upon whether the nomological conditional in question is of 'simple' or of 'causal' form, as we shall subsequently ascertain.

Let us assume that '$Kat^1 \cdot T^iat^1$' is appropriately interpreted as an eternal sentence describing a unique historical situation, rather than as a sentential function which might represent repeatable trials with such experimental arrangements. As a result, the sentence, '$(Kat^1 \cdot T^iat^1) \ni_n O^iat^1*$', is understood to assert that the historically unique situation described by '$Kat^1 \cdot T^iat^1$' possesses a dispositional tendency of strength n to *bring about* the historically unique outcome response described by 'O^iat^1*'. The claim that historically unique situations actually possess these singular causal tendencies, moreover, may be taken to be the fundamental tenet of the single case analysis of dispositional properties under consideration here; but it is most important to observe that, with respect to sentences interpreted in this way, the truth or falsity of the vast majority of (a) non-statistical (or, probabilistic) nomological conditionals and of (b) historical sentences true of **W** up to time t (in this case, t^1) is theoretically irrelevant to the truth of that specific nomological conditional. In order to capture this difference, therefore, let us assume that the truth of sentences of the form, '$p \ni_n q$', is not affected by the truth of any sentences in any worlds *except* the sentences 'p' and 'q' themselves as follows: Let us assume a set of p-worlds of the kind that has been described, where S is a non-decreasing sequence of finite sets of p-worlds, such that, if $i \leq j$, then $S_i \subseteq S_j$. Then the sentence '$p \ni_n q$' is true of **W** if and only if "almost every sequence" S of possible sets of p-worlds satisfies the following condition:

(III) $\lim_{k \to \infty}$ [the number of worlds in S_k in which 'q' is true]/ the number of worlds in $S_k = n$.

The notion of "almost every sequence" incorporated into these truth-conditions may likewise be made precise; for the sentence, '$p \ni_n q$', is true of **W** just in case the following (logically equivalent) condition is fulfilled:

(IV) $\lim_{k \to \infty}$ [the number of members of $\{(S^m) : m \leq k$ and (S^m) satisfies (III)$\}$]/the number of members of $k = 1$.

Since the specific sets of dispositional properties possessed by objects (or events) mentioned by 'p' are preserved form world to world, the p-worlds that are members of S include (but are not restricted to) Kyburg's "lawful future worlds". From this point of view, therefore, it seems clear that, in relation to the class of sentences under consideration, Kyburg's conditions (I) and (II) provide necessary but *not* sufficient conditions for the truth of 'causal' nomological conditionals of statistical (or, probabilistic) form. Only conditions (III) and (IV) appear to be logically adequate to represent the conception of dispositional properties as single case causal tendencies; for these reasons, therefore, conditions (III) and (IV) afford crucial ingredients for the formal calculus to follow.

A theoretically adequate characterization of non-extensional operators of this kind within an intensional language framework $\mathfrak{L}^*$, of course, necessitates some specification of the formal syntax and the model theory which define them. The introduction of such a formal system, i.e., the probabilistic causal calculus $\mathscr{C}$, shall be conducted in three phases, namely: first, the fork operator (or subjunctive conditional) will be added to a classical sentential calculus to provide the calculus of subjunctive conditionals $\mathscr{S}$; second, the u-fork (or universal strength causal conditional) will be added to calculus $\mathscr{S}$ to produce the calculus of causal conditionals of universal strength $\mathscr{U}$; and, third, the

n-fork (or probabilistic strength causal conditional) will be added to calculus $\mathscr{U}$ to provide the probabilistic causal calculus $\mathscr{C}$. Although the properties of each component of this formal system (concerning consistency, soundness, and completeness) which have been established will be specified, they shall not be proven here; instead, those who wish to pursue these points in detail should consult Fetzer and Nute [1979] and [1980] for discussions and proofs. The intended interpretation of this system will be elaborated in the section which follows, although its broad outlines should be familiar already; nevertheless, (a) each calculus reflects the assumptions underlying its construction (in particular, the thesis that true scientific conditionals do satisfy the requirement of maximal specificity for reference class descriptions); (b) each calculus has been constructed to validate the principles of infererence of scientific conditionals that obtain relative to nomically irrelevant, as well as nomically relevant, predicates and sentences; and, (c) as a result, a few of the axiom schemata incorporated in these calculi may seem unintuitive upon preliminary consideration which, upon further reflection, appear to be essential components of an adequate analysis of the kind undertaken here.

A PROBABILISTIC CAUSAL CALCULUS

Let us begin by assuming a familiar set of primitive signs and formation rules as the foundation for constructing a formal system which we may denote as the calculus of subjunctive conditionals $\mathscr{S}$ as follows, namely: (i) countably many sentence letters $A, A', ..., B, B', ...$, (ii) the logical signs '$-$' and '$\supset$', i.e., the negation sign and the material conditional; and, (iii) parentheses, braces, and brackets for punctuation, i.e., the primitive signs of a classical sentential calculus. The language of calculus $\mathscr{S}$ is produced by adding a new, primitive dyadic sentential operator, "$\ni\!\!-$", the fork, to determine the well-formed formulae of $\mathscr{S}$ as follows: an object x is a formula of $\mathscr{S}$ if, and only if, x belongs to every set S which contains every sentence letter and also contains '$-p$', '$p \supset q$', and '$p \ni\!\!- q$' whenever it contains 'p' and 'q'. Let us assume the standard contextual definitions for '$\cdot$', i.e., the conjunction sign; 'v', i.e., the disjunction sign; and, '$\equiv$', i.e., the material biconditional; and let us define the modal operators, '$\Box p$' and '$\Diamond p$' contextually as '$-p \ni\!\!- p$' and '$-(p \ni\!\!- -p)$', respectively. The rules of inference for calculus $\mathscr{S}$, moreover, are *modus ponens*, MP, and the *Gödel rule of necessitation*, NEC, i.e., if 'p' is a theorem, then '$\Box p$' is a theorem.

A formula of $\mathscr{S}$ is an axiom of $\mathscr{S}$ if and only if it is tautologous or has one of the following forms:

A1. $\Box(p \supset q) \supset (\Box p \supset \Box q)$;
A2. $\Box(p \supset q) \supset (p \mathrel{\Rightarrow} q)$;
A3. $(p \mathrel{\Rightarrow} q) \supset (p \supset q)$;
A4. $[p \mathrel{\Rightarrow} (q \supset r)] \supset [(p \mathrel{\Rightarrow} q) \supset (p \mathrel{\Rightarrow} r)]$;
A5. $[\Box(p \supset q) \cdot (q \mathrel{\Rightarrow} r)] \supset [\Diamond(p \cdot r) \supset (p \mathrel{\Rightarrow} r)]$;
A6. $[(p \mathrel{\Rightarrow} r) \cdot (q \mathrel{\Rightarrow} r)] \supset [(p \lor q) \mathrel{\Rightarrow} r]$.

$\mathscr{S}$ thus defined is an extension of a calculus which Nute [1975] has developed. For any well-formed formula 'p' and set Γ of wffs, '$\vdash_{\mathscr{S}} p$', '$\Gamma \vdash_{\mathscr{S}} p$', '$\Gamma$ is $\mathscr{S}$-consistent', and 'Γ is maximally $\mathscr{S}$-consistent' are defined in the usual ways. Since $\mathscr{S}$ is a normal extension of the familiar modal system $\mathfrak{M}$ (as are all the calculi presented here), a Lindenbaum lemma, a compactness theorem, a deduction theorem, and other familiar results likewise obtain for $\mathscr{S}$.

$\mathscr{S}$ resembles various calculi that have been recommended as analyses of the logical structure of ordinary language subjunctive conditionals in several respects; in particular, it is an extension of Lewis's $\mathscr{VW}$, although it is not an extension of Lewis's $\mathscr{VC}$ [1973], Stalnaker's $\mathscr{C}2$ [1968], or Pollock's $\mathscr{SS}$ [1976]. All of the latter systems lack A5 but contain

P1. $(p \cdot q) \supset (p \mathrel{\Rightarrow} q)$;

which, however, might serve a useful role in the analysis of ordinary language but would (as we have already observed) seriously obfuscate basic distinctions between lawlike and accidental generalizations. Both of these differences, of course, could be appropriate for their different contexts, insofar as A5 is a formal counterpart of the requirement of maximal specificity for subjunctive conditionals in scientific language, a requirement which ordinary language, in general, does not presuppose. But the propriety of P1 even in the context of the analysis of ordinary language subjunctive conditionals has been questioned by Bennett [1974]. Other calculi of subjunctive conditionals that differ significantly from $\mathscr{S}$ have been developed independently by Nute [1975a] and [1977]. These calculi not only lack A5 and P1 but contain

P2. $[(p \lor q) \mathrel{\Rightarrow} r] \supset [(p \mathrel{\Rightarrow} r) \cdot (q \mathrel{\Rightarrow} r)]$;

which is an intuitively attractive axiom schema for scientific conditionals. Further consideration of its merits will be reserved for later discussion. A number of preliminary results concerning calculus $\mathscr{S}$ may be given in summary:

SYNTAX, SEMANTICS, AND ONTOLOGY

THEOREM 1: the following are theorem schemata of $\mathscr{S}$:

a. $\vdash_s p \mathrel{\ni\mkern-6mu-} p$;
b. $\vdash_s \Box p \supset (q \mathrel{\ni\mkern-6mu-} p)$;
c. $\vdash_s \Box {-p} \supset (p \mathrel{\ni\mkern-6mu-} q)$;
d. $\vdash_s \Diamond p \supset [(p \mathrel{\ni\mkern-6mu-} q) \supset -(p \mathrel{\ni\mkern-6mu-} -q)]$;
e. $\vdash_s [(p \mathrel{\ni\mkern-6mu-} q) \cdot (q \mathrel{\ni\mkern-6mu-} p)] \supset [(p \mathrel{\ni\mkern-6mu-} r) \equiv (q \mathrel{\ni\mkern-6mu-} r)]$;
f. $\vdash_s [(p \cdot q) \mathrel{\ni\mkern-6mu-} r] \supset [p \mathrel{\ni\mkern-6mu-} (q \supset r)]$;
g. $\vdash_s [p \mathrel{\ni\mkern-6mu-} (q \cdot r)] \supset [(p \cdot q) \mathrel{\ni\mkern-6mu-} r]$;
h. $\vdash_s (p \mathrel{\ni\mkern-6mu-} -q) \vee \{[p \mathrel{\ni\mkern-6mu-} (q \supset r)] \supset [(p \cdot q) \mathrel{\ni\mkern-6mu-} r]\}$.

THEOREM 2: $\mathscr{S}$ is consistent.

A *model theory* for calculus $\mathscr{S}$ may be obtained as follows. An $\mathscr{S}$-*model* is an ordered triple $\langle I, f, [\,] \rangle$ such that for all wffs 'p', 'q', and 'r',

S1. $I \neq \emptyset$;
S2. $[\,]$ assigns to 'p' a subset $[p]$ of I;
S3. $[-p] = I - [p]$; $[(p \cdot q)] = [p] \cap [q]$; and so on, for all the other truth-functional connectives;
S4. f assigns to 'p' and each $i \in I$ a subset $f(p, i)$ of $[p]$;
S5. $[p \mathrel{\ni\mkern-6mu-} q] = \{i \in I : f(p, i) \subseteq [q]\}$;
S6. if $i \in [p]$, then $i \in f(p, i)$;
S7. if $f(p, i) = \emptyset$, then $f(q, i) \cap [p] = \emptyset$;
S8. if $f(q, i) \subseteq [r]$, $f(-(p \supset q), i) = \emptyset$, and $f(p \cdot r, i) \neq \emptyset$, then $f(p, i) \subseteq [r]$;
S9. $f(p \vee q, i) \subseteq f(p, i) \cup f(q, i)$.

Let us say that a wff 'p' is *true in an* $\mathscr{S}$-*model* $\langle I, f, [\,] \rangle$ just in case $[p] = I$. Let us say that a wff 'p' is $\mathscr{S}$-*valid* just in case 'p' is true in every $\mathscr{S}$-model.

The model theory for $\mathscr{S}$ is related to $\mathscr{S}$ in the usual ways; in particular $\mathscr{S}$ is both sound and complete; that is,

THEOREM 3: every theorem of $\mathscr{S}$ is $\mathscr{S}$-valid.

THEOREM 4: every $\mathscr{S}$-valid wff is a theorem of $\mathscr{S}$.

Notice from A2 and NEC that subjunctive conditionals are closely related to logical necessity; in fact, '$p \mathrel{\ni\mkern-6mu-} q$' is true whenever (but not only when) 'p' is a logically sufficient condition for 'q'. Quite a different relationship obtains between causal conditionals and logical necessity, however; for the set of *causally* sufficient conditions for 'q' and the set of *logically* sufficient conditions for 'q' are mutually exclusive. This circumstance, of course, strongly suggests that any attempt to analyze causal conditionals in terms of subjunctive conditionals cannot possibly succeed. In any case, no such analysis will be attempted here; instead, I shall introduce a new primitive symbol to represent the causal conditional of universal strength u and proceed to develop a formal calculus to define the logic of that connective.

The language of $\mathscr{U}$ is just the language of $\mathscr{S}$ augmented by an additional dyadic sentential operator, '$\ni_u$', the u-fork, to establish the well-formed formulae of $\mathscr{U}$ in the customary fashion. Whereas wffs of $\mathscr{S}$ are to be read as, 'if p were (tenselessly) the case, q would be (tenselessly) the case', these new wffs of $\mathscr{U}$ (of the form, '$p \ni_u q$') are to be read as, 'p would (tenselessly) invariably bring about q'. The rules of inference for $\mathscr{U}$ are MP and NEC as before. In addition to all tautologous wffs (of $\mathscr{U}$) and wffs of any one of the forms A1–A6, the axioms of $\mathscr{U}$ include all wffs of any of the following forms:

A7. $(p \ni_u q) \supset -\Box(p \supset q)$;
A8. $(p \ni_u q) \supset -(-q \ni_u -p)$;
A9. $(p \ni_u q) \supset (p \ni q)$;
A10. $[p \ni_u (q \supset r)] \supset [(p \ni_u q) \supset (p \ni_u r)]$;
A11. $[\Box(p \supset q) \cdot (q \ni_u r)] \supset \{\Diamond(p \cdot r) \supset [(p \ni_u r) \vee \Box(p \supset r)]\}$;
A12. $[(p \cdot q) \ni_u r] \supset [(p \ni_u q) \supset (p \ni_u r)]$;
A13. $\{[p \ni (q \ni_u r)] \cdot \Diamond(p \cdot r)\} \supset \{[(p \cdot q) \ni_u r] \vee \Box[(p \cdot q) \supset r]\}$;
A14. $[(p \ni_u q) \cdot \Box(q \supset r)] \supset [(p \ni_u r) \vee \Box(p \supset r)]$.

For any wff 'p' and set Γ of wffs, we define '$\vdash_u p$', '$\Gamma \vdash_u p$', 'Γ is $\mathscr{U}$-consistent', and 'Γ is maximally $\mathscr{U}$-consistent', in the usual ways.

The motivation for A7 has already been discussed. Many others have noted that the fork is not generally transposable; causal conditionals, however, are *never* transposable, a circumstance reflected by A8. A9–A11 are causal counterparts of axiom schemata of $\mathscr{U}$, while A12–A14 are causal counterparts of theorem schemata of $\mathscr{U}$. That the list of special axioms of the u-fork is considerably longer than the list for the fork is due primarily to the absence of a causal counterpart for A2; indeed, if such an axiom were a part of $\mathscr{U}$, all of A12–A14 would become derivable from other axioms of $\mathscr{U}$. A13, moreover, is of special importance as a partial implementation of the requirement of maximal specificity for causal conditionals.

Stronger versions of some of these axioms suggest themselves, as well as

SYNTAX, SEMANTICS, AND ONTOLOGY 63

other theses which are not included among the theorems of $\mathscr{U}$. Even without reviewing all of the candidates previously considered or the specific reason for rejecting some among them, it may be worthwhile to illustrate the situation by a few representative cases. At least two stronger versions of axiom A14, for example, may appear to pose attractive alternatives, namely:

P3. $[(p \ni_u q) \cdot (q \ni r)] \supset [(p \ni_u r) \lor \Box(p \supset r)]$;
P4. $[(p \ni_u q) \cdot \Box(q \supset r)] \supset (p \ni_u r)$.

To understand why P3 must be rejected, consider a case in which 'p' describes a necessary and sufficient causal condition for the outcome described by 'q', when 'r' in fact describes a necessary causal condition for the occurrence of p itself; for under these conditions, unless p and r cause one another, P3 is false. There do appear to be cases in which p and r do cause each other in just this fashion, namely: those of co-existence (or of co-variation), such as the mutual causation of the temperature and pressure of an ideal gas of a constant volume. Causal relations of this kind are possible within this calculus, whether or not their occurrence happens to be exceptional rather than the rule. To understand why P4 must likewise be rejected, consider the case in which 'r' is a theorem of $\mathscr{U}$. One final example is the "converse" of A13:

P5. $[(p \cdot q) \ni_u r] \supset [p \ni (q \ni_u r)]$.

Consider here a case where 'p' describes a causally sufficient condition for the outcome described by 'r' and 'q' is a contingent sentence which, although true, is nomically irrelevant to the occurrence of r. In a situation of this kind, while r would indeed occur, it seems excessively counterintuitive for q to qualify as "bringing about" r. Considerations such as these have affected the choice of axioms to include in the set of axiom schemata for calculus $\mathscr{U}$.

Once again, a number of preliminary results concerning $\mathscr{U}$ may be stated.

THEOREM 5: the following are some theorem schemata for $\mathscr{U}$:

a. $\vdash_u -\Box(p \ni_u q)$;
b. $\vdash_u (p \ni_u q) \supset (\Diamond p \cdot \Diamond -q)$;
c. $\vdash_u (p \ni_u q) \supset -(p \ni_u -q)$;
d. $\vdash_u (p \ni_u q) \supset -\Box(p \supset -q)$;
e. $\vdash_u [p \ni_u (q \supset r)] \supset [\Box q \supset (p \ni_u r)]$;
f. $\vdash_u [p \ni_u (q \supset r)] \supset [\Box(p \supset q) \supset (p \ni_u r)]$;
g. $\vdash_u (p \ni_u q) \supset [p \ni_u (p \cdot q)]$;
h. $\vdash_u [(p \ni_u q) \lor (p \ni_u r)] \supset \{[p \ni_u (q \lor r)] \lor \Box[p \supset (q \lor r)]\}$;
i. $\vdash_u [(p \ni_u q) \cdot (p \ni_u r)] \supset [p \ni_u (q \cdot r)]$.

THEOREM 6: $\mathscr{U}$ is consistent.

A model theory for calculus $\mathscr{U}$ may be obtained as follows. A $\mathscr{U}$-*model* is an ordered quadruple $\langle \text{I}, \text{f}, \text{g}, [\,] \rangle$ such that for all wffs 'p', 'q', and 'r',

U1. $\langle \text{I}, \text{f}, [\,] \rangle$ is an $\mathscr{S}$-model;
U2. g assigns to 'p' and each $i \in \text{I}$ a subset $g(p, i)$ of $[p]$;
U3. $[p \ni_u q] = \{i \in \text{I}: g(p, i) \subseteq [q]$ and $f(p \cdot -q, i) \neq \emptyset \}$;
U4. if $f(p, i) = \emptyset$, then $g(q, i) \cap [p] = \emptyset$;
U5. $[p \ni_u q] \cap [-q \ni_u -p] = \emptyset$;
U6. $f(p, i) \subseteq g(p, i)$;
U7. if $g(q, i) \subseteq [r]$, $f(p \cdot -q, i) = \emptyset$, and $f(p \cdot r, i) \neq \emptyset$, then $g(p, i) \subseteq [r]$
U8. if $g(p, i) \subseteq [q]$, then $g(p, i) \subseteq g(p \cdot q, i)$;
U9. if $f(p, i) \subseteq [q \ni_u r]$ and $f(p \cdot r, i) \neq \emptyset$, then $g(p \cdot q, i) \subseteq [r]$.

Let us say that a wff 'p' is *true in a $\mathscr{U}$-model* $\langle \text{I}, \text{f}, \text{g}, [\,] \rangle$ just in case $[p] = \text{I}$. Let us say that a wff 'p' is $\mathscr{U}$-*valid* just in case 'p' is true in every $\mathscr{U}$-model.

THEOREM 7: every theorem of $\mathscr{U}$ is $\mathscr{U}$-valid.

THEOREM 8: every $\mathscr{U}$-valid wff is a theorem of $\mathscr{U}$.

If the world is indeterministic, at least some true causal conditionals and true lawlike sentences are *not* of universal strength. It appears to be the case that, for some significant class of cases, the 'p' phenomenon *tends to bring about* the 'q' outcome, where nevertheless p is not a causally sufficient condition for q. I shall assume that the strength of this tendency for p to bring about q may be specified by a real number $n \in [0,1]$; and that any sentence of this kind may be represented as possessing the form, '$p \ni_n q$'. The language of $\mathscr{E}$, therefore, is just the language of $\mathscr{U}$ augmented by an additional dyadic sentential operator, '$\ni_n$', the n-fork, for each real number $n \in [0, 1]$, to establish the well-formed formulae of $\mathscr{E}$ in the usual way. Do note, however, that $\mathscr{E}$ has uncountably many wffs, while $\mathscr{S}$ and $\mathscr{U}$ have only countably many wffs. The new wffs of $\mathscr{E}$ (of the form, '$p \ni_n q$') are to be read as, 'p would (tenselessly) bring about q with propensity n' (or, more colloquially, 'p tends to bring about q with a strength equal to n'). The rules of inference for $\mathscr{E}$ are again MP and NEC. In addition to all tautologous wffs (of $\mathscr{E}$) and any wffs of any of the forms A1–A14, the axioms of $\mathscr{E}$ include all wffs of any of the following forms (where m, n, and $k \in [0, 1]$):

A15. $(p \ni_n q) \supset [-(p \ni q) \cdot -(p \ni -q)]$;
A16. $(p \ni_n q) \supset -(-q \ni_m -p)$;
A17. $[(p \ni_u q) \vee \square(p \supset q)] \supset \{[p \ni_n (q \supset r)] \supset (p \ni_n r)\}$;
A18. $(p \ni_1 q) \supset \{[p \ni_n (q \supset r)] \supset (p \ni_n r)\}$ for $n \neq 0$;
A19. $\{[\square(p \supset q) \cdot (q \ni_n r)] \cdot -(p \ni -r)\} \supset [(p \ni_n r) \vee (p \ni r)]$;
A20. $[(p \cdot q) \ni_n r] \supset \{[(p \ni q) \vee (p \ni_1 q)] \supset (p \ni_n r)\}$;
A21. $\{[p \ni (q \ni_n r)] \cdot -(p \ni -r)\} \supset \{[(p \cdot q) \ni_n r] \vee [(p \cdot q) \ni r]\}$;
A22. $(p \ni_n q) \equiv (p \ni_{1-n} -q)$;
A23. $(p \ni_n q) \supset -(p \ni_m q)$ for $m \neq n$;
A24. $[(p \ni_u q) \vee (p \ni_1 q)] \supset \{(p \ni_n r) \supset [p \ni_n (q \cdot r)]\}$;
A25. $\{[(p \ni_m q) \cdot (p \ni_n r)] \cdot [p \ni_k (q \cdot r)]\} \supset$
    $\{[p \ni_{(m+n)-k} (q \vee r)] \vee [p \ni (q \vee r)]\}$;
A26. $\square -(p \cdot q) \supset ([(r \ni_m p) \cdot (r \ni_n q)] \supset \{[r \ni_{m+n} (p \vee q)] \vee [r \ni (p \vee q)]\})$;

For any wff 'p' and set Γ of wffs, we define '$\vdash_{\mathscr{C}} p$', '$\Gamma \vdash_{\mathscr{C}} p$', '$\Gamma$ is $\mathscr{C}$-consistent', and 'Γ is maximally $\mathscr{C}$-consistent', in the usual ways.

A15–A21 are probabilistic counterparts of axioms adopted for the universal strength causal conditional – with some important exceptions. Just as universal strength causal conditionals and strict implications are mutually exclusive, so too the n-fork and the simple fork exclude one another. A15, A19, and A21 all reflect this relationship. A22–A26 represent the "probabilistic" character of the n-fork; in particular, A22 and A23 are "normalizing" axioms, while A24–A26 resemble familiar theorems of the calculus of probability. In contrast to the u-fork, the n-fork lacks probabilistic counterparts for A12 and A14, namely:

P6. $[p \ni (q \ni_n r)] \supset \{[(p \cdot q) \ni_n r] \vee \square [(p \cdot q) \supset r]\}$;
P7. $[(p \ni_n q) \cdot \square(q \supset r)] \supset [(p \ni_n r) \vee \square(p \supset r)]$.

In place of P6, $\mathscr{C}$ incorporates the weaker A21; and $\mathscr{C}$ lacks altogether a counterpart for P7. With respect to P6, consider a case in which 'r' not only is brought about by 'q' with a strength equal to n but also is brought about by 'p' with a strength equal to u. In this case, assuming that 'p' and 'q' are "independent causes" neither disjunct of the consequent of P6 would need to be true even if its antecedent were true; for since '$p \ni r$' and '$\Diamond [(p \cdot q) \cdot r]$' are true, so is '$(p \cdot q) \ni r$'. Consequently, '$(p \cdot q) \ni_n r$' must be false; and there seem no grounds for denying that '$\square [(p \cdot q) \supset r]$' might be false as well. With respect to P7, finally, consider the case in which 'q' is logically equivalent to '$r \cdot s$', '$p \ni_u r$' is true and '$p \ni_{1/2} (r \cdot s)$' is too.

A number of preliminary results concerning $\mathscr{C}$ may be stated here, namely:

THEOREM 9: the following are some theorem schemata for $\mathscr{C}$:

a. $\vdash_c (p \ni_n q) \supset [-\Box(p \supset q) \cdot -\Box(p \supset -q)]$;
b. $\vdash_c (p \ni_n q) \supset [-(p \ni_u q) \cdot -(p \ni_u -q)]$;
c. $\vdash_c (p \ni_n q) \supset (\Diamond p \cdot \Diamond -q)$;
d. $\vdash_c (p \ni_n q) \supset \Diamond (p \cdot q)$.

THEOREM 10: $\mathscr{C}$ is consistent.

A model theory for calculus $\mathscr{C}$ may be obtained as follows. For any finite set S, let $\#S$ denote the cardinality of S. Then a $\mathscr{C}$-model is an ordered quintuple $\langle I, f, g, s, [\,] \rangle$ such that for all wffs 'p', 'q', and 'r' and all j, k, m, and $n \in [0, 1]$,

C1. $\langle I, f, g, [\,] \rangle$ is a $\mathscr{U}$-model;

C2. s assigns to each k a function s^k which assigns to 'p' and to each $i \in I$ a non-decreasing sequence $s^k(p, i)$ of finite subsets of $g(p, i)$;

C3. $[p \ni_n q] = \{i \in I : \lim_{k \to \infty} [\#\{k : k \leq m \text{ and } \lim_{j \to \infty} \frac{\#(s^k(p,i)_j \cap [q])}{\#s^k(p,i)_j} = n\}/m] = 1$, $f(p, i) \not\subseteq [q]$, and $f(p, i) \not\subseteq [-q]\}$;

C4. if $f(p, i) \subseteq [q]$, then $(k)(\cup s^k(p, i)_j \subseteq [q])$;

C5. $[p \ni_n q] \cap [-q \ni_m -p] = \emptyset$;

C6. if $f(p \cdot -q, i) = \emptyset$ and $f(p, i) \not\subseteq [-r]$, then either

$(n)[\#\{k : k \leq m \text{ and } \lim_{j \to \infty} \frac{\#(s^k(q,i)_j \cap [r])}{\#s^k(q,i)_j} = n\}/m$ diverges];

or, $(En)(\lim_{m \to \infty} [\#\{k : k \leq m \text{ and } \lim_{j \to \infty} \frac{\#(s^k(q,i)_j \cap [r])}{\#s^k(q,i)_j} = n\}/m] = 1$

and $\lim_{m \to \infty} [\#\{k : k \leq m \text{ and } \lim_{j \to \infty} \frac{\#s^k(p,i)_j \cap [r]}{\#s^k(p,i)_j} = n\}/m] = 1)$.

C7. if $\lim_{m \to \infty} [\#\{k : k \leq m \text{ and } \lim_{j \to \infty} \frac{\#(s^k(p,i)_j \cap [q])}{\#s^k(p,i)_j} = 1\}/m] = 1$, then

SYNTAX, SEMANTICS, AND ONTOLOGY 67

$(En)(\lim_{m\to\infty} [\#\{k: k\leqslant m \text{ and } \lim_{j\to\infty} \frac{\#(s^k(p\cdot q,i)_j\cap[r])}{\#s^k(p\cdot q,i)_j} = n\}/m] = 1$

and $\lim_{m\to\infty} [\#\{k: k\leqslant m \text{ and } \lim_{j\to\infty} \frac{\#(s^k(p,i)_j\cap[r])}{\#s^k(p,i)_j} = n\}/m] = 1)$;

C8. if $f(p,i)\subseteq[-r]$, then $(k)[\text{if}(i')\{\text{if } i'\in f(p,i), \text{ then }$
$\lim_{j\to\infty} \frac{\#(s^k(q,i')_j\cap[r])}{\#s^k(q,i')_j} = n\} \text{ then } \lim_{j\to\infty} \frac{\#(s^k(p\cdot q,i)_j\cap[r])}{\#s^k(p\cdot q,i)_j} = n]$.

Let us say that a wff 'p' is *true in a* $\mathscr{C}$-model $\langle I, f, g, s, [\,] \rangle$ just in case $[p] = I$. Let us say that a wff 'p' is $\mathscr{C}$-*valid* just in case 'p' is true in every $\mathscr{C}$-model.

THEOREM 11: every theorem of $\mathscr{C}$ is $\mathscr{C}$-valid.

A completeness result for calculus $\mathscr{C}$, although highly desirable, is not presently available. Proofs of this kind typically rely upon recursive definitions of the wffs of the calculi under consideration, whereas $\mathscr{C}$ contains a nondenumerable set of wffs for which these effective procedures are not altogether sufficient. If and when the necessary techniques become available, moreover, they may ultimately lead to strengthening the syntax and weakening the model theory of this calculus.

ALTERNATIVE INTERPRETATIONS

The intended $\mathscr{C}$-model $\langle I, f, g, s, [\,] \rangle$ is one in which I is the set of all of those worlds that are logically possible, i.e., consistently describable, relative to a classical language framework $\mathfrak{L}$, i.e., that of a sentential calculus, and $[p]$ is the subset of those worlds in which the sentence 'p' is true. The three selection functions f, g, and s may then be explained in terms of a set of dispositional properties possessed by the given objects and events within a given world with special reference to the objects and events of this world, **W**. Let us assume that, for any object x such that x participates in the history of the world **W**, i.e., $x \in H(\mathbf{W})$, $(Ey)(x = y)$. Let us further assume that any such x may be described in at least three different ways, namely: (i) x may be described as identical to y when (a) $x \in H(\mathbf{W})$ and $y \in H(\mathbf{W})$, and (b) for all properties F, Fx if and only if Fy, where the values of the variable 'F' are allowed to range over (what we shall refer to as) *historical relations* as well as dispositional properties, i.e., in effect, every property 'F' is assumed to be the-having-of-a-dispositional-property-χ-at-a-certain-time-t; (ii) x may be described as identical to y when (a) $x \in H(\mathbf{W})$ and $y \in H(\mathbf{W})$, and (b) for all properties F and times t, Fxt if and only if Fyt, where the values of the variable F are restricted to *dispositional properties*; and, (iii) x may be described as identical to y when (a) $x \in H(\mathbf{W})$ and $y \in H(\mathbf{W})$, and (b) there exists some reference property K and some (dispositional or historical) property G such that x has K and G and for all objects z, if z has K and G then $z = x$ (since satisfying a definite description *is* a necessary and a sufficient condition for the introduction of a proper name within a language framework). Then, with respect to f, g, and s, the language 𝔏 and the world **W**:

For a wff 'p' of 𝔏 and a world i, $f(p, i)$ is intended to be the set of all those worlds in which 'p' is true and the permanent properties of any objects or events mentioned by 'p' are the same as they are in **W** — at least so far as the truth of 'p' will allow. It is important to remember that a subjunctive conditional, '$p \mathrel{\beth} q$', may be true in language 𝔏 supplemented by 𝒞, i.e., 𝔏*, relative to **W** just in the case either (i) 'p' strictly implies 'q', i.e., '$\Box(p \supset q)$' is a theorem of 𝒞; or (ii) 'q' describes a permanent dispositional property of the objects and events described by 'p', i.e., '$p \mathrel{\beth} q$' is a simple nomological conditional true of **W** in 𝒞; or (iii) 'p' describes a universal strength causal condition for the phenomenon described by 'q', i.e., '$p \mathrel{\beth}_u q$' is a causal nomological conditional that is true of **W** in 𝒞. Assuming the objects mentioned by 'p' are identified by means of a proper name or a definite description (as was suggested above), then every world in which those objects possess all the properties and events attributed to them by that maximally specific antecedent description will be a world in which the outcome phenomenon that is described by 'q' also occurs if and only if that subjunctive conditional is true. Thus, '$p \mathrel{\beth} q$' is true if and only if, for all $i \in I$, $f(p, i) \subseteq [q]$.

There are at least two important classes of subjunctive conditionals with false antecedents, namely: those in which the antecedent falsely attributes a dispositional property to some object otherwise identified ("If the Statue of Liberty were gold, etc.,...", for example); and those in which, although no object is falsely attributed any dispositional properties it does not possess, the antecedent falsely describes the occurrence of some test or trial to which that object has never been subjected ("If an atomic bomb had been detonated on the Statue of Liberty, etc., ...", for another). In cases of the first kind, f picks out all those worlds in which the object identified as "the Statue of Liberty" in 𝔏 satisfies its identifying description '$K \cdot G$' in 𝔏 and retains all of the permanent properties it actually has *except* for those that this change in its material bring about; while, in cases of the second kind, f picks out all those worlds in which the object thus identified retains all of the permanent properties it actually has but undergoes a trial that never actually occurred. A third class of sentences, of course, combine antecedents of those two kinds. These considerations, moreover, afford a foundation for our criticism of Kripke's conception of proper names as rigid designators; for the difference between rigid and non-rigid designation appears to be that non-rigid designators are only required to preserve the permanent properties of the objects they denote as instances of distinctive kinds of things, whereas rigid designators are implicitly supposed to preserve *both* the permanent *and* the transient properties of the objects they denote as distinctive individual things. Although non-rigid designation allows for counterfactual inference to proceed unimpaired (so long as the properties which conjointly specify both the kind of object and the unique characteristic thereby denoted are preserved from world to world) as *counterhistorical inference*, rigid designation apparently requires either the existence of at least one statistical law or the violation of at least one universal law in order to permit counterfactual inference to occur — as *counternomological inference*! [Cf. Reinhardt [1979] and Stern [1981] on aspects of this complex issue.]

Perhaps it should be emphasized that the difference between "rigid" and "non-rigid" proper names (or "designators") does not depend upon whether the proper name under consideration is or is not introduced by means of some definite description (rather than by means of a *non-linguistic* causal process), but upon whether or not merely transient properties are required to hold across possible worlds. As Kripke ([1979], pp. 10–11) has explained, there are alternative conceptions of *definite descriptions* themselves, as follows (where 'Ψ' should stand for '$K \cdot G$' as before):

(i) *non-rigid* definite descriptions of the form, '$(\imath x)\Psi x$', i.e., 'the (one and only) x such that Ψx', are satisfied within any world W by that object, if any, which happens to satisfy that description within world W; and,

(ii) *rigid* definite descriptions of the very same form, i.e., '$(\imath x)\Psi x$', are satisfied within any world W by that object, if any, which happens to satisfy that description within the actual world W , i.e., 'Ψx' denotes the same object, x, within every world W that it actually denotes within this world W .

Thus, non-rigid principles of identity for objects across possible worlds may be specified accordingly as follows (paralleling Putnam [1973], pp. 128–129), namely: for all W, all x and all y, x is identical to y in W if and only if x participates in the history of W, y participates in the history of W, and, there exists some property Ψ such that x has Ψ and, for all z, if z participates in the history of W and z has Ψ, then z is identical with x in W, and y has Ψ.

Analogously, rigid principles of identity for objects across possible worlds may be specified as follows (again paralleling Putnam [1973]), namely: for all W, all x and all y, x is identical to y in W if and only if x participates in the history of W, y participates in the history of W, and, there exists some property Ψ such that x has Ψ and, for all z, if z participates in the history of W and z has Ψ, then z is identical with x in W, and y has Ψ. [For additional discussion, see Nute [1978], [1978a], and [   ].]

In developing the calculus $\mathscr{S}$, consideration was given to a certain thesis recommended by a number of other authors for a logic of ordinary language subjunctive conditionals, namely:

P2. $[(p \vee q) \mathrel{\unicode{x22FB}} r] \supset [(p \mathrel{\unicode{x22FB}} r) \cdot (q \mathrel{\unicode{x22FB}} r)]$.

It was stated at that time that P2 is not an acceptable thesis for scientific conditionals, a rejection we are now in a position to explain. When '$p \mathrel{\unicode{x22FB}} q$' is true, all those worlds in which 'p' is true and in which all the permanent properties of those objects and events mentioned in 'p' are as similar to the actual dispositional properties of those things in W as 'p''s being true will allow are also worlds in which 'q' is true. But then these worlds have to be exactly the same as those worlds in which '$p \vee (p \cdot -q)$', etc., are true, for *these* sentences can be made true just by making 'p' true. Since '$p \mathrel{\unicode{x22FB}} q$' is true, a world in which '$p \cdot -q$' is true effects changes in the dispositional properties of objects mentioned in 'p' which a world in which only 'p' (but *not* '$p \cdot -q$', etc.) is true does not make, while our semantics requires that the fewest possible changes of this kind be made. So '$[p \vee (p \cdot -q)] \mathrel{\unicode{x22FB}} q$' must be true. But if we then accept P2, '$(p \cdot -q) \mathrel{\unicode{x22FB}} q$' is true, and it will be demonstrable that '$\square(p \supset q)$' is also true. Since this argument does not depend upon the particular sentences 'p' and 'q', if P2 is adopted then every simple subjunctive will logically entail the corresponding strict implication. Fortunately, the inference authorized by P2 is not permissible in $\mathscr{S}$; for the set of worlds which f picks out for '$p \cdot -q$' and the actual world cannot both be contained in the set of worlds f picks out for 'p' and the actual world when '$p \mathrel{\unicode{x22FB}} q$' is true.

The causal connective of universal strength is associated with function g. Since the u-fork is causal in character, this function turns out to be less restrictive than function f, since the only properties that must remain constant from world to world are those that are specific features of the situation 'p'. Instead of those worlds in which all of the permanent dispositions of all the objects and events mentioned by 'p' are preserved (so far as the truth of 'p' will allow), $g(p, i)$ is the set of worlds in which 'p' is true and in which the trial or test situation described by 'p' is preserved (so far as the truth of 'p' will permit). In the case of the antecedent, "If this match were burning, etc., ...", for example, we consider all those worlds in which *the match's burning* has all the dispositional properties it has in the actual world, i.e., the permanent dispositions of that object with respect to that trial event, in relation to whatever outcome phenomena they may display. This can include some worlds in which the match itself might have rather different properties than it actually has in relation to *other* test conditions. Thus, for example, a world included in the set selected by g might be such that a (wooden) match is a good conductor, a fragile object, and so on; but it could not be a world in which a burning match does not consume oxygen and generate heat, etc.

The function s, finally, selects sequences $s^k(p, i)_j$ of non-decreasing finite sets of p-worlds for any wff 'p' and world i. While the ontological rationale for function s is the same as that for function g, there are some important differences arising from the probabilistic character of these properties; in particular, their displays have the logical features of limiting frequencies rather than those of constant conjunctions. Since these worlds are subsets of worlds in $g(p, i)$ with respect to the nomically relevant properties of the situation described by 'p' influenced by universal dispositions, $\cup s^k(p, i)_j$ is contained in $g(p, i)$. I would emphasize the impossibility of constructing a *single sequence* $s^k(p, i)$ in order to display all of the *single case* dispositional properties of the situation described by 'p' by means of limiting frequencies, a matter which will receive further consideration very soon. It is also worth noting that this semantics distinguishes between dispositions of strength u and those of strength 1; for, if '$p \ni_1 q$' is true, there are -q-worlds in $g(p, i)$, but if '$p \ni_u q$' is true, there are *no* -q-worlds in $g(p, i)$. Both of these issues concern aspects of the theory of explanation; thus, the investigation of those topics will provide an appropriate context for additional elaboration on these important details.

Since the semantics offered here explicitly invokes *limiting frequencies* as (probabilistic) measures of dispositional properties of varying strengths, of course, it might be thought preferable to employ a purely *measure-theoretical* semantical approach by simply establishing that these statistical strength properties conform to the axioms of the calculus of probability, while letting the limiting frequencies look after themselves. Indeed, this general approach has recently been pursued by Suppes [1973] and by Giere [1976a]. Upon first consideration, moreover, a semantics of this kind would appear to offer certain theoretical advantages, in particular: (i) the differences between the semantics for probabilities as *propensities*, i.e., properties of repeatable conditions, and those for probabilities as *frequencies*, i.e., properties of trial sequences, might then seem far more obvious (cf. Gillies [1973], pp. 87–88); and, (ii) the corresponding model theory for probabilistic dispositions could then be substantially simplified by dispensing with its frequency relations. These apparent benefits, however, may be obtainable solely at the expense of acquiring hidden costs, as the following considerations will suggest.

Presumably, a measure-theoretical semantics with the appropriate features would assign to a world W and conditional antecedent 'p' a probability density function $\mu_{W,p}$ on the set of worlds such that $\int_{[p]} \mu_{W,p} \, d\mu_{W,p} = 1$. Thus, a sentence of the form, '$p \ni_n q$', is (plausibly) supposed to be true in W if and only if $\int_{[q]} \mu_{W,p} \, d\mu_{W,p} = n$, where (a) $g(p, W) \not\subseteq [q]$, and (b) $g(p, W) \not\subseteq [-q]$. But formulations of this kind are theoretically inadequate within this context, since they obscure the important distinction between *causal tendencies* (sometimes formalized by conditional probabilities) and *conditional probabilities* (which do not invariably reflect causal tendencies). Even if there were no causal connection between the situations described by 'p' and by 'q', for example, a conditional probability, defined as, say a degree of rational belief, '$Pr(q/p) = n$', might well be true and satisfy this conception *without* also satisfying conditions (III) and (IV), a prospect illustrated by Carnap's [1962]. The necessity to satisfy measure-theoretical conditions of adequacy, of course, is not at issue here, but rather that their satisfaction alone is not sufficient to establish the *kind* of interpretation intended. Indeed, the examination of "Humphreys' paradox" presented in Chapter 10 suggests as well that attempts to formalize (probabilistic) causal tendencies by means of conditional probabilities are, in principle, unlikely to succeed.

Nevertheless, the most important difference between the semantics that Kyburg has formulated, defined by (I) and (II), and the semantics being proposed, defined by (III) and (IV), is *essentially ontological*. The specific standard for selecting the members of the classes of possible worlds which display the statistical properties in question, under the dispositional interpretation, is determined by *the actual single case propensities of objects and events in the actual world* W; that is, whether or not the family of intended models is ever identified, in fact, the selection function s must select all and only those worlds which are characterized by the appropriate dispositional properties. Under the hypothetical frequency interpretation which Kyburg has proposed, by contrast, there is no corresponding principle that *selects* the intended models: although these sequences of sequences of trials are stipulated as exhibiting convergence properties similar to those exhibited by our 𝒞-models, they lack corresponding ontological determinants. Indeed, in this respect, Kyburg's proposed semantics for propensities seems to be essentially comparable with those advanced by van Fraassen [1977] and [1979], who maintains that, "the probability of event A equals the relative frequency with which it would occur, were a suitably designed experiment performed often enough under suitable conditions". But, until this desideratum is satisfied, there is no foundation, in principle, for selecting the intended models and thus *no determinate values* for those sequences to display. As a consequence, both Kyburg and van Fraassen require the conception of permanent dispositional properties, without which their semantics are not adequate.

PART II

EXPLANATION

4. STATISTICAL EXPLANATION AND STATISTICAL RELEVANCE

The ontological difference between the hypothetical frequency interpretations advanced by Kyburg [1974], [1978] and by van Fraassen [1977], [1979] and the single-case dispositional analysis advanced by Fetzer and Nute [1979], [1980], moreover, displays itself further in the kind of explanation they provide for the occurrence of singular outcomes and of sequential results as well. Under the dispositional interpretation, for example, both kinds of occurrences may be explained as manifestations of dispositional properties of the physical world; in particular, these probabilities as single-case statistical dispositions are equally applicable to the outcomes of singular or of multiple trials. Under their frequency interpretations, by contrast, the assignment of a "probability" to the occurrence of a certain result on a singular trial apparently is not literally meaningful, since "probability" itself is defined by means of limiting frequencies and relative frequencies over "long runs" of test trials. The fundamental difference which we have discovered to distinguish these interpretations from one another, therefore, is that the dispositional construction affords a theoretical conception of *single-case causal tendencies*, which these hypothetical frequency interpretations are unable to supply; but there remains yet another difference with respect to their explanatory capacity which merits additional consideration.

The important difference in the kind of explanation these interpretations provide emerges from viewing the world as a physical system whose history consists of an infinite number of temporally successive states. During the history of this sytem, certain statistical patterns may repetitiously occur; the problem, therefore, is to provide some explanation for the occurrence of these patterns. From this point of view, the difference between them is describable as follows: the dispositional interpretation provides a theoretical basis for accounting for these patterns in terms of the *system's initial conditions*, insofar as the occurrence of actual frequencies is explained by reference to the dispositional tendencies that generate them; while these frequency interpretations, by contrast, yield an empirical basis for accounting for these patterns in terms of the *pattern's ultimate configuration*, since the occurrence of actual frequencies is explained by reference to the hypothetical frequencies which control them. Consequently, the kind of explanation provided by a dispositional interpretation for the occurrence of actual frequencies during the course of the world's history is broadly *mechanistic* in character, while the kind of explanation afforded by these frequency constructions for those same occurrences is broadly *teleological* in character. To the extent to which the progress of science has been identified with a transition from teleological to mechanistic explanations, therefore, there even appear to be suitable inductive grounds for preferring the dispositional to the frequency approach.

Since the differences between these interpretations are rooted in differing degrees of ontological commitment, however, where the dispositional interpretation commits itself to the existence of single-case dispositional properties of objects and events in the actual world W which these hypothetical frequency interpretations completely eschew, perhaps their inadequacies could be overcome, at least in part, by an *actual frequency* construction, such as those advanced by von Mises [1964] or by Reichenbach [1949]. The decision procedure which Reichenbach advanced to contend with the problem of the single case, for example, was that of assigning singular occurrences to the narrowest reference classes for which reliable statistics can be compiled. Moreover, although Reichenbach himself apparently did not explicitly consider issues of explanation, the theories of explanation subsequently proposed by Wesley C. Salmon [1971] and by Carl G. Hempel [1968] may both be viewed as developments with considerable affinities to Reichenbach's position, which nevertheless afford distinct alternative solutions to the problems of single-case explanation. Perhaps at least some of the difficulties encountered with respect to the single case in particular and explanation in general on hypothetical frequency accounts might therefore be amenable to more satisfactory solution on the basis of principles which Reichenbach, Salmon, and Hempel have previously proposed, which we shall now consider.

REICHENBACH'S REFERENCE CLASSES

The theoretical foundation for Reichenbach's analysis of the single case is provided by the definition of "probability" itself as the limit of the frequency with which a certain attribute A actually occurs within a specific reference sequence S. Reichenbach himself assumes no properties other than the limiting frequency for A within S as necessary conditions for the existence of probabilities and thereby obtains an interpretation of the broadest possible generality. It is important to note, moreover, that Reichenbach

([1949], p. 72) envisions *finite* sequences as also possessing "limits" in the following sense:

> Notice that a limit exists even when only a finite number of elements x_i belong to S; the value of the frequency for the last element is then regarded as the limit. This trivial case is included in the interpretation and does not create any difficulty.

Indeed, one justification for the inclusion of these "limits" has been pointed out by Putnam [1951], namely: that when the sequence S contains only a finite number of members, those members may be counted repetitiously an endless number of times to generate trivial limiting frequencies.

As the basis for a theoretical reconciliation of these concepts, therefore, let us assume that a sequence S is *infinite* if and only if S contains at least one member and the description of its reference class R does not impose any upper bound upon the number of members of that class on syntactical or semantical grounds alone, within the appropriate language framework, $\mathfrak{L}zt$. Although "limits" may be properties of finite sequences on this construction, they are nevertheless not supposed to be properties of any single individual trials *per se* and may only be predicated thereof *as a manner of speaking*:

> I regard the statement about the probability of the single case, not as having a meaning of its own, but as representing an elliptical mode of speech. In order to acquire meaning, that statement must be translated into a statement about a frequency in a sequence of repeated occurrences. The statement concerning the probability of the single case thus is given a *fictitious meaning*, constructed by a *transfer of meaning from the general to the particular case*. (Reichenbach [1949], pp. 376–377)

Strictly speaking, therefore, "probabilities" are only properties of singular trials as members of reference classes *collectively*; yet, Reichenbach permits referring to "probabilities" as properties of such trials *distributively*, "not for cognitive reasons, but because it serves the purpose of action to deal with such statements as meaningful". Thus, for the purpose of distinguishing their meaning with respect to the occurrence of singular trials and "infinite" trial sequences, Reichenbach introduces a different term, i.e., "weight", for application to the single case.

In spite of this difference in meaning, however, the numerical value that should be assigned to attribute A as the outcome of a single trial T is determined, in principle, by the limiting frequency with which A occurs in an infinite trial sequence of kind R, where $T \in R$. The existence of a "single case weight" for the occurrence of attribute A therefore requires: (i) the existence of an infinite sequence of trials of kind R, (ii) with a limiting frequency for A equal to r, where (iii) it is not the case that there exists some other infinite sequence of kind R such that the limiting frequency for A is not equal to r; hence, '$P(Axt/Rxt) = r$' is true if and only if there is an infinite sequence of trials of kind R in which the limiting frequency for A is equal to r and, for all trial sequences S, if S is an infinite sequence of kind R, then the limiting frequency for outcomes of kind A is equal to r. In effect, therefore, every single individual trial T which happens to be a trial of kind R with respect to the occurrence of attribute A must be viewed as a member of a unique trial sequence R^*, consisting of all and only those singular trials that are trials of kind R with respect to the occurrence of attribute A; otherwise, violation of the uniqueness condition would generate explicit contradictions of the form, '$P(Axt/Rxt) \neq P(Axt/Rxt)$'.

Any singular individual trial, however, may be exhaustively described if and only if *every* property of that trial is explicitly specifiable, including the spatial and temporal relations of that instantiation of properties to all others. Let us assume that any singular individual trial T is a property of some object (or arrangement of objects) x, such that, for each such singular trial, 'Txt' is true if and only if '$F^1 xt \cdot F^2 xt \cdot \ldots$' is true, where '$F^1$', '$F^2$', ..., and so on are predicates designating distinct properties of that individual trial in $\mathfrak{L}zt$. Then the individual trial T is subject to exhaustive description, in principle, if and only if there exists some number m such that F^1 through F^m exhausts every property of that individual trial; otherwise, the individual trial T is not exhaustively describable, in principle, since there is no end to the number of properties which would have to be specified to provide an exhaustive description of that trial. The last time I turned the ignition to start my car, for example, was a single individual trial involving a 1970 Audi 100 LS 4-door red sedan, with a Kentucky license plate, a recorded mileage of 88,358.4 miles, that had been purchased in 1973 for $2,600.00 and driven to California during the summer of the same year, and so on, which was parked in the right-most section of a three-car wooden garage to the rear of a two-story building at 159 Woodland Avenue in Lexington, one a somewhat misty morning at approximately 10:30 A.M. on 12 March, 1976, on half-hour after I had drunk a cup of coffee, and so forth. As it happened, it would not start.

The significance of examples such as these, I surmise, has two different aspects. On the one hand, of course, it indicates the enormous difficulty of providing an exhaustive description of any such individual trial; in fact, it strongly suggests a *density principle for singular trial descriptions*, namely: that for any description of any singular trial T which occurs during the course of the world's history, say, '$F^1 xt \cdot \ldots \cdot F^m xt$', there exists some further description, '$F^1 xt \cdot \ldots \cdot F^n xt$', such that the set of predicates $\{'F^1', \ldots, 'F^m'\}$ is a proper subset of the set of predicates $\{'F^1', \ldots, 'F^n'\}$ — assuming there is an unlimited supply of predicates in the object language $\mathcal{L}zt$, of course. It is at least equally important to observe, on the other hand, that among all of the properties which have thus far been specified, only some, *but not all*, would be viewed as contributing factors, i.e., as relevant variables, in relation to the outcome attribute of starting or not, as the case happened to be; that is, even if this singular individual trial is not exhaustively describable, it does not follow that the *causally relevant* properties of this trial arrangement T are themselves not exhaustively describable; for the properties whose presence or absence contributed to bringing about my car's failure to start would include accumulated moisture in the distributor, perhaps, but would not include, presumably, my wearing a flannel overshirt at the time. It is at least logically possible, therefore, that among the infinity of properties which happened to attend this individual trial, only a finite proper subset would exhaust all of those exerting any influence upon that outcome attribute on that particular occasion.

The theoretical problem in general, therefore, may be stated as follows: for any singular individual trial T, (i) to determine the kind of trial R that T happens to be with respect to the occurrence of outcome attribute A; and (ii) to ascertain the limiting frequency r with which attribute A occurs within the infinite sequence S of trials of kind R, if such a sequence and such a limit n both happen to exist. The "problem of the single case", therefore, is that of determining the kind of trial R that any singular trial T happens to be with respect to the occurrence of an outcome attribute A, i.e., the problem of the selection of a unique reference sequence S for the assignment of a unique trial T. Consequently, Reichenbach's ([1949], p. 374) solution to this problem is central to his analysis of the single case:

We then proceed by considering *the narrowest class for which reliable statistics can be compiled*. If we are confronted by two overlapping classes, we shall choose their common class. Thus, if a man is 21 years old and has tuberculosis, we shall regard the class of persons of 21 who have tuberculosis [with respect to his life expectancy]. Classes that are known to be irrelevant for the statistical result may be disregarded. The class F is irrelevant with respect to the reference class R and the attribute class A if the transition to the common class $R \cdot F$ does not change the probability, that is, if $P(R \cdot F, A) = P(R, A)$. For instance, the class of persons with the same initials is irrelevant for the life expectancy of a person.

A property F is thus *statistically relevant to the occurrence of an attribute A with respect to a reference class R* (in Reichenbach's sense) if and only if,

(I) $P(Axt/Rxt \cdot Fxt) \neq P(Axt/Rxt)$;

that is, the limiting frequency for the outcome attribute A within the infinite sequence of $R \cdot F$ trials differs from the limiting frequency for that same attribute within the infinite sequence of R trials itself.

Since a unique individual trial T should be assigned to the narrowest reference class R "for which reliable statistics can be compiled", it seems clear that for Reichenbach, at least, a decision of this kind depends upon the state of knowledge $\mathcal{K}$ of an individual or collection of individuals z at some time t, i.e., the set of sentences $\mathcal{K}$ accepted or believed by z at t, no matter whether true or not (within the langauge framework $\mathcal{L}zt$). Reichenbach indeed maintains that, even though the "probability" for each single case does depend upon the state of our knowledge, this consequence does not obtain for probabilities as properties of classes, in general. The probability for particular attributes, such as death, for example, among 21 year old men reflects a frequency that obtains within the physical world, regardless of our knowledge and beliefs, which remains unaffected by the fact that the frequency for death might be different in narrower classes, such as that composed of 21 year old men with tuberculosis. Nevertheless, since it is theoretically impossible (in Reichenbach's opinion) to provide an independent interpretation for *single-case* probabilities, assignments of "weights" to singular events essentially depend upon states of knowledge $\mathcal{K}zt$. Reichenbach admits that statistical knowledge concerning a reference class R may be fragmentary and incomplete, where the problem is one of "balancing the importance of the prediction against the reliability available". However, as a general policy, Reichenbach recommends treating an individual trial T as a member of successively narrower and narrower reference classes R^1, R^2, ... and so on, where each class is specified by taking into account successively more and more statistically relevant properties F^1, F^2, ..., where $R^1 \supset R^2$, $R^2 \supset R^3$, and so on, $P(Axt/R^1xt) \neq P(Axt/R^2xt)$, $P(Axt/R^2xt) \neq P(Axt/R^3xt)$, etc.

Reichenbach observes that, strictly speaking, the choice of a reference class R is not identical to the choice of a reference sequence S, since the members of that class may be ordered in different ways, some of which could differ in probability for a particular attribute A. The intriguing question, however, is precisely how narrow the relevant class, in principle, should be, were our knowledge of frequencies complete (Reichenbach [1949], pp. 375–376):

According to general experience, the probability will approach a limit when the single case is enclosed in narrower and narrower classes, to the effect that, from a certain step on, further narrowing will no longer result in noticable improvement. It is not necessary for the justification of this method that the limit of the probability, respectively, is = 1 or is = 0, as the hypothesis of causality [i.e., the hypothesis of determinism] assumes. Neither is this necessary *a priori*; modern quantum mechanics asserts the contrary. It is obvious that for the limit 1 or 0 the probability still refers to a class, not to an individual event, and that the probability 1 cannot exclude the possibility that in the particular case considered the prediction is false. Even in the limit the substitute for the probability of a single case will thus be a class probability, ...

The evident reply on Reichenbach's analysis, therefore, is that the appropriate reference class R relative to knowledge context $\mathscr{K}zt$ containing every sentence that is true of the world and no sentence that is false (within the language $\mathfrak{L}$) would be some reference class R^i where $T \in R^i$ and, for every class R^j such that $T \in R^j$ and $R^i \supset R^j$, it is not the case that $P(Axt/R^ixt) \neq P(Axt/R^jxt)$; that is, with respect to attribute A and trial T, R^i should be an *ontically homogeneous reference class*, in the sense that (i) $T \in R^i$; (ii) $P(Axt/R^ixt) = r$; and, (iii) for all subclasses R^j of R^i to which T belongs, $P(Axt/R^jxt) = r = P(Axt/R^ixt)$.

Notice that the class R^i itself is not necessarily unique with respect to the set of properties $\{F^i\}$ specified by the reference class description for R, since any class R^j such that $R \supseteq R^j$ and $R^j \supseteq R$ (when R is an ontically homogeneous reference class with respect to attribute A and trial T) will likewise be an ontically homogeneous reference class with respect to attribute A and trial T, even though the set of properties $\{F^j\}$ specified by the reference class description for R^j differs from that for R, i.e., $\{F^i\} \neq \{F^j\}$. Consequently, although any classes which happen to be such that $R \supseteq R^j$ and $R^j \supseteq R$ will have to have all and only the same trial members and will generally yield the same frequencies for the same specific attributes with respect to those same trial members, their reference class descriptions, nevertheless, will not invariably coincide. From this perspective, therefore, resolving the single-case reference class problem by assigning a single case to an ontically homogeneous reference class does not provide a unique *description* solution but rather a unique *value* solution. On Reichenbachian principles, however, it may be argued that there *is* a unique description solution as well as a unique value solution, namely: *the class whose description includes specification of the largest set of properties of that individual trial* "for which reliable statistics can be compiled".

The largest set of properties of an individual trial T that might be useful for this purpose, moreover, need not be logically equivalent to the set of all the properties of that individual trial; for any singular trial T may be described by means of predicates which violate the requirement that reference classes not be logically limited to a finite number of members on syntactical or semantical grounds alone. Let us therefore assume that *a predicate expression is logically impermissible for the specification of a reference class description* if (a) that predicate expression is necessarily satisfied by nothing at all or (b) that predicate expression is necessarily satisfied by at most N objects or events during the course of the world's history. Predicates which happen to be satisfied by only one member during the course of the world's history, therefore, are permissible predicates for reference class descriptions so long as their extensions are not finite on logical grounds alone (such as might occur with any predicate expression essentially requiring proper names for its explicit definition or whose satisfaction by some proper name would yield a logical truth, an issue which Hempel [1968], pp. 123–129, has discussed in detail). Let us further assume that no predicate expression which logically entails the attribute predicate (or, its negation) is permissible, on the basis of previous considerations.

On these conditions, the largest set of permissible predicates to describe a single trial T will not be logically equivalent to the set of all the properties of that individual trial; but, nevertheless, it will remain the case that, in general, such "narrowest" reference class descriptions are satisfied by only one individual object or event during the course of the world's history, where this result occurs *as a matter of logical necessity*. For if a "narrowest" reference class description were satisfied by more than one such object or event, there would be some predicate expression, say, 'F^*', such that one of them satisfies 'F^*' and any other does not; otherwise, they would not be different objects or events. Since this conclusion follows from the principles of identity for objects and events (elaborated in Chapter 2), moreover, it may be properly characterized as "a matter of logical necessity". It should be noted, however, that a reference class description could be satisfied by no more than one distinct object or event without being a "narrowest" such description; thus, for example, an ontically homogeneous reference class might have only one member, where its reference class description is neither logically limited to a finite number of members nor includes every permissible predicate describing that one member.

These considerations suggest an insuperable objection to the applicability, in principle, of the frequency interpretation of probability; for if it is true that each individual trial T is describable, in principle, by some set of predicates such that (i) every member of that set is a permissible predicate for the purpose of a reference class description and (ii) that reference class description itself is satisfied by that individual trial alone, then the indispensable criterion of statistical relevance is systematically inapplicable for the role it is intended to fulfill. For under such circumstances, each individual trial T is the solitary member of a reference class R^* described by a reference class description consisting of the conjunction of some set of permissible predicates 'F^1', ..., 'F^n', i.e., $\{`F^n\text{'}\}$, where, lacking any information concerning the statistical relevance or irrelevance of any property F^i of any singular trial — other than that the attribute A did occur (or, did not occur) on that singular trial — it is systematically impossible to specify which of those properties of trial T are statistically relevant and which are not; and the occurrence of its outcome, whether A or not-A, in principle, *has to be attributed to the totality of properties present at that individual trial*. Reliable statistics, after all, are only as reliable as the individual statistics upon which they are based; so *if the only statistical data that may be ascertained, in principle, concerns the occurrence of outcome attributes on singular trials where each singular trial is the solitary member of a reference class, there is no basis for accumulating the "reliable statistics" necessary for the applicability of the frequency criterion.* Of each individual trial T^1, T^2, ..., it is possible in principle to specify a homogeneous reference class description R^{*1}, R^{*2}, ...; but since each trial is the sole member of its particular reference class, it is impossible to employ the frequency criterion to ascertain which properties, if any, among all of those present at each such singular trial, were statistically irrelevant to its outcome.

The theoretical problem of the single case, let us recall, requires, for any single individual trial T, (a) to determine the kind of trial R that T happens to be with respect to the occurrence of outcome attribute A; and, (b) to ascertain the limiting frequency r with which attribute A occurs in some infinite sequence of trials of kind R, when such a sequence and such a limit both happen to exist. With respect to any single individual trial T that actually occurs in the course of the world's history, however, these conditions are, in effect, automatically satisfied; for, (i) any such trial T may be described by some set of permissible predicates $\{`F^n\text{'}\}$ specifying a kind of trial R^* of which T is the solitary member where, nevertheless, (ii) the reference class R^* is not logically limited to any finite number of members and Reichenbach's limit concept for infinite sequences of this kind is trivially satisfied by 0 or 1. Any other such singular trial T^i, after all, may likewise be described by some set of permissible predicates, i.e., $\{`F^i\text{'}\}$, where $\{`F^n\text{'}\} \neq \{`F^i\text{'}\}$, a condition of individuation distinct events must surely satisfy. With respect to each such singular trial T, therefore, outcome attribute A either occurs or does not occur. Assume that T belongs to reference class R for which the probability

86 EXPLANATION

of A is $\neq 0$ and $\neq 1$; i.e., (a) $T \in R$; and, (b) $P(Axt/Rxt) = r$, where $0 \neq r \neq 1$. Then there necessarily exists a subclass R^* of R such that $P(Axt/R^*xt) = 0$ or $= 1$, namely: the class specified by some set of permissible predicates $\{'F^n{'}\}$ of which trial T is the solitary member. Hence, since $P(Axt/R^*xt) \neq P(Axt/Rxt)$ and $R \supset R^*$, the properties distinguishing R^* and R must be statistically relevant with respect to the occurrence of attribute A, necessarily, by the frequency criterion; moreover, for any other reference class R such that (a) $T \in R$; and, (b) $P(Axt/Rxt) = r$, where $0 \neq r \neq 1$, it is theoretically impossible that R is a homogeneous reference class for trial T relative to attribute A, in Reichenbach's sense. Indeed, on the basis of the frequency criterion of statistical relevance, such an assumption is invariably false.

SALMON'S STATISTICAL RELEVANCE

It is interesting to observe that Reichenbach himself tended to focus upon the problems involved in the *prediction* of singular events rather than those involved in their *explanation*. The differences that distinguish explanations from predictions, however, may figure in significant ways within the present context; for if it happens to be the case that the purpose of a prediction is to establish grounds for *believing that* a certain sentence (describing an event) is true, but the purpose of an explanation is to establish grounds for *explaining why* an event (described by a certain sentence) occurs, i.e., if predictions are appropriately interpreted as *reason-seeking* why-questions, while explanations are appropriately interpreted as *explanation-seeking* why-questions (Hempel [1965], pp. 334–335), it might well turn out that at least part of the problem with the frequency criterion of statistical relevance is that it is based upon an insufficient differentiation between explanation and prediction contexts. Consider the following: (a) Reason-seeking why-questions are relative to a specific epistemic context, i.e., a knowledge context $\mathscr{K}zt$ within a language $\mathfrak{L}zt$, where, with respect to an hypothesis H whose truth is not known and a sentence S in $\mathfrak{L}zt$ (where neither S nor $-S$ is entailed by $\mathscr{K}zt$), some *requirements of evidential relevance* should be employed according to which (in part) S is *inductively relevant* to the truth of H, relative to $\mathscr{K}zt$ and $\mathfrak{L}zt$, provided that (i) S is not deductively relevant to the truth of H, i.e., S entails neither H nor $-H$ in $\mathfrak{L}zt$; and, (ii) the measure of evidential support ES for H, relative to $\mathscr{K}zt \cdot S$, differs from the measure of evidential support for H, relative to $\mathscr{K}zt \cdot -S$, in $\mathfrak{L}zt$; that is,

(II) $ES(H/\ \mathscr{K}zt \cdot S) \neq ES(H/\mathscr{K}zt \cdot -S)$;

where, in effect, any sentence S within $\mathscr{K}zt$ whose truth or falsity makes a difference to the evidential support for an hypothesis H within $\mathfrak{L}zt$ must be taken into consideration in determining the evidential support for H, relative to $\mathscr{K}zt$. (b) Explanation-seeking why-questions are relative to a certain ontic context, i.e., the nomological regularities and particular facts of the actual world, $\mathbf{W}$, where, with respect to the explanandum-event e described by an explanandum-sentence E within the language $\mathfrak{L}zt$, some *requirements of explanatory relevance* are properly employed according to which (in part), for any predicate 'F' and reference predicate 'R' in $\mathfrak{L}zt$, 'F' is nomologically *relevant* to the occurrence of e as described by 'E', relative to the reference predicate 'R', provided that (i) neither 'E' nor '$-E$' is entailed by 'F' or by 'R' in $\mathfrak{L}zt$; and, (ii) the degree of nomic expectability NE of 'E', relative to '$R \cdot F$', differs from the degree of nomic expectability of 'E', relative to '$R \cdot -F$' (as well as $\mathbf{W}$ and $\mathfrak{L}zt$); that is,

(III) $NE(E/Rxt \cdot Fxt) \neq NE(E/Rxt \cdot -Fxt)$;

where, in effect any property whose presence or absence, relative to a certain reference property R, makes a difference to the nomic expectability for the explanandum-event e (as described by 'E' in $\mathfrak{L}zt$) must be taken into consideration in formulating an adequate explanation for 'E', relative to $\mathbf{W}$ and $\mathfrak{L}zt$.

If predictions belong to the epistemic reason-seeking context, then it is entirely plausible to take into account any property G whose presence or absence changes the evidential support for an hypothesis H with respect to a knowledge context $\mathscr{K}zt$; indeed, the requirement of evidential relevance would demand that any such property G of any individual x which is inductively relevant to H within the knowledge context $\mathscr{K}zt$ has to be taken into account when calculating the evidential support for H, when the sentence G^* attributing G to x is such that,

(IV) $ES(H/\mathscr{K}zt \cdot G^*) \neq ES(H/\mathscr{K}zt \cdot -G^*)$;

where, as it might be expressed, the statistical relationship between property G and the attribute property A described by the hypothesis H may be merely one of statistical correlation, rather than one of causal connection, i.e., it is not necessary that G be a property whose presence or absence is explanatorily relevant to the occurrence of attribute A. For the presence or absence of the property G might be *predictively significant*, whether or not it contributes to bringing about outcomes of that kind.

It is not at all obvious, however, that the frequency criterion provides any theoretical latitude for differentiating between statistical relations of these distinctive kinds; on the contrary, it appears as though any factor with respect to which frequencies differ is on that account alone evidentially (or, explanatorily) relevant to an hypothesis H (or, an explanandum E), indifferently, without reflecting any theoretical distinction between the explanation-seeking and the reason-seeking situations themselves. If every member of the class of twenty-one year old men R were also a member of the class of tuberculous persons F, for example, then the property of being tuberculous would be statistically irrelevant to the outcome attribute of death D. But even though this property might reasonably be ignored for the purpose of prediction (as a matter of statistical correlation), it would not be reasonable to ignore this property for the purpose of explanation (when it makes a causal contribution to such an individual's death). If there is a significant difference between these kinds of situations, therefore, then those principles appropriate for establishing relevance relations within one of these contexts may be theoretically inappropriate for establishing relevance relations within the other. Perhaps the crucial test of the utility of Reichenbach's criterion, therefore, may be found within the context of explanation, rather than that of prediction; for an examination of the "statistical relevance" explication of explanation advanced by Salmon should provide additional opportunities to evaluate the criticisms of that principle thus far advanced.

Salmon diverges from Reichenbach's formulations, not in departing from the frequency interpretation of probability, but rather in assigning a single case to the *broadest* instead of to the *narrowest* relevant class, as follows:

If every property that determines a place selection is statistically irrelevant to A in R, I shall say that R is a *homogeneous reference class* for A. A reference class is homogeneous if there is no way, even in principle, to effect a statistically relevant partition without already knowing which elements have the attribute in question and which do not. ... The aim in selecting a reference class to which to assign a single case is not to select the narrowest, but the widest, available class. ... I would reformulate Reichenbach's method of selection of a reference class as follows: choose the broadest homogeneous reference class to which the single event belongs. (Salmon [1971], p. 43)

Precisely because Salmon retains the frequency interpretation of probability, his own formulations encounter difficulties analogous to those reported above; but the introduction of the concept of a partition and of the "screening-off" rule may be viewed as significant contributions to the frequency construction of explanation, as follows:

(i) On Salmon's analysis, *a partition of a reference class R* is established by a division of that class into a set of mutually exclusive and jointly exhaustive subsets by means of a set of properties $F^1, F^2, ..., F^m$ and their complements where each ultimate subset of that class $R \cdot C^1, R \cdot C^2, ..., R \cdot C^n$ is homogeneous with respect to the outcome attribute A. This procedure should be envisioned as introducing a refinement in the formulation of Reichenbach's criterion of statistical relevance, since it implicitly assumes that a property F is *statistically relevant* to the occurrence of an attribute A with respect to a reference class R if and only if,

(I*) $P(Axt/Rxt \cdot Fxt) \neq P(Axt/Rxt \cdot -Fxt)$;

that is, the limiting frequency for the outcome attribute A within the infinite sequence of $R \cdot F$ trials differs from the limiting frequency for that same outcome within the infinite sequence of $R \cdot -F$ trials.

(ii) Furthermore, a property F *screens-off* a property G with respect to an outcome attribute A within the reference class R if and only if,

(V) $P(A/R \cdot F \cdot G) = P(A/R \cdot F \cdot -G) \neq P(A/R \cdot -F \cdot G)$;

where the equality between the limiting frequency for A within the classes $R \cdot F \cdot G$ and $R \cdot F \cdot -G$, on the one hand, establishes that the property G is not statistically relevant with respect to the attribute A within the reference class $R \cdot F$; and the inequality between the limiting frequency for A within the class $R \cdot -F \cdot G$ and the class $R \cdot F \cdot G$, in particular, on the other hand, establishes that the property F is statistically relevant with respect to the attribute A within the reference class $R \cdot G$. Salmon suggests, for example, that the presence of a low-pressure area F screens-off sudden drops in barometer readings G within the reference class of temporal periods in specified vicinities R, relative to the occurrence of stormy weather A, where the low-pressure area F should take precedence over the falling barometer G with respect to that attribute A: a contrived drop in the barometer reading, after all, would not provide a reliable indication of the onset of inclement weather, while the presence of such a low-pressure area generally would. (Salmon [1971], pp. 54–55)

The subtle shift from Reichenbach's explicit formulation (I) to Salmon's implicit formulation (I*), moreover, warrants further elaboration, especially insofar as it tacitly solves a problem for extensional frequency formulations. According to formula (I), of course, a property F is statistically relevant to an attribute A with respect to a reference class R if and only if the limiting frequency for A, given R-and-F, differs from the limiting frequency for A, given R, i.e., given R-and-F and R-and-not-F. In cases in which *every member* of a reference class R happens to possess a property F, however, the class of R-

and-not-F will be the null class; consequently, the limiting frequency for A within R cannot possibly differ from the limiting frequency for A within R-and-F, since R-and-F and R are then identical, necessarily. Any property F that every member of a reference class R happens to possess, therefore, cannot possibly be statistically relevant to any attribute A with respect to that same reference class R, which may appear appropriate in the case of *predictions* but appears excessively perplexing in the case of *explanations*. To the extent to which statistical relevance relations are intended to provide a theoretical foundation for establishing explanatory relevance relations, therefore, it is important to observe that, in cases of this kind, a property F may be explanatorily relevant to attribute A within reference class R *even when* $P(Axt/Rxt \cdot Fxt) = P(Axt/Rxt)$! As a result, it seems reasonable to conclude that either formulations such as (I) must be interpreted *intensionally*, i.e., as applicable to hypothetical rather than to historical sequences (which entails abandoning purely extensional conceptions) or else they establish sufficient but *not* necessary conditions of explanatory relevance (which motivates a preference for alternative formulations). The employment of principle (I*) within this context, therefore, overcomes at least one difficulty otherwise encountered by principle (I).

Salmon consolidates these ingredients as the foundation for his account of explanation on the basis of the principle that screening-off properties F should take precedence over those properties $G^1, G^2, \ldots$ which they screen-off within an explanation situation. According to Salmon, therefore, an explanation of the fact that x, a member of the reference class R, is a member of attribute class A as well, may be provided by establishing the following conditions:

(1) $R \cdot C^1, R \cdot C^2, \ldots, R \cdot C^n$ is a homogeneous partition of R relative to A;
(2) $P(A/R \cdot C^1) = r^1$; $P(A/R \cdot C^2) = r^2$; ...; $P(A/R \cdot C^n) = r^n$;
(3) $r^i = r^j$ only if $i = j$; and,
(4) $x \in R \cdot C^m$ (where $m \in \{1, 2, \ldots, n\}$).

Consequently, the appropriate reference class to specify in order to explain an outcome A for a trial T is the ontically homogeneous reference class $R \cdot C^m$ such that (a) $T \in R \cdot C^m$; (b) $P(A/R \cdot C^m) = r^m$; and, (c) for all other reference classes $R \cdot C^1, R \cdot C^2, \ldots, R \cdot C^n$, which are ontically homogeneous relative to attribute A, $r^i = r^m$ only if $i = m$, which presumably, is intended to guarantee that there is one, and only one, reference class to which T may appropriately be assigned, namely: the *broadest* one of that kind. From this point of view, therefore, Salmon provides a unique description solution as well as a unique value solution to the problem of the single case.

Salmon's conditions of adequacy, let us note, are sufficient for their intended purpose *only if* broadest homogeneous reference classes may be described by means of (what may be referred to as) *disjunctive properties*, i.e., predicate expressions of the form, '$R \cdot F^1 \vee F^2 \vee \ldots \vee F^n$', where a sentence of the form, '$P(A/R \cdot F^1 \vee F^2 \vee \ldots \vee F^n) = r$', is true if and only if '$P(A/R \cdot F^1) = P(A/R \cdot F^2) = \ldots = P(A/R \cdot F^n) = r$' is true, as Salmon himself [1975] has acknowledged. Otherwise, Salmon's conditions would be theoretically objectionable, insofar as it might actually be the case that there were an infinite set of reference classes, $\{R \cdot C^i\}$, whose members satisfy conditions (1) and (2) only if they do not satisfy condition (3), and conversely. The difficulty with this maneuver as a method of preserving Salmon's conditions (1)–(4) as sufficient conditions of explanatory adequacy, however, is that it entails the adoption of (what may be referred to as) *a degenerating explanation paradigm*; for the occurrence of an attribute A on trial T may properly be explained by identifying that trial as a member of a successively more and more *causally heterogeneous* reference class under the guise of the principle of reference class homogeneity. For if condition (3) is retained, then if, for example, the division of the class of twenty-one year old men R by means of the properties of having tuberculosis F^1 or heart disease F^2 or a brain tumor F^3 establishes a subclass $R \cdot C^m$ such that, with respect to the attribute of death D, $P(D/R \cdot F^1 \vee F^2 \vee F^3) = r$ is an ultimate subset of the homogeneous partition of that original class, i.e., $\{R \cdot F^1 \vee F^2 \vee F^3\} = \{R \cdot C^m\}$, then the explanation for the death of an individual member i of the class R resulting from a brain tumor, perhaps, is only explainable by identifying that individual as a member of the class $R \cdot C^m$, i.e., as a member of the class of twenty-one year old men who have either tuberculosis or heart disease or a brain tumor.

The significance of this criticism appears to depend upon how seriously one takes (what may be referred to as) *the naive conception of scientific explanation for the occurrence of singular events*, namely: that the occurrence of an outcome attribute A on a singular trial T should be explained by citing *all and only* those properties of that particular trial T which contributed to bringing about that specific outcome A, i.e., that a property F is *explanatorily relevant* to attribute A on trial T if and only if F is a *causally relevant* (or, more broadly, a *nomically relevant*) property of trial T, relative to outcome A. From this perspective, moreover, Salmon's explication of explanation is theoretically objectionable for at least two distinct reasons, as follows:

First, *statistically relevant* properties are not necessarily *causally relevant* (or, *nomically relevant*) properties, and conversely (as Coffa [1974] has also observed). If it happens that the limiting frequency r for the attribute of death D among twenty-one year old men R differs in relation to those whose initials are the same F when that class is subject to a homogeneous partition, then that property is explanatorily relevant, necessarily, on the basis of the frequency criterion of statistical relevance; and it might even happen that a property of this kind screens-off another property G, such as having tuberculosis, in spite of the fact that property G is causally relevant to attribute D, while property F is not. Indeed, Salmon himself elsewhere suggests that, "relations of statistical relevance must be explained on the basis of relations of causal relevance" (Salmon [1975], p. 400), where these relations of statistical relevance apparently fulfill the role of evidential indicators of relations of causal relevance (Salmon [1975a], pp. 121–129 and pp. 141–145).

Second, the admission of disjunctive properties for the specification of an ultimate subset of a homogeneous partition of a reference class R does not satisfy the desideratum of explaining the occurrence of an outcome A *on* trial T by citing *all and only* properties of that particular trial, whether causally relevant, nomically relevant, statistically relevant, or otherwise. This difficulty, however, appears to be less serious, in principle, since a modification of Salmon's conditions serves for its resolution, namely: Let us assume that a reference class description, '$R+$', is *stronger than* another reference class description, 'R', if and only if '$R+$' entails 'R', but not conversely. Then let us assume as new condition (3*) Salmon's old condition (4) together with the addition of new condition (4*) in lieu of Salmon's old condition (3) as follows:

(4*) $R \cdot C^m$ is a strongest homogeneous reference class;

that is, the reference class description, '$R \cdot C^m$', is stronger than any other reference class description '$R \cdot C^j$' (where $j \in \{1, 2, ..., n\}$) such that $x \in R \cdot C^j$. The explanation of the death of an individual member i of a reference class R resulting from a brain tumor, therefore, is then only explainable by identifying i as a member of the strongest class, $R \cdot C^m$, i.e., as a member of the class of twenty-one year old men who have brain tumors, which is a broadest homogeneous reference class of the explanatorily relevant kind under this modification of the frequency conception of statistical explanation.

It is important to observe, however, that none of these considerations mitigates the force of the preceding criticism directed toward the frequency criterion of statistical relevance itself. For it remains the case that any singular trial T that belongs to a reference class R for which the probability of the occurrence of attribute A is $\neq 0$ and $\neq 1$ will likewise belong to innumerable subclasses $R \cdot C^j$ of R; and, indeed, T itself will necessarily belong to some subclass R^* of R such that $P(A/R^*) = 0$ or $= 1$, which, on Salmon's own criteria, requires that trial T be assigned to R^*, provided R^* is a broadest (or, a strongest) ultimate subclass of a homogeneous partition of the original class R. In effect, therefore, those properties that differentiate trial T as a member of class R from all other trial members of that class establish the basis for effecting a homogeneous partition of that class, a partition for which the probability for attribute A within every one of its ultimate subclasses (whether strongest or not) must be $= 0$ or $= 1$, necessarily. And this result itself should be viewed as reflecting a failure to distinguish those principles suitable for employment within the context of explanation from principles suitable within the context of prediction, promoted by (what appears to be) the mistaken identification of statistical relevance with explanatory relevance.

It should not be overlooked that the properties *taken to be* statistically relevant to the occurrence of attribute A on trial T relative to the knowledge context $\mathcal{K}zt$ are not necessarily those that actually *are* statistically relevant to that attribute within the physical world. For this reason, Salmon's account emphasizes the significance of the concepts of *epistemic* and of *practical* homogeneity, which, however, on Salmon's explication, actually turn out to be two different kinds of *inhomogeneous* reference classes, where, for reasons of ignorance or of impracticality, respectively, it is not possible to establish ontically homogeneous partitions for appropriate attributes and trials (in Salmon [1971], p. 44). Salmon's analysis thus does not provide an explicit characterization of the conception of reference classes that are believed to be homogeneous whether or not they actually are; nevertheless, there is no difficulty in supplementing his conceptions as follows:

Let us assume that a reference class R is *epistemically homogeneous* with respect to an attribute A and trial T within the knowledge context $\mathcal{K}zt$ if and only if the set of sentences $\mathcal{K}$ accepted or believed by z at t, considerations of truth all aside, logically entails some set of sentences, $\{S\}$, whose members assert (a) that $T \in R$; (b) that $P(A/R) = r$; (c) that for all subclasses R^j of R to which T belongs, $P(A/R^j) = r = P(A/R)$; and, (d) that it is not the case that there exists some class $R^i \supset R$ such that $P(A/R^i) = P(A/R)$. The satisfaction of conditions (a), (b) and (c), therefore, is sufficient to fulfill the epistemic version of the conception of an *ontically homogeneous* reference class R for trial T with respect to attribute A, while the satisfaction of (d) as well is sufficient to fulfill the epistemic version of the conception of a *broadest homogeneous* reference class R for trial T with respect to attribute A, within the knowledge context $\mathcal{K}zt$. In order to differentiate these conceptions from Salmon's original formulations, therefore, we may refer to these as, say, *supplementary definitions* of epistemic homogeneity, without denying the potential utility of both sets of concepts.

HEMPEL'S MAXIMAL SPECIFICITY

The conclusions that emerge from the preceding investigation of Salmon's own analysis of statistical explanation support the contention that, relative to the frequency criterion of statistical relevance, statistical explanations are only *statistical* relative to a knowledge context $\mathcal{K}zt$, i.e., as the matter might be expressed, "God would be unable to construct a statistical relevance explanation, not as a limitation of His power but as a reflection of His omniscience". It is therefore rather intriguing that Salmon himself has endorsed this very conclusion as a criticism of *Hempel's* account of statistical explanation (Salmon (1974], p. 165); for, as the considerations which follow are intended to display, the fundamental difference between their analyses is that Hempel's explicit relativization of statistical explanations to the knowledge context $\mathcal{K}zt$, as it were, affirms *a priori* what, under Salmon's explication, is *a posteriori* true, namely: that for probabilities r such that $0 \neq r \neq 1$, the only homogeneous reference classes are epistemically homogeneous. The issue which Salmon's criticism fails to make clear, however, is that this difficulty arises as a necessary consequence of reliance upon the frequency criterion of statistical relevance for *any* non-epistemic explication, including, of course, Salmon's *own* ontic explication. As we shall also discover, there are at least two additional points with respect to which these analyses are distinctive, in spite of their initial appearance of marked similarity.

In order to establish the soundness of these claims, therefore, let us consider the three principal ingredients of Hempel's epistemic explication. First, Hempel introduces the concept of an *i-predicate in* $\mathcal{K}zt$ which, in effect, is any predicate 'F^m' such that a sentence '$F^m i$' asserting the satisfaction of 'F^m' by the individual i belongs to $\mathcal{K}zt$, regardless of the kind of property that may be designated thereby. Hempel then defines "statistical relevance" as follows:

'F^m' will be said to be *statistically relevant* to 'Ai' in $\mathcal{K}zt$ if (1) 'F^m' is an i-predicate that entails neither 'A' nor '$-A$' and (2) $\mathcal{K}zt$ contains a lawlike sentence '$P(A/F^m) = r$'

specifying the probability of 'A' in the reference class characterized by 'F^m'. (Hempel [1968], p. 131)

Insofar as sentences of the form, '$P(A/F^m) = r$', are supposed to be *lawlike*, it is clear that, on Hempel's analysis, (a) their reference class descriptions must be specified by means of permissible predicates and (b) these sentences are to be understood as supporting subjunctive and counterfactual conditionals. What is *not* clear, however, is the theoretical justification for ascribing counterfactual or subjunctive force to probabilities interpreted as frequencies under either hypothetical or extensional constructions, which is highly problematic, as we have previously ascertained.

The condition that the knowledge context $\mathcal{K}zt$ contain a set of probability sentences, moreover, has been subjected to criticism by Kyburg [1970a] on the grounds that it requires $\mathcal{K}zt$ to contain an enormous number of these sentences. So long as a *normative* conception of knowledge prevails, however, surely such an objection has no force at all, since innumerable consequences will invariably attend the acceptance of any particular proposition 'p' by z at t, given (CR-1) and (CR-2) alone. Even on the basis of a *descriptive* conception, in fact, there appear to be reasons for questioning this criticism, namely:

(i) the sentences belonging to $\mathcal{K}zt$ are accepted or believed by z at t, i.e., they represent what z takes to be the case at t;

(ii) this set of sentences, therefore, may be believed to be exhaustive, relative to attribute A and trial T, whether that is really true or not; and,

(iii) presumably, an explication of this kind is intended to specify the conditions that should be fulfilled, in principle, in order to provide an adequate explanation for the occurrence of a particular event, relative to some knowledge context $\mathcal{K}zt$, without presuming that these conditions are always or even generally satisfied.

Second, Hempel defines the concept of *a maximally specific predicate 'M' related to 'Ai' in $\mathcal{K}zt$*, where 'M' is such a predicate if and only if (a) 'M' is logically equivalent to a conjunction of predicates that are statistically relevant to 'Ai' in $\mathcal{K}zt$; (b) 'M' entails neither 'A' nor '$-A$'; and, (c) there is no predicate expression stronger than 'M' which satisfies (a) and (b); i.e., if 'M' is conjoined with a predicate that is statistically relevant to 'Ai' in $\mathcal{K}zt$, then the resulting expression entails 'A' or '$-A$' or else it is logically equivalent to 'M'. Thus, for outcome A on trial T, 'M' is intended to provide a description of that trial as a member of a reference class R which is specified by the conjunction of every predicate 'F^m' that is statistically relevant to that outcome on that trial, i.e., 'M' is a conjunction of *all and* *only* the statistically relevant *i*-predicates (or their negations) satisfied by trial T in $\mathcal{K}zt$.

Third, Hempel formulates the *revised requirement of maximal specificity*, as follows: An argument

(VI) $\dfrac{\begin{array}{c} P(A/F) = r \\ Fi \end{array}}{Ai}$ $[r]$,

where r is close to 1 and all constituent sentences are contained in $\mathcal{K}zt$, qualifies as an explanation of the explanandum-phenomenon described by the explanandum-sentence 'Ai' (or of the fact that i is a member of the attribute class A) within the knowledge context $\mathcal{K}zt$, only if,

> (RMS*) For any predicate, say 'M', which either (a) is a maximally specific predicate related to 'Ai' in $\mathcal{K}zt$ or (b) is stronger than 'F' and statistically relevant to 'Ai' in $\mathcal{K}zt$, the class $\mathcal{K}$ contains a corresponding probability sentence, '$P(A/M) = r$', where, as in (VI), $r = P(A/F)$. (Hempel [1968], p. 131)

Since a predicate expression 'F^j' is stronger than a predicate expression 'F^i' if and only if 'F^j' entails but is not entailed by 'F^i', moreover, a predicate which is logically equivalent to the conjunction of 'F' with *any other permissible i-predicate* 'F^k' such that $\mathcal{K}zt$ contains the corresponding probability sentence, '$P(A/F \cdot F^k) = r$', will satisfy condition (b) of the requirement of maximal specificity, provided that, as in (VI), $r = P(A/F)$.

The motivation for Hempel's introduction of this requirement, we should recall, was his discovery that *statistical* explanations suffer from a species of explanatory ambiguity arising from the possibility that an individual trial T might belong to innumerable different reference classes, $R^1, R^2, \ldots$, for which, with respect to a specific attribute A, the probabilities for the occurrence of that outcome could greatly differ, i.e., the reference class problem for single case explanation. In particular, Hempel has displayed concern for the possibility of the existence of alternative explanations consisting of alternative explanans which confer *high probabilities* upon both the occurrence and the *non*-occurrence of an attribute A, relative to the physical world **W** or a knowledge context $\mathcal{K}zt$, a phenomenon which cannot arise in the case of explanations involving universal, rather than statistical, lawlike sentences. It is important to observe, however, that there are *two* distinct varieties of explanatory ambiguity, at least one of which is not resolved by Hempel's requirement of maximal specificy. For although Hempel provides a unique *value* solution to the single case problem (which entails a resolution of the difficulty that arises with conflicting explanations which confer high probabilities upon their explanandum sentences), Hempel's approach does not provide a unique *description* solution for the reference class problem, a difficulty that still continues to afflict his conditions of adequacy for explanations invoking universal *or* statistical lawlike sentences within an epistemic *or* an ontic context. Hempel's explication, therefore, appears to afford only a restricted resolution of one species of statistical ambiguity, which, however, does not provide a solution to the general problem of explanatory ambiguity for explanations of either kind. For explanations involving universal laws as well as those involving statistical laws continue to suffer from the difficulties that arise from a failure to contend with the problem of providing a unique description solution for single case explanations.

Hempel resolves the problem of statistical ambiguity, in effect, by requiring that, within a knowledge context $\mathcal{K}zt$, the occurrence of the outcome attribute A on a trial T is adequately explained by identifying a reference class R such that (a) $T \in R$; (b) $P(A/R) = r$; and, (c) for all subclasses R^j of R to which T belongs, $P(A/R^j) = r = P(A/R)$; but he does not also require that (d) it is not the case that there exists some other class R^i containing R such that $P(A/R^i) = P(A/R)$. Thus, on the basis of condition (b) of (*RMS**), if, for example, the probability for the outcome of death D within the class of twenty-one year old men who have tuberculosis R is equal to r and the probabilities for that same outcome within the classes $R \cdot F^1$, $R \cdot F^1 \cdot F^2, \ldots$, of twenty-one year old men who have tuberculosis R and who have high blood pressure F^1, or who have high blood pressure F^1 and have blue eyes $F^2, \ldots$ are likewise equal to r, then the explanation for the occurrence of death for an individual who belongs not only to class R but also to class $R \cdot F^1$, to class $R \cdot F^1 \cdot F^2$, and so on, will be adequate, *regardless of which of these reference classes is specified by the explanans for that single case*. Consequently, on Hempel's explication, there is not only no unique explanation for the occurrence of any such explanandum-event but, if the density principle for single trial descriptions is sound, *there may be an infinite number of adequate explanations for any one such explanandum.* And this surprising result applies alike for explanations invoking universal laws, since if salt R dissolves in water A as a matter of physical law (within a knowledge context or without), then table salt $R \cdot F^1$ dissolves in water, hexed table salt $R \cdot F^1 \cdot F^2$ dissolves in water, ... as a matter of physical law as well; so if a single trial T involves a sample of hexed table salt, an adequate explanation for its dissolution in water may refer to any one of these reference classes or to any others to which it may happen to belong, providing only that, for all these properties F^i, it remains the case that every member of $R \cdot F^i$ possesses the attribute A as a matter of physical law.

Comparison with Salmon's explication suggests at least two respects in which Hempel's analysis involves theoretically objectionable conditions of adequacy. The first, let us note, is that Hempel's requirement of maximal specificity (RMS^*) does not incorporate any appropriate relevance criteria that would differentiate statistically relevant from statistically irrelevant properties in the sense of principle (I*). For Hempel has defined his conception of statistical relevance so broadly that any property F^i at all such that there exists some probability r for which a sentence of the form, '$P(A/F^i) = r$', is either true or believed to be true, qualifies that property as "statistically relevant" to the attribute A, independently of any consideration for whether or not there might be some class R such that $R \supset F^i$ and $P(A/R) = r = P(A/F^i)$; in other words, *Hempel's conception of statistical relevance is not a relevance relation of the appropriate kind* (a point which Salmon [1971], pp. 7–12 and p. 35 has also emphasized). In order to contend with this difficulty, therefore, general revision of Hempel's definition is required along the following lines:

> 'F^m' will be said to be *statistically relevant* to 'Ai' relative to 'R' in $\mathscr{K}zt$ if and only if (1) 'F^m' and 'R' are i-predicates that entail neither 'A' nor '$-A$'; (2) $\mathscr{K}zt$ contains the lawlike sentences, '$P(A/R \cdot F^m) = r^i$' and '$P(A/R \cdot -F^m) = r^j$'; and, (3) the sentence, '$r^i \neq r^j$', also belongs to $\mathscr{K}zt$.

This alternative, in effect, represents the exchange of Reichenbach's criterion of relevance (I) for Salmon's criterion of relevance (I*).

The second is that Hempel's account of explanation is not capable of yielding a unique description solution for the problem of the single case because it is logically equivalent to the *supplementary definition of epistemic homogeneity*, in particular, to the epistemic conception of an *ontically homogeneous* reference class, rather than to the epistemic conception of a *broadest homogeneous* reference class. There appears to be no reason, in principle, however, which precludes the reformulation of Hempel's own requirements so as to incorporate the conditions necessary for fulfilling the desideratum of providing a unique description solution (for arguments having the form (VI) previously specified) as follows:

(RMS**) For any predicate, say 'M', (a) if 'M' is weaker than 'F' and is statistically relevant to 'Ai' in $\mathcal{K}zt$, then the class $\mathcal{K}$ contains a corresponding probability sentence asserting that $P(A/M) \neq r$; and, (b) if 'M' is stronger than 'F' and is statistically relevant to 'Ai' in $\mathcal{K}zt$, then the class $\mathcal{K}$ contains a corresponding probability sentence asserting that $P(A/M) = r$; where, as in (VI), $r = P(A/F)$.

Condition (*RMS***), therefore, not only requires that any property of trial T that is statistically related but nevertheless is not statistically relevant to 'A' must be excluded from an adequate explanation of that outcome on that trial in $\mathcal{K}zt$, but also requires that trial T must be assigned to the *broadest* homogeneous reference class of which it is a member, in the sense that 'F' must be the *weakest* predicate statistically relevant to 'Ai' in $\mathcal{K}zt$ which satisfies Hempel's original (*RMS**), i.e., 'F' itself has to be a maximally specific predicate relative to 'Ai' in $\mathcal{K}zt$. Moreover, on (what appears to be) the reasonable presumption that the definition of a maximally specific predicate precludes the specification of reference classes by *non-trivial* disjunctive properties, i.e., disjunctive predicates which are not logically equivalent to some non-disjunctive predicate, these conditions actually require that that trial be assigned to the *strongest* homogeneous reference class of which it is a member. From this point of view, therefore, the reformulation of Hempel's requirement appears to provide a (strongest) unique description solution as well as a unique value solution to the single case problem within the spirit of Hempel's explication.

The revised formulation of Hempel's requirements [incorporating appropriate relevance conditions and (*RMS***) as well] and the revised formulation of Salmon's requirements [incorporating his original conditions (1), (2) and (3*), together with new condition (4*) as well], of course, both accommodate the naive conception of scientific explanation to the extent to which they satisfy the desideratum of explaining the occurrence of an outcome A on the singular trial T by citing *all and only relevant properties of that specific trial*. Neither explication, however, fulfills the expectation that a property F is *explanatorily* relevant to an outcome A on trial T if and only if F is *causally* relevant (or, *nomically* relevant) to outcome A on trial T, so long as they remain wedded to the frequency criterion of statistical relevance. Nevertheless, precisely because Hempel's explication *is* epistemic, i.e., related to a knowledge context $\mathcal{K}zt$ which could contain sentences satisfying the specified conditions (no matter whether true or not), it is not subject to the criticism that the only adequate explanations are *non-statistical*, i.e., those for which the probability $r = 0$ or $= 1$. On the other hand, it *is* subject to the criticism that there are no *non-epistemic* adequate explanations for which the probability $r \neq 0$ and $\neq 1$, i.e., there are no *ontic* (or, *true*) statistical explanations. As it happens, this particular criticism has been advanced by Alberto Coffa [1974], who, while arguing that Hempel's epistemic explication is necessitated by (1) implicit reliance upon the frequency interpretation of physical probability, in conjunction with, (2) implied acceptance of a certain "not unlikely" reference class density principle, unfortunately neglects to emphasize that Salmon's own *ontic* explication is similarly afflicted, precisely because the fatal flaw is not to be found in the epistemic-ness of Hempel's conception but rather in *reliance on the frequency criterion of statistical relevance for any ontic explication.*

If Coffa's reconstruction of Hempel's rationale were sound, then it would be important to observe that, provided (1) is satisfied and (2) is true, there are, *in principle*, no non-epistemic adequate explanations for which probability $r \neq 0$ and $\neq 1$; that is, an epistemic explication is then not only *the only theoretically adequate kind* of an explication, but an explication remarkably similar to Hempel's specific formulations would appear to be *the only theoretically adequate explication*. So if it is not the case that an epistemic explication of the Hempel variety is the only theoretically adequate construction, then either (1) is avoidable or (2) is not true. Intriguingly, the density principle Coffa endorses, i.e., the assumption that, for any specific reference class R and outcome A such that trial $T \in R$, there exists another subclass R^i of R such that, (i) $T \in R^i$ and (ii) $P(A/R) \neq P(A/R^i)$, is demonstrably false, since for all subclasses R^i such that $R \supseteq R^i$ and $R^i \supseteq R$, this principle does not hold, i.e., it does not apply to any homogeneous reference class R, whether or not T is the only member of R. Insofar as every distinct trial T^n is describable, in principle, by some set of permissible predicates, $\{`F^n`\}$, such that trial T^n is the solitary member of the kind R^* thereby defined, however, it is evident that Coffa's density principle is not satisfied by even *one single trial* during the course of the world's history. Coffa's principle is plausible, therefore, only so long as there appear to be *no* homogeneous reference classes; for once the existence of reference classes of the R^* kind has been theoretically identified, it becomes obvious that this density principle is false. Nevertheless, another density principle in lieu of Coffa's principle does generate the same conclusion, namely: the density principle for singular trial descriptions previously introduced. Coffa has therefore advanced an unsound argument for a true conditional conclusion, where, as it happens, by retaining one of his premises and replacing the other that conclusion does indeed follow, albeit on different grounds.

In his endorsement of Coffa's contentions, Salmon [1974] explicitly agrees that, on Hempel's explication, there are no non-epistemic adequate explanations for which the probability $r \neq 0$ and $\neq 1$; thus he remarks, "There are no homogeneous reference classes except in those cases in which *either* every member of the reference class has the attribute in question *or else* no member of the reference class has the attribute in question". With respect to his own explication, by contrast, Salmon asserts, "The interesting question, however, is whether under any other circumstances R can be homogeneous with respect of A — e.g., if one-half of all R are A". It is Salmon's view, in other words, that Hempel's position entails an *a priori* commitment to determinism, while his does not. But the considerations adduced above demonstrate that determinism is a consequence attending the adoption of the frequency criterion of statistical relevance itself, i.e., determinism is as much a logical implication of Salmon's position as it is of Hempel's. This result, moreover, underlines the necessity to draw a clear distinction between *statistical* relevance and *explanatory* relevance as such; for, although it is surely true that any two distinct events are describable, in principle, by different sets of permissible predicates, it does not follow that they are necessarily not both members of some *causally* (or, *nomically*) *homogeneous* reference class R for which, relative to attribute A, it is the case that $P(A/R) = r$, where $0 \neq r \neq 1$, even though, as we have ascertained, they may *not* both be members of some *statistically homogeneous* reference class R^* for which, with respect to attribute A, $P(A/R^*) = r$, where $0 \neq r \neq 1$. Thus, although it is not logically necessary *a priori* that the world be deterministic, the adoption of the frequency criterion of statistical relevance for the specification of explanatory relevance relations is nevertheless a sufficient condition for establishing determinism's truth.

When the reformulated versions of Salmon's and Hempel's explications which entail assigning each singular trial T to the "strongest" homogeneous reference class R of which it is a member (with respect to attribute A) are compared, the revised Hempel explication provides, in effect, a meta-language formulation of the revised Salmon object-language explication, with the notable exception that the Hempel-style explication envisions explanations as arguments for which some *high probability* requirement is appropriate, while the Salmon-style explication does not. The issue of whether or not explanations should be construed as arguments is somewhat elusive, of course, since there appears to be no problem, in principle, with separating *explanation-seeking* and *reason-seeking* varieties of (let us say) "statistical arguments", i.e., as sets of sentences divided into premises and conclusions, where the premises provide the appropriate grounds, reasons, or evidence for their conclusions. But however suitable a high probability requirement might be relative to the reason-seeking variety of statistical argument, there appear to be no suitable grounds for preserving such a requirement relative to the explanation-seeking variety of statistical argument in the face of the following consideration, namely: that *the imposition of a high probability requirement between the explanans and the explanandum of an adequate statistical explanation renders the adequate explanation of attributes that occur only with low probability logically impossible, in principle* (as Salmon [1971] has emphasized). It is notable that Hempel himself has withdrawn this requirement (in the "Postscript 1976" to the German edition of his [1965], listed here as Hempel [1976]) for reasons of this kind.

Although we shall return to the question of whether or not explanations should be construed as arguments in the chapters that follow, let us observe that each of the species of fequency interpretation encounters difficulties of its own. In particular, the *hypothetical frequency* interpretations, which Kyburg [1974], [1978] and van Fraassen [1977], [1979] formulate, appear to be capable of providing support for counterfactual and subjunctive conditionals, but only at the expense of embracing *teleological* principles of explanation. The *actual frequency* interpretations advanced by Reichenbach [1949] and von Mises [1964], moreover, overcome such commitments to teleology, but they are unable to establish support for counterfactual and subjunctive conditionals, on the one hand, and entail the *trivialization* of principles of explanation, by virtue of an implicit commitment to ontological determinism, on the other. Although we have not examined the *finite frequency* interpretations endorsed by Russell [1948] and by Sklar [1970] in detail, it should be obvious that they are incapable of overcoming the fundamental inadequacies which attend all of the varieties of frequency interpretation; indeed, their inadequacies are similar to those afflicting the actual frequency interpretations we have already considered. None of these interpretations, of course, affords the basis for an adequate theoretical solution to the problems of single case explanation. The satisfactory resolution of these significant difficulties thus appears to require an interpretation that is both *ontological* and *indeterministic*, i.e., the conception of probabilities as single-case dispositional properties. For an epistemic explication of the Hempel variety is not the only theoretically adequate construction of statistical explanations, precisely because reliance upon frequency interpretations of physical probability and frequency criteria of statistical relevance *are* avoidable theoretical alternatives. Contrary to Salmon's ([1971], p. 82) allegation, therefore, the choice between the propensity and the frequency conceptions within the context of explanation does not appear to be a matter of "philosophical preference", after all.

5. A SINGLE CASE THEORY OF CAUSAL EXPLANATION

In an article entitled, 'Are Statistical Hypotheses Covering Laws?', Levi has presented the provocative argument that a covering law account of statistical explanation encounters a dilemma arising from the theoretical necessity to reconcile the following desiderata (Levi [1969], p. 299), namely:

On the one hand, statistical explanation is to be viewed as explanation by law; and, on the other hand, an account of statistical explanation must be based upon an acceptable interpretation of statistical probability.

Levi asserts that these desiderata cannot be simultaneously satisfied and that, as a result, the function of covering laws in Hempel's explication had better be envisioned as fulfilled by material rules of inference. The force of Levi's argument, of course, rests upon his contention that there is no acceptable interpretation of statistical probability which succeeds in converting statistical hypotheses into covering laws (Levi [1969], p. 303). Levi admits that he is unable to provide an impossibility theorem here but claims that, "a review of the more obvious candidates ought, at a very minimum, to place the onus of proof very squarely on the shoulders of those who believe such statements can be constructed". The purpose of this chapter is to demonstrate that, contrary to Levi's claim, the single-case dispositional interpretation succeeds in fulfilling this objective.

Although Levi's argument will turn out to be ill-founded (so long as the following arguments are sound), the desiderata that he adopts in the course of his inquiry appear to be worthy of preservation; thus, for example, he tends to endorse the requirements that, for any lawlike sentence S (of universal or statistical form), S must be (a) capable of supporting counterfactual and subjunctive conditionals, (b) logically unrestricted to any finite number N of instances during the course of the world's history, and, (c) capable of serving the function of a covering law in explanatory arguments of an appropriate form. Moreover, to whatever extent Levi intends to be understood as dealing with *extensional* constructions, rather than *intensional* explications, of lawlike sentences, his position appears to be defensible. The difficulty, however, is that Levi himself does not make this matter clear: such phrases as, "extensional language" or "first-order extensional language", for example, do not even occur as elements of his exposition. Nevertheless, as we have found, extensional conceptions of lawlike sentences, whether statistical or not, are, in principle, unable to overcome obstacles such as these.

The serious candidates for the interpretation of lawlike sentences which remain, therefore, are all *intensional* explications; indeed, so long as teleological analyses are excluded from further consideration, they are all variations on *dispositional* constructions. Surprisingly or not, the development of an adequate conception of statistical dispositions, largely stimulated by the publication of Popper's [1957] and [1959], has fostered a plethora of interpretations exhibiting diverse degrees of family resemblance to one another. The discussion which follows, therefore, should not be entertained as dealing with every possible alternative conception of dispositional properties but as examining the principal variations upon this dispositional theme which appear to warrant review within the present context. Levi's own views in relation to these issues, advanced, for example, in Levi [1967] and [1973] (which have led others, such as Sklar [1974] and Kyburg [1974], to regard his position as within the dispositional family, an interpretation that Levi [1977] now vehemently rejects), in particular, require attention elsewhere. A number of significant alternative positions will receive consideration here, including, for example, the "long run" dispositional conception that Hempel [1965] and Hacking [1965] have both endorsed, and the "single case" dispositional interpretations which Mellor [1969], [1971] and Giere [1973], [1973a], and [1976a] have recommended. The author shall contend that *none of these* analyses does justice to problems such as those Levi has cited and that, in their place, an account is required of the kind that Fetzer [1971], [1974], and [1974d], initially proposed, and that Fetzer [1976a] and Fetzer and Nute [1979], [1980] subsequently developed to the form in which it appears in Chapter 3.

"LONG RUN" DISPOSITIONAL CONCEPTS

Popper [1957], Hempel [1965], and Hacking [1965] all advance or endorse formulations of (what we shall refer to as) the "long run" propensity interpretation, according to which physical probabilities are statistical dispositions to display particular patterns of response under appropriate test conditions. As Popper ([1957], p. 67), for example, explains it,

Every experimental arrangement is *liable to produce*, if we repeat the experiment very often, a sequence with frequencies which depend upon this particular experimental arrangement. These virtual frequencies may be called probabilities. But since these probabilities turn out to depend upon the experimental arrangement, they may be looked upon as *properties of this arrangement*. *They characterize the disposition or the propensity* of the experimental arrangement to give rise to certain characteristic frequencies *when the experiment is often repeated.*

Hempel ([1965], p. 378) cites Popper's paper with approval, elaborating upon the example of attributing a probability of ¼, say, to the outcome of landing on side III as the result of tosses with a homogeneous regular tetrahedron:

> This disposition might be characterized by means of a subjunctive conditional phrase: if the tetrahedron were to be tossed a large number of times, it would yield the result III in about one-fourth of the cases. Implications in the form of counterfactual and subjunctive conditionals are thus hallmarks of lawlike statements both of strictly universal and of statistical form.

Hacking ([1965], p. 10) expresses remarkably similar sentiments in comments upon the meaning of the phrase, "frequency in the long run", such as these:

> "Frequency in the long run" is all very well, but it is a property of the coin and tossing device, not only that, in the long run, heads falls more often than tails, but also that this would happen even if in fact the device were dismantled and the coin melted. This is a dispositional property of the coin: what the long run frequency is or would be or would have been.

Although Hacking [1965] follows Braithwaite [1953] in endeavoring to develop the significance of this conception more thoroughly through the elaboration of principles of acceptance and of rejection for hypotheses of this kind, we may assume that all three here embrace essentially the same "long run" conception.

From an historical perspective, the "long run" propensity interpretation was systematically anticipated by Peirce (for example, Buchler [1955], pp. 164-173); indeed, Peirce himself was explicit in explaining the necessity for endless trials to display this "would-be" property of coins and dice and tosses:

> ... to define the die's "would-be", it is necessary to say how it would lead the die to behave on an occasion that would bring out the full consequence of that "would be"; and this statement will not of itself imply that the "would-be" of the die *consists* in such behavior. Now in order that the full effect of the die's "would-be" may find expression, it is necessary that the die should undergo an endless series of throws from the dice box, the result of no throw having the slightest influence upon the result of any other throw, or, as we express it, the throws must be *independent* each of every other. (Buchler [1955], p. 169)

CAUSAL EXPLANATION 107

According to this conception, therefore, an unrestricted generalization that attributes a permanent "long run" statistical disposition to every member of a reference class K (under an appropriate description 'K') within a language $\mathfrak{L}^*zt$, relative to test conditions T^i and outcome response O^i, would exhibit the following form,

(I) $\quad (x)(t)[Kxt \ni\!\!\!- (T^i xt \ni\!\!\!-^*_n O^i xt^*)]$;

which asserts, ⌜For all x and all t, if x were K at t, then the strength of the dispositional tendency for T^i-ing x at t to bring about O^i-ing x at t^*, *over the long run*, is $\overline{n}$⌝; or, say, ⌜For all x and all t, if x were K and were subjected to an infinite number of trials of kind T^i over a period of time t, then x would display an outcome response of kind O^i with limiting frequency $\overline{n}$⌝. Thus, in effect, the "long run" propensity interpretation attributes *a universal strength dispositional property to display specific limiting frequencies n^i over infinite sequences of appropriate trials T^i to every object or arrangement of kind K, with respect to outcomes of kind O^i*.

Upon preliminary consideration, therefore, a construction of this kind may appear to be somewhat anomalous conceptually, until one considers that, in this respect, especially, *hypothetical frequency* interpretations are exactly the same as *"long run" propensity* interpretations: both, in effect, attribute invariable limiting frequencies to the distribution of particular outcomes over infinite sequences of trials of appropriately specified kinds. The fundamental difference between them, moreover, remains their respective ontological foundations; for the "long run" propensity interpretation, like its "single case" counterpart, is committed to the existence of *real causal connections between events:* consequently, once again, the dispositional interpretation exhibits the broadly mechanistic character which distinguishes dispositional interpretations from their hypothetical frequency counterparts. The principal semantical property differentiating between these dispositional interpretations themselves, therefore, is that "single case" interpretations *permit* what "long run" interpretations *prohibit*, namely: the possibility of divergence among these generating probabilities and their frequency displays. In order to present a formal analysis of the semantical properties of "long run" propensities in some fashion parallel to that employed in Chapter 3 for "single case" propensities, let us once again invoke the concept of *p-worlds* understood as before and proceed as follows:

We shall continue to assume that '*p*' is appropriately interpreted as an eternal sentence describing a logically possible unique historical situation, rather than as an occasion sentence representing a class of different trials. The truth of sentences of the form, '$p \ni\mathrel{-}\!\!*_n q$', of course, is presumed to be unaffected by the truth of any sentences in any worlds except the sentences 'p' and 'q' themselves, where, in particular, specific sets of dispositional properties possessed by objects (or events) mentioned by 'p' are preserved from world to world. Let us further assume that S is a non-decreasing sequence of finite sets of p-worlds, i.e., if $i \leq j$, then $S_i \subseteq S_j$. Then a sentence of the form, '$p \ni\mathrel{-}\!\!*_n q$', will be true of the actual world $\mathbf{W}$ if and only if,

(II) $\quad \lim\limits_{k \to \infty} \dfrac{\text{the number of worlds in } S_k \text{ in which '}q\text{' is true}}{\text{the number of worlds in } S_k} = n.$

Thus, in the case of the "long run" conception, a single sequence of trials suffices to exhibit the frequency displays that require an infinite set of infinite sequences of trials for its "single case" counterpart. Condition (II), moreover, affords the foundation for an alternative model theory for the probabilistic causal calculus $\mathscr{C}$, as follows:

For any finite set S, let $\#S$ denote the cardinality of S. Then a $\mathscr{C}$*-model is an ordered quintuple $\langle \mathrm{I, f, g, s^*, [\,]} \rangle$ such that for all wffs 'p', 'q', and 'r' and all k, m, and $n \in [0, 1]$:

C*1. $\langle \mathrm{I, f, g, [\,]} \rangle$ is a $\mathscr{U}$-model;

C*2. s* assigns to 'p' and each $i \in \mathrm{I}$ a non-decreasing sequence $\mathrm{s}^*(p,i)_k$ of finite subsets of $g(p, i)$;

C*3. $[p \ni\mathrel{-}\!\!*_n q] = \{i \in \mathrm{I} \colon \lim\limits_{k \to \infty} \dfrac{\#(\mathrm{s}^*(p,i)_k \cap [q])}{\#\mathrm{s}^*(p,i)_k} = n,\; f(p,i) \not\subseteq [q],$
and $f(p, i) \not\subseteq [-q] \}$;

C*4. if $f(p, i) \subseteq [q]$, then $\cup \mathrm{s}^*(p,i)_k \subseteq [q]$;

C*5. $[p \ni\mathrel{-}\!\!*_n q] \cap [-q \ni\mathrel{-}\!\!*_m -p] = \emptyset$;

C*6. if $f(p \cdot -q, i) = \emptyset$ and $f(p,i) \not\subseteq [-r]$, then either $\dfrac{\#(\mathrm{s}^*(q,i)_k \cap [r])}{\#\mathrm{s}^*(q,i)_k}$
diverges or $\lim\limits_{k \to \infty} \dfrac{\#(\mathrm{s}^*(q,i)_k \cap [r])}{\#\mathrm{s}^*(q,i)_k} = \lim\limits_{k \to \infty} \dfrac{\#(\mathrm{s}^*(p,i)_k \cap [r])}{\#\mathrm{s}^*(p,i)_k}$;

C*7. if $\lim\limits_{k \to \infty} \dfrac{\#(\mathrm{s}^*(p,i)_k \cap [q])}{\#\mathrm{s}^*(p,i)_k} = 1$, then $\lim\limits_{k \to \infty} \dfrac{\#(\mathrm{s}^*(p \cdot q, i)_k \cap [r])}{\#\mathrm{s}^*(p \cdot q, i)_k} =$
$\lim\limits_{k \to \infty} \dfrac{\#(\mathrm{s}^*(p,i)_k \cap [r])}{\#\mathrm{s}^*(p,i)_k}$;

C*8.   if $\lim_{k \to \infty} \frac{\#(s^*(q,i)_k \cap [r])}{\#s^*(q,i)_k} = n$ for all $j \in f(p,i)$ and $f(p,i) \subseteq [-r]$,

then $\lim_{k \to \infty} \frac{\#(s^*(p \cdot q, i)_k \cap [r])}{\#s^*(p \cdot q, i)_k} = n$.

Let us say that a wff 'p' is *true in a* $\mathscr{C}$*-model* $\langle I, f, g, s^*, [\ ]\rangle$ just in case $[p] = I$. Let us say that a wff 'p' is $\mathscr{C}$*-valid* just in case 'p' is true in every $\mathscr{C}$*-model.

THEOREM 11*: every theorem of $\mathscr{C}$* is $\mathscr{C}$*-valid.

Since more than one sequence (or more than one sequence of sequences) of worlds may display the propensities actually possessed by things, there can be no *single* intended $\mathscr{C}$*-model (or $\mathscr{C}$-model). Instead, there will be a family of intended $\mathscr{C}$*-models (or $\mathscr{C}$-models), such that each member of that family is exactly like every other member except for the fourth component; that is, they will be exactly alike with respect to I, f, g, and [ ]. Furthermore, for each intended $\mathscr{C}$*-model and each intended $\mathscr{C}$-model, these *pairs of models* will have exactly the same first, second, third, and fifth components. As a result, we may say that the alternative model theories for the calculus $\mathscr{C}$ are *extensionally equivalent* in the sense that, for any wff 'p' of $\mathscr{C}$, for any intended $\mathscr{C}$-model M and any intended $\mathscr{C}$*-model M*, the set of worlds in M at which 'p' is true will be exactly the same as the set of worlds in M* at which 'p' is true. In particular, 'p' will be true at the actual world W in M if and only if 'p' is true at the actual world W in M*. Even though this extensional similarity obtains between these model theories, however, the *intensions* of wffs containing probabilistic causal conditionals nevertheless differ between them, as the differences between the semantical conditions C3 and C*3 clearly demonstrate. Thus, the theory of $\mathscr{C}$-models *identifies* the strength of a propensity with the limit of a long run of repetitions of the same historically unique individual trial; the theory of $\mathscr{C}$-models, by contrast, permits these long run limits to diverge from the strength of such a propensity, but asserts that it is *overwhelmingly probable* that any such long run of repetitions will converge upon the value of that propensity as its limit. As a consequence, the theory for $\mathscr{C}$*-models is a theory of long run propensities, while the theory of $\mathscr{C}$-models is a theory of single case propensities.

All of this, of course, can be expressed somewhat less formally as well. Thus, to recognize the logical difference between them, notice that the hold which single case probabilities have on long run results differs categorically from the grip exercised upon them by long run probabilities. Under these long run constructions, the relative frequencies for particular outcomes displayed by an infinite sequence of trials *necessarily* equal their probabilities. This results from the fact that long run probabilities are just what the long run frequencies are or would be or would have been. Under a single case construction, however, the relative frequency for a specific outcome within such an infinite sequence only *probably* equals its probability. This results from the fact that these long run outcomes must be computed on the basis of Bernoulli's theorem — even when the sequence itself consists of an infinite number of trials. Consequently, although a long run probability permits the prediction of long run results with the force of logical necessity, a single case probability only permits the prediction of long run results with a probability of 1. It would therefore be mistaken to assume that the hypothetical frequencies which would be generated during an infinite sequence of trials *necessarily* equal their probabilities; for although this is indeed the case for long run probabilities, it is only overwhelmingly probable for single case probabilities. As a matter of fact, of course, the sequences within the range of our experience are always finite; consequently, this logical difference is important in principle but not in practice.

The fundamental ontological difference between them, of course, is that one interprets probabilities as long run dispositional properties, while the other interprets them as single case dispositional properties. An analysis which takes long run probabilities as basic, however, may succeed in developing an acceptable long run construct but must still contend with the problem of accounting for the assignment of "probabilities" to singular events. The force of this consideration, moreover, becomes apparent within the context of single case explanations; for under a long run interpretation, a property F *will be explanatorily relevant with respect to the occurrence of an outcome O relative to a reference property K*, let us say, provided that,

(III) $P^*(Oxt/Kxt \cdot Fxt) \neq P^*(Oxt/Kxt \cdot -Fxt)$;

that is, the long run propensity for the outcome attribute O with arrangements of kind $K \cdot F$ differs from the long run propensity for that same attribute with arrangements of kind $K \cdot -F$ (where the asterisk identifies an intensional explication of the probability relation). The issue that arises, therefore, is the connection between these long run results and particular single case outcomes; for, as Hacking ([1965], p. 50) has emphasized, what holds in "the long run" does not necessarily matter to each single case: even if each singular trial *were* characterized by such "long run" dispositional tendencies, how are they explanatorily relevant to accounting for the outcomes of each *singular* trial? Reichenbach, of course, sought to resolve this difficulty by introducing the concept of *weight* as the predictional value of a sentence pertaining to a single case, without notable success. The point, however, is not to reaffirm the inadequacies of Reichenbach's maneuver but to notice that, in order to account for single case outcomes, even advocates of the *long run* propensity construct are theoretically obligated to introduce a corresponding *single case* analysis.

The most important benefit of the "single case" approach, therefore, is that it not only accounts for the meaning of *single case* probabilities but also solves the problem of *long run* probabilities; for, given the values of the relevant single case probabilities, calculations of long run probabilities for various combinations of outcomes over various lengths of trials may be made on the basis of the mathematical principles for statistical probabilities. Thus, the fundamental advantage of the single case interpretation is that it yields a construct which is theoretically significant for *both* the long run *and* the single case — and it achieves this effect without introducing any new and difficult problems for the theory of science. With respect to Levi's desiderata, therefore, namely: that an adequate interpretation of statistical hypotheses as covering laws must provide probabilistic generalizations which (a) support counterfactual and subjunctive conditionals, (b) are logically unrestricted to any finite number N of instances during the course of the world's history, and (c) can function as covering laws in explanatory arguments of the appropriate form, the "long run" propensity interpretation encounters no difficulties with either (a) or (b). For, with respect to condition (b), statistical hypotheses under this interpretation concern every experimental arrangement of a certain kind K (or, $K \cdot F$), where, even though there might be only a finite number of instantiations of these arrangements during the course of the world's history, as a matter of fact these hypotheses are not logically limited to any finite number of those arrangements. Moreover, in fulfillment of condition (a), this interpretation explicitly construes these statistical hypotheses as attributing statistical dispositions to experimental arrangements of the appropriate kind, where these probabilistic properties themselves reflect the hypothetical frequencies that would be generated if arrangements of that kind were subjected to long runs of trials, an explicitly subjunctive construction.

The difficulties encountered on the "long run" propensity interpretation, therefore, concern condition (c); for although statistical hypotheses interpreted as "long run" propensities are perfectly capable of functioning as covering laws in explanatory arguments of explanandum sentences for "long run" outcomes, they are *not* capable of serving that function for "single case" results. Thus, even the long run propensity interpretation apparently succumbs to the force of Levi's criticism. Before turning to consideration of its "single case" counterparts, however, perhaps it should be observed that, in comparison with alternative "long run" interpretations, the long run *propensity* interpretation represents a significant advance over long run *frequency* interpretations (regardless of their detailed features), insofar as the propensity conception, but not its frequency counterparts, affords an elegant solution to the problem of explicating the *requirement of randomness*; for the construct appropriate to frequency analyses is based upon the limit properties of certain subsequences of particular sequences of trials, while the construct appropriate to propensity analyses (in *either* their long run *or* their single case varieties) concerns the strength of the tendency for various outcomes from trial to trial; that is,

(A) Under frequency constructions, a sequence of trials of kind F within a reference sequence of kind R will be *normal* if and only if (i) its outcomes are *free from aftereffect*, i.e., the frequencies for the O^i outcomes of various trials do not depend upon the outcomes of the preceding trials; and, (ii) its outcomes are *insensitive to ordinal selection*, i.e., the frequency for O outcomes within the entire sequence is equal to the frequencies for outcomes of those O kinds within any subsequence selected by taking every k*th* trial;

where this definition reflects Reichenbach's ([1949], pp. 143–144) conception. By contrast,

(B) Under propensity constructions, a sequence of trials of kind F with experimental arrangements of kind K will be *random* if and only if (i) the strength of the tendency for outcomes of kind O is the same from trial to trial, i.e., these propensities are *equal*; and, (ii) the strength of the tendency for an outcome of kind O^i on trial T^m and an outcome of kind O^j on trial T^n is equal to the product of the strength of the tendency for O^i on trial T^m times the strength of the tendency for O^j on trial T^n, i.e., they are also *independent*.

Thus, it is possible to demonstrate, quite generally, that a trial sequence is *random* if and only if the experimental arrangements generating it are ontically homogeneous (in the propensity sense), i.e., their *nomically relevant* properties are the same from trial to trial. For if any such trial were conducted with an arrangement whose propensity to produce an outcome of kind O on a trial of that type differed from that of any other member of that sequence, then the propensities for each of the outcomes would not be equal and independent; and, conversely.

A corresponding claim on behalf of *normality*, of course, will obtain just in case the trial sequence S itself is ontically homogeneous (in the frequency sense), i.e., their *statistically relevant* properties are the same from trial to trial. Thus, the demonstration in terms of frequencies is similarly plain, on the convenient assumption that trials with arrangements of kind $R \cdot F$ may be introduced at regular intervals such that they form a subsequence selected by taking every kth trial. If this assumption is regarded as *too* convenient, however, since temporal relations, say, ordinarily establish their order, then the demonstration that a sequence is normal if and only if it is also homogeneous may be viewed with suspicion — although, in any case, the propensity proof that *a sequence of trials is random if and only if the nomically relevant properties affecting each of the possible outcomes do not vary from trial to trial* remains entirely unimpaired. Perhaps this difficulty underlies those reservations concerning the use of the concept of normality (or, randomness) with respect to a physical sequence expressed, for example, by Coffa [1974a] and Kyburg [1974a], which Salmon [1977], especially has endeavored to overcome. So far as I have been able to discern, however, this difficulty is a problem solely for the frequency construction, being logically irrelevant to the propensity approach.

The differences between randomness and normality, of course, reflect the difference between the conception of probabilities as properties of repeatable conditions, the mathematical framework for which was formalized by Kolmogorov [1956], and the conception of probabilities as properties of trial sequences, formalized, for example, by von Mises [1964]. As Popper ([1957], p. 89) has observed, the transition from the frequency to the propensity interpretation thus represents an exchange of mathematical frameworks. It may therefore be worthwhile to review the basic differences Gillies [1973] has cited, namely:

first, as we have already noted, von Mises introduces "probabilities" as attributes of (infinite) trial sequences, while Kolmogorov envisions them as attributes of (repeatable) fixed conditions;

second, von Mises explicitly defines "probability" by means of limiting frequencies within (more or less) normal sequences, while Kolmogorov only restricts its meaning as satisfying his axioms; and,

third, although von Mises expends great efforts in attempting to provide an adequate analysis of "normality" (or, "randomness"), this problem does not appear to require Kolmogorov's attention.

As Gillies ([1973], p. 87) has remarked, these differences may be less crucial than they appear, since "normal" trial sequences would be generated (with high probability) through the repetition of "random" conditions. Nevertheless, the difference in attitude regarding "randomness" may not be difficult to explain, namely: a sequence can be *normal* only if it happens to be infinite (since it will otherwise be impossible to satisfy condition (A) (ii) for large k), while a sequence may be *random* regardless of whether it happens to be finite or infinite — indeed, it could consist of only a *single trial* of fixed conditions! Consequently, the propensity approach automatically resolves a major problem attending the frequency conception (which Humphreys [1977] seeks to solve).

These definitions are also important, of course, insofar as they specify sufficient conditions for the applicability of Bernoulli's theorem; for it is not logically necessary under either interpretation for *finite* frequencies to vary if and only if certain probabilities likewise vary. Indeed, the principles of (let us say) "practical certainty" which they imply for finite trial sequences are identical, to wit:

> *The Short Run Principle*: Let S_n be a sequence of normal or random trials with, say, arrangements of kind R or kind K; then if the probability for an outcome of kind O is equal to r, then if S_n is a "relatively long" sequence, it is *practically certain* that the relative frequency f of O-outcomes will be approximately equal to r; and,
>
> *The Single Case Principle*: Let T be a single trial member of a sequence of normal or random trials of kind R or kind K; then if the probability for an outcome of kind O is very close to one (or to zero), then it is *practically certain* that outcome O will (or will not) occur.

The Short Run Principle itself (as well as its corollary), moreover, has been advanced by no less a figure than Cramér ([1946], pp. 149–150) as his formulation of the *frequency interpretation of probability hypotheses*. Nevertheless, these principles should not be confused with either interpretation of probability, since they are implied by them both (which perhaps may account for the extent to which the practice of statistical inference proceeds without the necessity to resolve these thorny foundational problems).

It is important to recognize, of course, that — theoretical differences notwithstanding — a property F would be assumed to be statistically relevant (or, nomically relevant) to the occurrence of an outcome O on either interpretation, in general, *under appropriate test conditions*, if for suitable n,

(IV) $\mathscr{F}_n(Oxt/Rxt \cdot Fxt) \neq \mathscr{F}_n(Oxt/Rxt \cdot -Fxt)$;

that is, when an appropriate frequency distribution $\mathscr{F}_n$ for outcomes of kind O over n trials of kind F with arrangements of kind R differs from the frequency distribution for outcomes of that O kind over trials of kind $-F$ with arrangements of that R kind. Statistical differences of this sort, of course, provide no logical guarantee of statistical (or, nomic) relevance; nevertheless, as Cramér's *Short Run Principle* declares, there is a *probabilistic* relationship of this very kind. In order to appraise the significance of any frequency distribution, therefore, it is necessary to ascertain whether or not it records a *random sample*, i.e., whether or not that data actually describes the results of n trials of kind F with arrangements of kind R or K, a problem to which we shall subsequently return.

ALTERNATIVE "SINGLE CASE" CONCEPTS

The desideratum that a probabilistic hypothesis S has to be capable of serving as a covering law in explanatory arguments of the appropriate form, of course, guarantees that Levi's three conditions will be jointly sufficient, but it does so in an open-ended way, insofar as those additional requirements that must be met, over and beyond (a) and (b), are not explicitly formulated. This quality should not necessarily be supposed to be a deficiency, unless a more definite set of adequacy conditons is already at hand, which is not the case or, at least, not obvious. Indeed, we have just discovered a subtle but important difference in the application of probabilistic hypotheses as statistical laws which depends upon *the character of the explanandum*, in particular, upon whether the explanandum describes a singular occurrence or some infinite trial sequence; for these sentences, under the "long run" propensity interpretation, appear suitable for explaining the occurrence of limiting frequencies over infinite trial sequences, but not for explaining the specific outcomes of each singular trial. Even interpretations that are deliberately contrived for application to the single case, however, may or may not succeed, even if they are dispositional conceptions, as Mellor's [1969], [1971] and

Giere's [1973], [1973a], and [1976a] shall now exhibit. In order to establish (what I take to be) their theoretical limitations, therefore, I will recommend four desiderata which an adequate interpretation of statistical probability must satisfy; that is, such a conception *must* envison these probabilities (i) as single case statistical dispositions (ii) attributing non-extensional causal connections (iii) to appropriately specified experimental arrangements, where (iv) a distinction is drawn between dispositional tendencies of statistical strength 1 and dispositional tendencies of universal strength u. The following arguments are intended to defend these desiderata and to establish that Mellor's interpretation fails to satisfy conditions (i) and (ii), while Giere's interpretation fails to satisfy conditions (iii) and (iv), which warrants their rejection.

Mellor claims that both Popper and Hacking have been misleading in comparing the propensity of an experiment with the fragility of a glass, on the grounds that the results of trials with a "chance set-up", i.e., a statistical arrangement, are not strictly comparable with the results of dropping a fragile glass (Mellor [1969], p. 26):

> The main point is that it cannot be the *result* of a chance trial that is analogous to the breaking of a dropped glass. It is true that the breaking may be regarded as the result of dropping the glass, but to warrant the ascription of a disposition, it must be supposed to be the *invariable* result. Other things being equal, if a glass does not break when dropped, that suffices to show that it is not fragile. But if propensity is to be analogous to fragility, the result of a chance trial is clearly not the analogue of the glass breaking since, in flat contrast to the latter, it must be supposed *not* to be the invariable result of any trial on the set-up. If it were so, the trial simply would not be of a chance set-up at all. Other things being equal, if a chance trial does not have any given result, that does not suffice to show that the set-up lacks the corresponding propensity.

Thus, in Mellor's view, the results of testing a "disposition" are always uniform, i.e., invariably the same, necessarily; while the results of trials on a "chance set-up", by contrast, cannot possibly be invariably the same. If "propensity" is supposed to be a dispositional property like "fragility", therefore, then the feature of a propensity which is analogous to the breaking of a glass when it is dropped cannot possibly be the result of a trial on a chance set-up: "If propensity, then, is a disposition of a chance set-up, its display, analogous to the breaking of a fragile glass, is not the result, or any outcome, of a trial on the set-up. The display is the chance distribution over the results, ..." (Mellor [1969], p. 26). Thus, by identifying the "display" of a propensity with its chance distribution, Mellor believes that he has succeeded in developing the propensity concept of probability as a dispositional construct perfectly analogous to fragility.

It is evident that Mellor's objective is the explication of a propensity construct which is "dispositional" in a strong sense. This requires that the presence of that property be "displayed" on each and every trial to which it is subjected; otherwise, it would fail to be completely analogous to fragility, which Mellor takes as his dispositional paradigm. By identifying the "display" of a propensity with its chance distribution, therefore, Mellor has indeed succeeded in obtaining a conception which is "dispositional" in this strong sense. It is also evident, however, that Mellor obtains this "success" at considerable expense. In the first place, the claim that the "display" of a propensity always accompanies its trial has now become semantically tautologous. For, if there were no chance distribution, there would certainly be no propensity: a propensity, after all, *is* a chance distribution. In the second place, although Mellor may call its chance distribution the "display" of a propensity, this is obviously an empirically vacuous claim. For, while its chance distribution is necessarily present upon each and every trial with an experimental arrangement, this "display", as a statistical property of those singular trials, is not experientially accessible.

The basic difficulty with Mellor's analysis, therefore, is that it relies upon an ambiguous concept of display. Normally, of course, a property is "displayed" by its observable manifestations. In Mellor's scheme, however, a property is also "displayed" by its mere theoretical presence. It seems far preferable, after all, to envision the "display" of a propensity as the relative frequencies it generates during long and short runs of trials. Thus, it is apparently Mellor, rather than Popper and Hacking, who is confused upon this point. These considerations, moreover, suggest the underlying cause of his troubles, namely: Mellor has mistaken an analogy for an isomorphism. These properties, after all, may perfectly well be alike in certain respects without being alike in all. A propensity, for example, is a statistical disposition; fragility, by contrast, is a universal disposition. These properties, therefore, are alike in being dispositions, yet unlike insofar as one is of statistical, the other of universal, strength. In adopting fragility as the paradigm for his analysis, therefore, Mellor placed himself in a conceptual dilemma.

Although Mellor therefore fails to take seriously the notion of a single case statistical disposition, it is important to observe that he further endeavors to preserve a Humean conception of these causal connections by means of principles of inference regarding particular "coherent betting quotients":

The sense thus given to the gambler's claim, that an A-event has a chance p of being attended by a B-event, should be entirely acceptable to a Humean. It is that the gambler can know that after some number of repeated bets at CBQ p he will break even. His knowledge here is of nothing but facts about occurrences of A-events and B-events, since these suffice to determine whether he does break even. No non-Humean connections in nature are needed to establish the fact that this A-event, by virtue of being of kind A and of the truth of the statistical law, is such that it is reasonable to prefer a partial belief of degree p to any other on its being attended by a B-event. And that is, on the propensity theory, just what it is to say that p is the chance – the objective, empirical, single case probability truly ascribable to the situation. (Mellor [1971], p. 164)

Thus, by attempting to relate single case propensities and coherent betting quotients, Mellor believes that he has succeeded in developing a construct which does not violate hoary Humean principles. But there are at least two important reasons for doubting the success of his endeavor, as we are about to discover.

The first, of course, is that an extensional interpretation of lawlike sentences, universal or otherwise, affords no foundation, in principle, for supporting counterfactual and subjunctive conditionals, a feature essential to the adequacy of any conception of statistical hypotheses as covering laws. Admittedly, Mellor himself does not appear to be altogether comfortable with this aspect of his position, acknowledging, in fact, that, "It is debatable whether a Humean can so distinguish lawlike from accidental universals", if these sentences cannot support counterfactuals (Mellor [1971], p. 164). But awareness of this problem does not provide a suitable substitute for its solution; indeed, it appears to be an irony of his position that Mellor forfeits what perhaps qualifies as the *single greatest strength* of a dispositional construction, namely: that dispositional properties, properly understood, supply the ontological justification for subjunctive and counterfactual attributions. Indeed, in light of this aspect of his analysis, it should come as no surprise that Mellor's views on dispositions, in general, as set forth in Mellor [1974], appear to provide a measure of understanding which is less than completely comprehensive, as I have sought to explain in detail in Fetzer [1978a].

The second, moreover, is that the rationale which Mellor does provide for his identification of single case propensities and coherent betting quotients, i.e., CBQ's, itself appears to be incapable of withstanding critical scrutiny, as recent discussions by Giere [1973], by Sklar [1974], and by Salmon [1979a] all suggest. As Giere, in particular, has emphasized, "there is an important technical difficulty with the argument concerning the definition of 'breaking even'. It is true that the *ratio* of successful bets approaches the betting rate with probability one. But the *difference* between numbers of success and failure is not so constrained. It may simply grow ever larger." Thus, since success or failure to a gambler with a finite stake is a matter of the difference, rather than of the ratio, between his actual number of successful bets and his actual number of failures, Mellor has failed to establish a coherent rationale for identifying the numerical values of single case propensities p with the numerical values of chance betting quotients q. As Giere suggests, it may be no accident that Mellor finds himself compelled to resort to arguments such as these; indeed, a more adequate defense of his conception might even be a theoretical impossibility.

Although Mellor's analysis thus fails to satisfy conditions (i) and (ii) and should therefore be rejected, our examination of his approach provides an opportunity to distinguish several senses in which propensities may or may not be envisioned as "invariable", i.e., as physical constants. These senses pertain to (a) their strength distributions, (b) their hypothetical long run displays, and (c) their actual single case outcomes.

(a) Although a propensity may be regarded as either (i) the strength of the dispositional tendency for a particular outcome, or (ii) the strength distribution over all the possible outcomes, in both cases propensities are viewed as invariable properties of appropriately specified experimental set-ups.

(b) These properties, moreover, generate characteristic hypothetical long run relative frequencies for their various outcomes, which may be calculated by means of Bernoulli's theorem, where these relative frequencies, like the propensities they measure, are also regarded as invariable.

(c) But while the propensity to produce a particular outcome on a singular trial with an experimental set-up is a constant property of that set-up, the results of trials actually conducted with such set-ups are not at all invariable, unless, of course, the property involved happens to be a universal, rather than a statistical, disposition.

Since Giere interprets propensities as single case statistical dispositions which attribute non-extensional causal connections to experimental arrangements, his analysis represents an enormous improvement over that advanced by Mellor. Not the least of the benefits of Giere's [1973] explication, for example, is a clarification of the relationship between degree of belief and single case propensity, namely: "If a set-up has a tendency of strength ½ to produce outcome A on a particular trial, then ½ is the correct degree of belief in A on that trial because this is the value for which the subjective expectation fits the objective expectation". The difficulties which afflict Giere's interpretation, therefore, are subtle features that might very easily be regarded as "minor points of detail" or as "matters of slight consequence". The author, on the contrary, shall contend, in the arguments to follow, that these features, however subtle, are precisely those that distinguish Giere's analysis from an adequate explication of the properties under consideration. The first of these concerns the solution Giere recommends to the problem of the single case, which we will now consider.

According to Giere [1973a], Popper's [1959] explication of propensities as dispositional properties does not succeed for at least two (superficially quite different) reasons: first, he offers (what Giere takes to be) a "long run", rather than a "single case", analysis; and, second, the analysis that he does provide fails to solve the single case reference class problem. As I explained in Fetzer [1971], Popper's own account appears to be ambiguous, insofar as certain of his remarks support a "long run" construction, while other remarks suggest a "single case" interpretation. More interesting to contemplate, therefore, is the force of Giere's second criticism, which he formulates, with respect to the issue of what should count as a repetition of "the same experiment", under Popper's explication (which he believes to be defective), as follows:

> Thus Popper fails to solve the problem of the single case for the old-fashioned reason that he provides no solution to the problem of the reference class. By contrast, a single-case propensity interpretation, which makes no essential reference to any sequence (virtual or real), automatically avoids the problem of the single case and the problem of the reference class as well. The propensities associated with a trial of a setup are no more relative to any description than any other physical characteristics, e.g., mass, specific gravity, or electrical conductivity. The prima facie objections to the single-case propensity interpretation are entirely different. (Giere [1973a], p. 473)

Although the foundation for Giere's optimism is not altogether obvious, it appears to be rooted in his adoption of the following conception: "we will call a system stochastic if, and only if, for every possible initial state, it has a probability distribution defined over its final events" (Giere [1979], p. 442). Indeed, Giere reiterates essentially the same conception in a number of other places (including Giere [1975], p. 218; Giere [1976], pp. 70–71; and especially Giere [1977], pp. 43–45).

Admittedly, there is a certain aura of plausibility in Giere's straightforward declaration that a single case propensity "is a characteristic of a particular trial of a designated system – no reference to a repeatable *kind* of system is necessary" (Giere [1977], p. 44). But, plausible or not, this claim is either trivial or false. If Giere intends to maintain that properties of this variety are *single-case properties* which distinguish each individual trial, then he may preserve the truth of his contention, but only at the expense of its triviality; for, as Quine ([1974], p. 5) has also noted, *in the absence of principles for the specification of trials by their kinds, the dispositional proper-*

ties, *statistical or otherwise, which are thought to attend the occurrence of these individual trials*, $T^1, T^2, \ldots,$ *and so on, must be attributed to the totality of properties present at each such trial*. Thus, unless Giere is willing to concede the theoretical necessity for each trial to be characterized as one of *a system of a specific kind K*, then all of the arguments which have been considered with respect to the frequency interpretation will apply, *mutatis mutandis*, to Giere's version of the single case propensity interpretation. Consequently, of any particular outcome O^j and trial T^i, since that outcome either occurs or does not occur, it will be theoretically impossible, without such principles, to hypothesize or to conjecture which properties, if any, among all those present at each such singular trial, were causally (or, nomically) irrelevant to its outcome. But the adoption of such principles for the appropriate specification of experimental arrangements is *logically equivalent* to the characterization of those systems by their repeatable kinds. Consequently, Giere fails to solve the problem of the single case for the old-fashioned reason that he provides no solution to the problem of the reference class.

Another fascinating aspect of Giere's explication is his desire to assimilate dispositions of the statistical strength 1 to dispositions of the universal strength u as exceptionless uniformities. Formally speaking, he proposes to define "system necessity" for a stochastic system, "not simply (as) unit propensity, but, so to speak, (as) unit propensity in all possible worlds" (Giere [1979], p. 444). As Giere explains, the only stochastically necessary result, for stochastic systems in general, is the complete set of possible outcomes for that stochastic system; nevertheless, it seems to him desirable to identify *stochastic necessity*, in general, with physical necessity, and *stochastic impossibility*, i.e., propensity zero, with physical impossibility. Perhaps the principal motivation for this maneuver, moreover, appears to be the preservation of a parallel with "inductive probabilities", where, according to familiar Bayesian schemes, an hypothesis may be assigned an "inductive probability" of 1 if, and only if, it is a logically necessary proposition. The problem with this approach, however, is that, in the case of *infinite sets of possible outcomes* (e.g., as would occur with an enormous dartboard and extremely fine darts in relation to each point of impact upon that board: each specific point will be the point of impact with propensity zero, while one or another point must be the point of impact with propensity one), a paradox arises (which Giere attributes to Skyrms), namely: that it is physically necessary that *some* specific outcome occur, while it is physically impossible for any *specific* outcome to occur; consequently, with physical necessity, some physically impossible outcome must occur (Giere [1979], p. 447

and note 7). Giere concludes with uncertainty over the resolution of this dilemma, since current physical theories for stochastic systems suggest there *are* physically possible outcomes that should be assigned probabilities of measure zero, which conflicts with the philosophical desideratum of identifying physical probabilities with propensities and zero propensities with physical impossibilities (Giere [1979], p. 449).

Although Giere's discussions of the propensity interpretation tend to focus upon examples from the quantum domain, the problems that are involved have more general importance; as Hacking ([1973], p. 486) has observed,

> ... we constantly use propensity models in every kind of gross affair: agriculture, meteorology, medicine. We need urn models of epidemics to discover whether we have an epidemic and should quarantine the neighbors. It is mere gossip to suggest we know some connection between quanta and infection. Even if current microphysics turned out all wrong and were replaced by a deterministic theory we should still use urn models. For the propensity theorist to rest his arguments on quantum theory is to retain an equivocal joker while handing all the aces to the Bayesian personalists.

Indeed, the considerations which have gone before strongly support the view that the single case propensity interpretation provides the only account of physical probability which is capable of converting statistical hypotheses into covering laws; consequently, there are excellent grounds for *not* abandoning the conception of probabilities as propensities. Moreover, as Hacking reminds us, the theoretical utility of this interpretation, within statistics and without, appears to be substantial. If recent physical theories, quantum mechanical or otherwise, suggest that some physically possible outcomes *must* be assigned the measure zero, therefore, abandoning propensities hardly seems to be a reasonable theoretical option. Dispensing with the identification of zero propensities and physical impossibilities, however, would appear to be both rational and responsible, especially in the wake of Skyrms' paradox.

All of this, of course, makes very good sense from the perspective of the formal semantics for single case causal tendencies advanced in Chapter 3; for the arguments presented there supported the necessity for exhibiting the frequency relations that obtain over infinite trial sequences in order to formalize the features of single case statistical dispositions, in particular. Giere [1976a], we discovered there, attempts to do *without* these frequency relations in an endeavor to draw the distinction between probabilities as frequencies and probabilities as propensities more forcefully. The consequences attending this maneuver that have been uncovered here, however, when combined with those displayed by our previous investigations, strongly suggest that *it is not theoretically desirable and may even be logically inconsistent to identify propensities of measure one with physical necessities.* In the absence of this identification, moreover, there will be circumstances in which an accumulation of propensities of measure zero will equal measures of one, and conversely; however, the mathematical properties of these conditions are not therefore mysterious, since they are those of standard measure theory, in general (Kolmogorov [1956], for example).

According to standard measure theory, let us note, it is appropriate for physical impossibilities to be assigned measures of zero, but not conversely; that is, measures of zero are not assigned solely to physical impossibilities. As Kolmogorov has remarked,

To an impossible event (an empty set) corresponds, in accordance with our axioms, the probability $P(\emptyset) = 0$, but the converse is not true: $P(A) = 0$ does not imply the impossibility of A. When $P(A) = 0$, from principle (b) [i.e., *The Single Case Principle*] all we can assert is that when the conditions $\mathscr{S}$ are realized but once, event A is practically impossible. It does not at all assert, however, that in a sufficiently long series of tests the event A will not occur. On the other hand, one can deduce from the principle (a) [i.e., *The Short Run Principle*] merely that when $P(A) = 0$ and n is very large, the ratio m/n will be very small (it might, for example, be equal to $1/n$). (Kolmogorov [1956], p. 5)

Thus, these considerations are reflected by condition C3 of the model theory for the probabilistic causal calculus $\mathscr{C}$; for the condition that $f(p, i) \nsubseteq [q]$ is superfluous except for the case in which (the strength) n equals one, and the condition that $f(p, i) \nsubseteq [-q]$ is superfluous except for the case in which (the strength) n equals zero.

Conditions such as these are quite essential in order to distinguish between '$p \ni_u q$' and '$p \ni_1 q$', on the one hand, and '$p \ni_u -q$' and '$p \ni_0 q$', on the other. For the truth of '$p \ni_1 q$' must allow that 'q' could be false even though 'p' were true, i.e., that '$p \ni_u q$' (and '$p \ni q$') are false; and the truth of '$p \ni_0 q$' must allow that 'q' could be true even though 'p' were true, i.e., that '$p \ni_u -q$' (and '$p \ni -q$') are also false. These ontological differences, of course, have relatively straightforward logical and epistemic consequences. Moreover, since the strength of single case propensities has to be displayed through relative frequencies, it should be both obvious and intuitive that measures of zero and of one are logically compatible with arbitrarily large numbers of exceptions. Thus, to paraphrase Peirce, *these frequency relations appear to be theoretically indispensable to bring out the full consequences of these single-case "would-be"s*. Although Giere's attitude toward "description-relativity" will receive further attention in Chapter 6, therefore, there are ample grounds for doubting the adequacy of the conception he recommends, especially since it violates conditions (iii) and (iv) together.

A SINGLE-CASE THEORY OF EXPLANATION

The conception that remains envisions these probabilities as single case statistical dispositions attributing non-extensional causal connections to appropriately specified experimental arrangements, where a distinction is drawn between dispositional tendencies of statistical strength 1 and dispositional tendencies of universal strength u, i.e., the single case construction which is presented in detail in Chapter 3. As we have previously ascertained, this construction affords an interpretation of statistical hypotheses as probabilistic generalizations which (a) support counterfactual and subjunctive conditionals and (b) are logically unrestricted to any finite number of instances during the course of the world's history; the issue that remains, therefore, is whether these probabilistic generalizations (c) can function as covering laws in explanatory arguments of the appropriate form. This conception will be an adequate interpretation, relative to Levi's desiderata, therefore, provided that it affords a suitable foundation for an appropriate explication of the logical structure of causal explanations for singular events. Pursuit of this objective, however, presupposes several preliminary formulations, which we shall now consider.

Notice that the specification of an adequate explanation is necessarily relative to the selection of some appropriate explanandum sentence E, which delineates those specific features of the explanandum-phenomenon that are of particular interest within that explanation situation, i.e., the choice of an explanandum sentence is *pragmatically determined*. This pragmatical component appears to be ineliminable, in principle, since any specific outcome response that occurs at some time t^* during the course of the world's history, say, O, will be a singular event which — by virtue of the infinity of properties and relations it bears to other events, provided the density principle for such event descriptions is correct — cannot even be exhaustively described; consequently, no event is explainable in every respect, but only as *an event of a certain kind* or, more precisely, as *an event under a certain description*, as Hempel especially has emphasized. Consideration should therefore be directed to the appropriate form for explanandum-descriptions.

Careful readers may have observed that Hempel and Salmon adopt somewhat different conceptions of the appropriate form of explanandum-sentences, since Hempel's explication is relative to the explanandum-phenomenon described by the explanandum-sentence "Ai" (or of the fact that the individual i is a member of the attribute class A), while Salmon's explication is relative to the explanandum-phenomenon described by the explanandum-sentence "$Ri \cdot Ai$" (or of the fact that i, a member of the reference class R, is also a member of the attribute class A). Since the specific object or event under consideration may be specified by means of a proper name or a definite description, which serves to identify, let us say, "the individual i", while the corresponding reference class R might be specified misleadingly or not suggested at all by particular formulations of explanation-seeking why-questions, let us assume that Hempel's construction is preferable to Salmon's construction, with the provision that these descriptions are understood as *eternal sentence* forms, such as, '$Oct*$', of corresponding *occasion sentences*, such as, 'Oc' (where, as before, 'c' denotes an individual constant rather than an ambiguous name).

A degree of caution must be exercised, of course, in distinguishing between the various conceptions of maximal specificity we have heretofore considered; in particular, (a) the requirement of maximal specificity for true scientific conditionals (as elaborated in Chapter 3) must be distinguished from (b) Hempel's revised requirement of maximal specificity ($RMS*$) for adequate statistical explanations (as elaborated in Chapter 4), especially insofar as the recommended reformulation of (b) previously proposed, ($RMS**$), must not be confused with the condition of adequacy for causal explanations that will now be proposed, which we may refer to as (let us say) (c) *the requirement of strict maximal specificity (RSMS)*, where this requirement, unlike those based upon *frequency criteria* of statistical relevance, depends upon *dispositional criteria* of nomic (or, causal) relevance (as elaborated in Chapter 3). The purpose of this condition, moreover, is to exclude nomically (or, causally) irrelevant predicates from the lawlike premise(s) of adequate explanations for the occurrence of an explanandum-phenomenon.

Since nomically (or, causally) irrelevant predicates possess no explanatory significance with respect to explaining why a member of the reference class K (under a description 'K' within a language $\mathfrak{L}zt$) possesses a certain property χ, on the one hand, or to explaining why a member of the reference class $K \cdot T^i$ (under a corresponding description, once again) possesses a certain other property O^i, on the other, an adequate conception of explanation should incorporate a rule for selecting the appropriate lawlike sentence to include in the explanans for the explanation of all such explananda, namely:

> *The Requirement of Strict Maximal Specificity*: An explanation of why an explanandum event — the possession of a property χ by an individual c or an outcome response O^i — occurs is adequate *only if* every property described by the antecedent condi-

tion(s) of any lawlike sentence S in the explanans of that explanation is nomically relevant to the occurrence of its attribute property χ or outcome response O^i, within $\mathfrak{L}zt$.

An alternative but equivalent formulation of this requirement (that some may find formally more revealing), moreover, may be advanced as follows: if 'S' is a lawlike sentence which occurs in the explanans for an explanandum sentence 'E', 'K' is the reference class description of 'S' and 'G' is a predicate which is *not* nomically relevant to the occurrence of e (as described by 'E' in $\mathfrak{L}zt$), then neither '$(x)(t)(Kxt \supset Gxt)$' nor '$(x)(t)(Kxt \supset -Gxt)$' is a logical truth in $\mathfrak{L}zt$. Since the requirement of maximal specificity requires that any lawlike sentence must include *every* predicate nomically relevant to its attribute property (when it is true), the requirement of *strict* maximal specificity requires further that any scientific explanation should include *only* predicates nomically relevant to its explanandum phenomenon within its lawlike premises (when it is adequate).

We may assume that explanations are properly supposed to be answers to explanation-seeking why-questions, which are susceptible to formulation with appropriate clarity by employing meta-linguistic conditions of the Hempelian variety, rather than object-language conditions of the Salmonian variety, as should be apparent from the importance of the language framework $\mathfrak{L}zt$ for the description of the explanandum-phenomenon, for the characterization of conditions of nomic (or, causal) relevance, and so forth. Consequently, we shall interpret explanations as *arguments*, i.e., as sets of sentences divided into two parts, explanans and explanandum, where the explanans is intended to provide appropriate grounds for explaining why the explanandum phenomenon e (as described by the explanandum sentence 'E') has occurred at the time t^*; however, we shall subsequently consider whether there are any outstanding objections, such as Stegmüller's [1973] so-called "paradox of the explanation of the improbable", which successfully undermine such a conception.

On the basis of these considerations, it is possible to advance a single criterion for the adequacy of a (*nomically significant*) *causal explanation* for the occurrence of a singular event, regardless of whether that event occurs as a manifestation of a universal or of a statistical nomological generalization:

(V)   A set of sentences S, known as the "explanans", provides *an adequate nomically significant causal explanation of the occurrence of a singular event described by another sentence E, known as its explanandum, relative to the language framework* $\mathfrak{L}zt$, if and only if:

(a) the explanandum is either a deductive or a probabilistic consequence of its explanans;

(b) the explanans contains at least one lawlike sentence of (universal or statistical) 'causal' form that is actually required for the deductive or probabilistic derivation of the explanandum from its explanans;

(c) the explanans satisfies the requirement of strict maximal specificity (*RSMS*) with respect to its lawlike premise(s); and,

(d) the sentences constituting the explanation — both the explanans and its explanandum — are true, relative to the language framework $\mathfrak{L}zt$.

The reference class description of any lawlike premise that occurs in such an explanation, therefore, must include *all and only* those predicates nomically (or, causally) relevant to the occurrence of that explanandum-phenomenon. A set of sentences satisfying conditions (a) and (b), from this point of view, may be regarded as a "potential" explanation, while those which also satisfy (c) and (d) are "adequate", within the language framework $\mathfrak{L}zt$. [The precise significance of the qualifying phrase, "nomically significant", incidentally, will receive consideration in Chapter 10.]

This criterion supports the conception that, from the logical point of view, there are two kinds of causal explanation, depending upon whether the general law(s) invoked in the explanans are essentially universal or essentially statistical in kind, i.e., depending upon whether they concern a dispositional tendency of universal or of statistical strength. If the law(s) invoked in the explanans are essentially universal, the logical properties of the relationship between the sentences constituting an explanans and its explanandum will be those of complete (deductive) entailment; while if they are essentially statistical, this relationship will be that of only partial (deductive) entailment. The logical relation, in either case, is *strictly deductive*; in particular, contrary to Hempel's previous explications, it is not one of logical probability in Carnap's sense, i.e., it is *not* a relation of *evidential support*. Consequently, these two kinds of explanation may be called "universal-deductive" and "statistical-probabilistic", respectively.

An example of a *universal-deductive* explanation for the occurrence of an explanandum event would be the following:

(CL)   For all x and all t, if x were gold at t, then heating x to 1063°C at t would invariably bring about its melting at t*;
(C1)   Jan's bracelet b is made of gold at t;
(C2)   Jan's bracelet b is heated to 1063°C at t;
(ES)   Jan's bracelet b melts at t*.

For surely the explanans invoking the covering law (CL) together with the initial conditions (C1) and (C2) fulfills the criteria for an adequate explanation of its explanandum-sentence (ES) (provided, of course, that the requirement of strict maximal specificity has been fulfilled and these sentences are true). This explanation, moreover, exemplifies the characteristics of (nomically significant) universal-deductive causal explanations in general and may schematized as follows:

(VI) $$\frac{(x)(t)[Gxt \ni (Hxt = 1063°C \ni_u Mxt^*)]}{Gbt \cdot Hbt = 1063°C} \quad [u]$$
$$Mbt^*$$

where the single line separating the explanans from its explanandum is intended to indicate that the logical properties of the relationship between this explanans and its explanandum are those of complete, rather than partial, entailment; and the symbol in brackets indicates that the degree of nomic expectability for an outcome of this kind on a trial of this kind is of universal, as opposed to statistical, strength.

An example of a *statistical-probabilistic* explanation for another explanandum event might likewise be the following:

(CL)   For all x and all t, if x were polonium^{218} at t, then a time trial of three minutes duration at t would bring about the loss of nearly half the mass of x by t + 3 minutes with strength .9.
(C1)   Jones brought a sample s of 8 grams of polonium^{218} into the lab for analysis at t;
(C2)   Jones weighed the sample s three minutes after bringing it in;
(ES)   The mass of s at t + 3 minutes was approximately 4 grams total.

For surely the explanans invoking the covering law (CL) together with the initial conditions (C1) and (C2) fulfills the criteria for an adequate explanation of its explanandum-sentence (ES) (provided, once again, that the requirement of strict maximal specificity has been fulfilled and these sentences are true). This explanation similarly exemplifies the characteristics of (nomically significant) statistical-probabilistic causal explanations in general and may be schematized as follows:

(VII) $\quad (x)(t)[Mxt = 8 \text{ gms} \mathrel{\supset\!\!\!-} (Txt = 3 \text{ min} \mathrel{\supset\!\!\!-}_{.9} Mxt^* = 4 \text{ gms})]$

$$\frac{Mst = 8 \text{ gms} \cdot Tst = 3 \text{ min}}{Mst^* = 4 \text{ gms}} \quad [.9]$$

where the double line separating the explanans from its explanandum is intended to indicate that the logical properties of the relationship between this explanans and its explanandum are those of partial, as opposed to complete, entailment; and the number in brackets indicates the degree of nomic expectability for an outcome of this kind on a trial of this kind which is, of course, of statistical, rather than universal, strength.

Although Hempel assumed as recently as [1965] that the number in brackets is an inductive probability in Carnap's [1962] sense, Hempel [1968] has exhibited less certainty over the propriety of that interpretation; indeed, he has expressed the opinion that Reichenbach's policy of assigning singular events to "the narrowest reference class" should now be understood as Reichenbach's version of the *requirement of total evidence* and that *this requirement is not an explanatory relevance condition* (Hempel [1968], pp. 121–122). From the present perspective, moreover, Hempel's own revised requirement of maximal specificity (*RMS**), when employed in conjunction with frequency criteria of statistical relevance, may appropriately be entertained as applying within the *context of prediction*, even though, as we have discovered, it may not serve its intended purpose within the *context of explanation*. Thus, for this specific purpose, Hempel's original definition of statistical relevance would appear to be theoretically unobjectionable, provided, of course, that these probability sentences are no longer required to be *lawlike*. These implications, of course, make it overwhelmingly unlikely that Hempel's thesis of the symmetry of explanation and prediction could possibly be sustained.

Since Hempel's [1965] rationale for interpreting the bracketed numbers as inductive probabilities in Carnap's [1962] sense does not apply at all on the present explication, let us consider this issue in some detail. Under the single case dispositional construction of lawlike sentences, these numbers are properly envisioned as "deductive", rather than as "inductive", in character, in the sense of being the meta-language analogue of the nomological relation in question, i.e., they both convey similar content, but they do so from different points of view, namely: as the *strength of the dispositional tendency* for arrangements of that kind to produce outcomes of that kind on any singular trial, in the case of n (or, u); and, as a description of the relevant initial conditions and lawlike generalizations, because of which that precise *degree of nomic expectability* obtains on that specific individual trial, in the case of [n] (or, [u], as it happens to be). Thus, the single case dispositional rationale for equating the numerical values of n and [n] is decidedly different than the standard "long run" (frequency or propensity) rationale; indeed, it even differs from Mellor's [1971] conception, where single case probabilities serve as "coherent betting quotients".

Nevertheless, there is an important intersection between these different approaches to understanding the significance of such bracketed numbers; for the single case dispositional analysis provides a justification for why these numbers may *also* be envisioned as "fair betting quotients". This connection is clearly reflected by Reichenbach's [1949] *logical interpretation*, according to which "probability" serves as a meta-language operator applying to sentences as its objects. On this interpretation, probability sentences represent not the frequency with which a particular outcome occurs within a certain sequence of trials, but rather the frequency with which particular *sentences* describing those outcomes happen to be true relative to a sequence of *sentences* describing those trials. If the statistical probability for an ace as the outcome within some sequence of trials consisting of tosses with a fair die and tossing device happens to be 1/6, for example, then it will necessarily be the case that the logical probability with which sentences asserting, "This outcome is an ace", accompany the members of the sequence of sentences asserting, "This trial consists of a toss with a fair die and tossing device", equals 1/6 as well (since these assertions will be true if and only if the events they describe occur). For, under this conception, it would be possible to claim that a particular explanandum, regarded as a "conclusion", follows from a particular explanans, regarded as its "premises", with a certain logical probability, whose numerical value represents the frequency of the truth of that conclusion, given the truth of its premises (theoretically assuming, of course, that all of the relevant properties of each such trial have been specified, i.e., that all of the relevant premises are available).

Since these logical probabilities are defined by limiting frequency relations, it might be suggested that, relative to explanations of forms (VI) and (VII), even on "single case" interpretations, these arguments merely explain their explanandum events by showing that they are to be expected with certain limiting frequencies "over the long run" given the specification of initial conditions included in their explanans. But that would be a faulty inference, for the value of [n] is intended to characterize the strength of

CAUSAL EXPLANATION   131

an explanans relative to its explanandum — where the event to be explained is *the occurrence of a specific outcome on a singular trial*. There is no doubt that the long run constructions exercise a grip upon long run results which differs categorically from the hold exerted by the single case construction. But relative to the explanation of singular events, the grip exerted by single case propensities is not matched by the hold exercised by long run probabilities. For the issue is *not* one of the limiting frequency with which an explanandum sentence (such as, "This outcome is an ace") accompanies a description of *initial conditions* (such as, "This trial consists of a toss with a fair die and tossing device"), but rather one of the connection between an entire *explanans*, i.e., the description of initial conditions together with the relevant general laws, and its respective *explanandum*.

The single case dispositional construction, therefore, establishes the "fair betting quotient" for a particular outcome O, under conditions of the specified kind, not only over the long run but also for the single case; for the numerical value of $[n]$, viewed as a "degree of nomic expectability", not only functions as an *estimation* of the limiting frequency with which sentences describing outcomes of that kind would be true over a long sequence of trials (with the force of overwhelming probability) but also functions as a *designation* of the degree of entailment with which such an explanandum follows from such an explanans on any singular trial (with the force of logical necessity). Although the long run constructions are capable of providing a *long run justification* for identifying (long run) probabilities with single case wagers, therefore, only the single case construction is able to provide a *single case justification* for identifying (single case) probabilities with single case wagers. My motivation for characterizing $[n]$ as the "degree of nomic expectability", therefore, is not only to avoid the possibly misleading connotations that might attend its unqualified endorsement as a "degree of partial entailment" (relative to Carnap's explication), but also to focus attention upon the theoretical aspects of this "single case" conception.

One important measure of the acceptability, in principle, of these conditions of adequacy for (nomically significant) causal explanations of singular events, of course, is the extent to which they provide intuitively satisfactory theoretical resolutions of (otherwise) *problematical cases*. Among those of special interest with respect to the present explication, for example, are presumptive instances of "universal-deductive" explanations such as Kyburg [1965] has proposed, an illustration of which is the following:

(CL)   For all x and all t, if x were a sample of table salt which had a

dissolving spell cast upon it at t, then dropping x into water at t would invariably bring about its dissolving at t^*;
(C1) This is a sample s of table salt at t;
(C2) This sample s has had a dissolving spell cast upon it at t;
(C3) This sample s is dropped into water at t;
(ES) This sample s dissolves at t^*.

Although this "explanation" satisfies all of Hempel's conditions of adequacy for "deductive-nomological" explanations (Hempel [1965]), it still appears to be unsatisfactory, insofar as the dissolving spell, especially, is not causally relevant to the occurrence of the explanandum event which is described by the explanandum sentence (ES). The question therefore arises, "Does such an 'explanation' qualify as *adequate* relative to conditions (a)–(d) above?"

The answer, of course, is not difficult to discern, since, in particular, condition (c), i.e., the requirement of strict maximal specificity, precludes the presence of nomically (or, causally) irrelevant predicates G in the reference class description K (or, $K \cdot T^i$) of any lawlike premise actually required to derive that explanandum from its explanans. If the property of having had a dissolving spell cast upon it *is* nomically (or, causally) irrelevant to the occurrence of that explanandum event, therefore, then an adequate (nomically significant) *universal-deductive* causal explanation for such an explanandum could exhibit the following form:

(VIII) $$\frac{(x)(t)[TSxt \mathrel{\ni\mkern-5mu-} (Wxt \mathrel{\ni_u} Dxt^*)]}{TSst \cdot Hst \cdot Wst} \quad [u]$$
$$\overline{Dst^*}$$

in which the occurrence of the sample's dissolving at t^*, i.e., Dst^*, would be explained by its being table salt at t, i.e., $TSst$, and by its being submerged in water at t, i.e., Wst (provided, of course, that the requirement of strict maximal specificity ($RSMS$) has been fulfilled and that these sentences are true). The property of having had a dissolving spell cast upon it, i.e., Hst, serves the role of a superfluous premise, which is not required for the derivation of the explanandum from its explanans and consequently possesses no explanatory significance within this specific explanation situation. In order to appreciate the function of ($RSMS$) even further within this context, notice that, if the property of being *table salt TS*, rather than *salt S,* is not nomically (or, causally) relevant to the occurrence of that explanandum, then that property must also be excluded from the required lawlike sentence.

Examples of this kind, of course, are not exclusively restricted to "deductive-nomological" explanations, since Salmon [1965] has offered parallels which satisfy Hempel's conditions of adequacy for "inductive-statistical" or "probabilistic" explanations (Hempel [1965], [1968]). Thus, a corresponding illustration would be the following:

(CL) For all x and all t, if x has a cold at t and x takes vitamin C at t, then a rest period of a week's duration would bring about recovery from that cold by $t + 7$ days with, say, strength .75;
(C1) Smith i has a cold at t;
(C2) Smith i takes vitamin C at t;
(C3) Smith i has a rest period of a week's duration at t;
(ES) Smith i recovers from his cold by $t + 7$ days time.

Once again, of course, although this "explanation" satisfies all of Hempel's conditions of adequacy for explanations of this kind, it appears to be unsatisfactory, so long as taking vitamin C, in particular, is not causally relevant to the occurrence of the explanandum event described by the explanandum sentence (ES). An adequate *statistical-probabilistic* explanation for such an explanandum could exhibit the corresponding form:

(IX) $$\frac{(x)(t)[Cxt \mathbin{\ni} (Txt = 7 \text{ days} \mathbin{\ni_{.75}} Rxt^*)]\quad Cit \cdot Vit \cdot Tit = 7 \text{ days}}{Rit^*}\ [.75]$$

in which the subject's recovery from his cold by t^*, i.e., Rit^*, is explained by his contracting the cold at time t, i.e., Cit, and by his having a period of rest of a week's duration thereafter, i.e., $Tit = 7$ days (provided, once again, that the requirement of strict maximal specificity ($RSMS$) is fulfilled and that these sentences are true). The property of consuming vitamin C at t, i.e., Vit, of course, occurs here as a superfluous premise; but whether or not that property actually possesses *no* explanatory significance within this specific explanation situation obviously depends upon the dispositional properties and nomological regularities of the actual world W. Likewise, the assumption that all of the explanatorily significant properties relevant to that specific explanandum event have been taken into account depends upon the truth of each of the sentences which is actually required for the probabilistic deduction of the explanandum from its explanans including, especially, its lawlike premise.

Another important measure of their acceptability, of course, is the extent to which they provide intuitively satisfactory resolutions of (alleged) *explanatory paradoxes*. Perhaps the instance of greatest relevance, moreover, is

Stegmüller's [1973] "paradox of the explanation of the improbable", which both Krüger [1976] and van Fraassen [1978] have discussed. Van Fraassen has formulated this "paradox" with special clarity; consequently, his discussion will be considered here. Stegmüller advances a principle which he refers to as *the Leibniz condition*, which asserts that an adequate explanation for an explanandum outcome E must afford some *justification* for expecting the occurrence of E rather than of an incompatible outcome $-E$. Van Fraassen claims that this requirement is telling with respect to Hempel's explication, but not to Salmon's, whereas the opposite is true. The "paradox" itself comes to this:

> Consider the question why E occurred, rather than F, where F is some event incompatible with E. Let us require of the answer that it *provide a reason why E occurred rather than F*. We may take this to imply, though perhaps not be implied by, the condition that the information adduced in that answer *gives us a reason for believing that E occurs rather than F*. (van Fraassen [1978], p. 162)

Stegmüller presses this point by emphasizing that, if an explanation for the occurrence of an *improbable* outcome E cannot satisfy this condition, *neither* can the explanation for the occurrence of its *probable* alternative $-E$. Since van Fraassen is unable to discern a solution to this "paradox", he concludes that, "it puts into a nutshell the central difficulties for the philosophical account of explanation" (van Fraassen [1978], p. 163).

The heart of the matter, of course, appears to be profound confusion between "explanation-seeking" and "reason-seeking" why-questions, which Hempel [1965] and [1968] especially has contributed to distinguishing. If Stegmüller's requirements were rigorously enforced, they would entail even more disastrous consequences than did Hempel's own *high probability* requirement, namely: *that the explanation of events which occur only with some probability would be logically impossible, in principle.* Thus, desiderata of this kind might possibly serve some purpose within the context of *prediction* (or, perhaps more broadly, within the context of *induction*) but surely cannot be tolerated within the context of *explanation*. The strongest arguments for such a conception, no doubt, find their ultimate foundation in an *a priori* commitment to determinism; certainly no one who takes seriously the possibility of ontic indeterminism, i.e., the theoretical possibility of *ultimately irreducible* statistical explanations, should be disturbed by Stegmüller's "paradox". From the perspective of a dispositional explication of statistical explanation, moreover, the situation is not at all perplexing; for, so long as the explanandum outcome E is explained by taking into account *all and only* the nomically (or, causally) relevant properties of that explanation situation, its explanans will be adequate, in principle, even when outcomes of that kind occur with a strength between zero and one. Indeed, provided the conditions (a) through (d) are satisfied, the only "paradox" that remains would appear to be *to explain what more could possibly be desired of an adequate explanation*, ontic determinism notwithstanding. Although van Fraassen disparages Krüger's [1976] preliminary efforts to clarify the conception that *A tends to produce E*, the considerations which have led us to this juncture strongly suggest that Krüger's conceptions, unlike those van Fraassen embraces, are, in principle, right-headed.

Stegmüller's "paradox of the explanation of the improbable", therefore, certainly provides no basis for doubting the acceptability of a dispositional conception of (nomically significant) causal explanations for single events; indeed, this explication appears to illuminate and to resolve that "paradox", at least to the extent to which it does not hinge upon *a priori* commitments to determinism. When measured against the problematical cases (advanced by Kyburg and by Salmon), on the one hand, and the explanatory paradoxes (supported by Stegmüller and by van Fraassen), on the other, there appear to be ample grounds for drawing the conclusion that, under the single case dispositional interpretation, probabilistic hypotheses *are* capable of fulfilling the role of covering laws. Consequently, it appears reasonable to conclude, in the absence of contrary evidence, that the conception of probabilities as single case statistical dispositions actually succeeds in satisfying Levi's [1969] conditions (a), (b), and (c) for its acceptability. Since there may be other reasons for its rejection, however, including, for example, issues of testability within an epistemic context, let us simply assume that this interpretation qualifies as *prima facie adequate* and consider these matters further in chapters to come.

Since the development and elaboration of this conception has required careful, painstaking, and detailed dissection of alternative constructions, including, of course, the "long run" frequency and propensity explications, it is remarkable to discover that Levi himself has condemned this approach virtually as a matter of *philosophical methodology*. He castigates attempts to provide possible-world formal semantics for these interpretations, alleging that they obfuscate more than they clarify:

These obscurantist diversions are supplemented by controversies concerning whether probability dispositions are long run relative frequency dispositions or single case dispositions and whether or not the truth of chance statements precludes determinism. By insisting on the construal of chance predicates as primitive predicates (*vis à vis* descriptions of test behavior) in the manner outlined previously, I mean to emphasize the gratuitous, diversionary and obscurantist character of such "interpretations" and to insist on the fundamental importance of providing an account of direct inference to the understanding of the conception of chance or objective probability. (Levi [1977], p. 447)

Now, while it would be foolish to deny that probabilistic hypotheses represent extremely important problems with respect to questions of epistemology which an acceptable analysis of the scope and limits of scientific knowledge must surely resolve, it would be silly to suppose that nothing else matters. Although Levi maintains that, "the controversy over whether chances are long run dispositions or single case dispositions is another example of a dispute which gives propensity interpretations a bad name" (Levi [1977], p. 454), it is difficult to imagine how the problems with which we have been preoccupied — including the investigation of dispositional properties, lawlike sentences, causal connections and statistical explanations — could possibly be subject to successful resolution in the *absence* of consideration of that specific issue. Whether or not Levi rests content with the conception of *chance predicates as primitive predicates*, i.e., as formally undefined, it would be narrow-minded, indeed, to arbitrarily restrict the course of inquiry in different directions. Even though Levi ([1979], p. 342) persists in his assault upon alternative conceptions, therefore, let us bear in mind that the merits of a philosophical position are properly measured by the problems which it resolves: the dilemma he posed in Levi [1969] could not possibly have been dissolved, were our resources restricted to those he prescribes in Levi [1977] and in Levi [1979].

6. THE DISPOSITIONAL CONSTRUCTION OF THEORIES

Since the conception of (nomically significant) causal explanations for singular events advanced in Chapter 5 depends in part upon the construction of *scientific explanations as deductive arguments*, i.e., as always deductive and as sometimes probabilistic, the adequacy of that conception requires the acceptability of an interpretation of this kind. Although we have discovered that Stegmüller's "paradox of the explanation of the improbable" poses no insurmountable obstacles to this conception, perhaps there are other important reasons for doubting its adequacy; in particular, Salmon [1977a] and [1978] has suggested that deductive arguments possess certain characteristics which are *not* appropriate for scientific explanations, which therefore dictate the necessity to abandon this interpretation. In order to defend the propriety of the dispositional construction, therefore, we shall consider Salmon's views on this and related matters; for, as will soon become apparent, the position which he has begun to develop in these recent publications is not only inconsistent with his previous conception of scientific explanation (as he himself candidly concedes), but also incompatible with the distinction between causal and non-causal scientific explanations which emerges from a dispositional interpretation of lawlike sentences, as we shall subsequently ascertain.

Salmon's fundamental objection to the conception of explanations as arguments revolves about the role of *irrelevant considerations*; in particular, he believes that irrelevant properties undermine scientific explanations, while irrelevant premises do not undermine deductive arguments:

Inference, whether inductive or deductive, demands a requirement of total evidence — a requirement that *all* the relevant evidence be mentioned in the premises. This requirement, which has substantive importance for inductive inferences, is automatically satisfied for deductive inferences. Explanation, in contrast, seems to demand a further requirement — namely, that *only* considerations relevant to the explanandum be contained in the explanans. This, it seems to me, constitutes a deep difference between explanations and arguments. (Salmon [1977a], p. 151)

Thus, as Salmon points out, even in the relevance logic proposed by Anderson and Belnap [1975], the deductive principle, "if '$p \cdot q$' is true, then 'p' is true", is a valid schema; yet a parallel situation in the case of explanation could undermine the acceptability of the corresponding explanans. Insofar as the conditions of adequacy (a) through (d) that have been proposed do exclude explanatory irrelevancies from the *lawlike premises* of adequate explanations, as examples (VIII) and (IX) exemplify, perhaps these elements should also be excluded from any occurrence within the *explanans as a whole*, in which case, of course, (VIII) and (IX) would not exemplify adequate explanations, after all. Thus, Salmon's conception might require abandoning the interpretation of scientific explanations as deductive arguments completely; but it could also be entertained as supporting an additional condition of adequacy, as follows.

In order to appreciate this theoretical possibility, let us consider the prospects for reformulating the conception of explanation previously proposed by stipulating that the appropriate explanans, S, for a proffered explanation of the occurrence of a singular event described by another sentence, E, ought to qualify as "adequate" *only if S consists exclusively of sentences that are actually required for the deductive or probabilistic derivation of the explanandum from its explanans*. Although there appears to be nothing objectionable about such a requirement, in principle, it is not entirely obvious either that its presence is required or that, if adopted, it would serve its purpose well. Salmon's concerns, of course, are undoubtedly alleviated by the requirement of strict maximal specificity (*RSMS*), which has no analogue among Hempel's conditions of explanatory adequacy for deductive or statistical explanations. But this condition, although theoretically well-founded, is completely compatible with the conception of explanations as arguments. Moreover, even though such an additional condition would exclude those sentences not actually required for the derivation of the explanandum from its explanans, they would not also automatically exclude explanatory irrelevancies from explanandum-sentences as well: that a certain gold object happens to be *Jan's bracelet* or that 8 grams of polonium[218] happens to be *Jones's sample*, for example, introduce features of their explanation situations which may be pragmatically determined but are nevertheless explanatorily irrelevant to their explanandum-phenomena, *per se*.

Difficulties such as these, of course, arise for Salmon's explication of explanation, *mutatis mutandis*, just as they arise for Hempel's. Consequently, they appear to have no special force with respect to the question of whether or not explanations should be envisioned as arguments. In the absence of further evidence, therefore, it seems reasonable to conclude that, to the extent to which Salmon's rejection of the conception of explanations as arguments is theoretically well-founded, it does not count against its intended objective; and, conversely, to the extent to which Salmon's rejection of the conception of explanations as arguments counts against its intended objective, it is not theoretically well-founded. Although the addition of such a condition would generate *sufficient* conditions of explanatory adequacy, its incorporation is not also *necessary;* for, we have found no grounds here for thinking that the conditions (a) through (d) specified already are not both necessary and sufficient or for rejecting the implied conception of explanations as arguments. Indeed, the evidence in support of this conception has not been exhausted by our discussion here.

CAUSAL AND NON-CAUSAL EXPLANATIONS

All of the explanations which we have considered heretofore have shared an important but rather subtle feature, namely: their explanandum-phenomena have all involved the coming-to-have-a-property-at-a-time by a particular individual thing, in which *some change of properties occurs*. The explanations of why Jan's bracelet b melts at t^*, of why Jones's sample s has 4 grams mass at t^*, and of why Smith i recovers from his cold at t^*, for example, are all cases in which the thing under consideration (bracelet b, sample s, person i) gains or loses by t^* some property it had at t [its solidity (in the case of b), its 8 grams mass (in the case of s), its cold (in the case of i)], where each such case illustrates the inadequacy, in principle, of the definition of identity by means of the following formulation,

(A) $(x)(y)[(x = y) \equiv (F)(Fx \equiv Fy)]$;

since (with the possible exception of b, which may no longer be a *bracelet* at all) these cases satisfy the following conditions,

(B) $(c \text{ at } t = c \text{ at } t^*) \cdot (EF)(Fct \cdot -Fct^*)$;

where, as before, 'c' ranges over individual constants as its values. Thus, we again discover the necessity for the formulation,

(C) $(x)(y)[(x = y) \equiv (F)(t)(Fxt \equiv Fyt)]$;

according to which the names or descriptions 'x' and 'y' name or describe the same object if and only if *the objects 'x' and 'y' name or describe instantiate the same ordered sets of dispositions in the same sequence of historical events*, which is the dispositional conception.

The case of Jan's bracelet, of course, illustrates the role of definite descriptions in the introduction of proper names; for, although bracelets are not normally the bearers of names at all, certainly there would be nothing amiss from a logical point of view in introducing the name, '*b*', shall we say, on the basis of (a) some underlying reference class *R* (such as jewelry which belongs to Jan, i.e., *Jxjt*) of which (b) that object is supposed to be a uniquely different member Ψ (such as her only bracelet, her only gold jewelry, or her only gold bracelet, i.e., *Gxt·Bxt*). Thus, the corresponding definite desciption which describes that object to the exclusion of all others, namely:

(D) $(Ex)\{(Jxjt \cdot Gxt \cdot Bxt) \cdot (y)[(Jyjt \cdot Gyt \cdot Byt) \supset (x = y)]\}$;

would warrant the introduction of the proper name '*b*' within the language $\mathfrak{L}zt$ [where '$(\imath x)(Gxt \cdot Bxt)$' stands for 'the gold bracelet (belonging to Jan at *t*)', which serves as an abbreviation for (D) within $\mathfrak{L}zt$] as follows,

(E) $b = (\imath x)(Gxt \cdot Bxt)$;

where the object named by '*b*' exists so long as and only so long as the object described by the description '$(\imath x)(Gxt \cdot Bxt)$' exists (or, continues to endure). Notice, in particular, moreover, that insofar as the name '*b*' is introduced by means of the description '$(\imath x)(Gxt \cdot Bxt)$' within the language framework $\mathfrak{L}zt$, a sentence of the form, '*Gbt*', will be a logical truth, relative to $\mathfrak{L}zt$, as long as (E) retains the status of a conventional assumption (or, shall we say, of a "naming postulate") within $\mathfrak{L}zt$ [which remains possible so long as (D) is true].

It is important to observe, therefore, that not all explanations involve the exchange of some property *F* at *t* for some other property *G* at *t**; indeed, some explanations do not involve the coming-to-have-a-property-at-a-time by a particular individual thing at all. Consider, for example, the explanation of why Jan's bracelet *b* has the melting point of 1063°C or of why Jones's sample has a half-life of 3.05 minutes. In cases of this kind, by comparison, the having-of-a-property-at-a-time by a particular individual thing is explained, where *no change of properties occurs*. Thus, the explanation for *b* having a melting point of 1063°C at *t*, informally expressed as follows,

(CL)   For all *x* and all *t*, if *x* were gold at *t*, then *x* would have a melting point of 1063°C at *t*;
(C1)   Jan's bracelet *b* is made of gold at *t*;
(ES)   Jan's bracelet *b* has a melting point of 1063°C at *t*;

exemplifies the characteristics of (nomically significant) non-causal explanations in general and may be schematized as follows:

(X) $(x)(t)(Gxt \ni Pxt = 1063°C)$
     Gbt
     ───────────────
     $Pbt = 1063°C$

where the single line separating the explanans from its explanandum is intended to indicate that the logical properties of the relationship between this explanans and its explanandum are those of complete, rather than partial, entailment; and the absence of any symbol in brackets indicates that this explanation is non-causal, rather than causal, where the explanandum-phenomenon invariably accompanies the instantiation of its explanans.

Whether the explanandum-phenomenon involves a dispositional property of universal or of statistical strength, of course, the logical structure of (nomically significant) non-causal explanations remains the same. Consequently, there are (at least) two fundamentally different varieties of scientific explanation, namely:

(1) *non-causal* explanations, which explain why particular things have specific properties at certain times (where *no* change of properties occurs) by invoking 'simple' lawlike sentences; and,

(2) *causal* explanations, which explain why particular things have specific properties at certain times (where *some* change of properties occurs) by invoking 'causal' lawlike sentences;

where the distinction between lawlike sentences of 'simple' and of 'causal' form is elaborated in Chapter 3. Moreover, even though non-causal explanations exhibit a "simpler" form than do causal explanations, they are not therefore immune from the problems of explanatory irrelevance that afflict their "more complex" causal counterparts: an explanation of why Jan's bracelet b has a melting point of $1063°C$ at t, for example, which allowed among its explanatorily relevant properties that b is a bracelet at t, i.e., Bbt, that b has had a melting spell cast upon it at t, i.e., Sbt, and so forth, therefore, would be just as undesirable as a non-causal explanation as are its causal explanation counterparts.

Accordingly, the appropriate conditions of adequacy for (*nomically significant*) *non-causal explanations* for the occurrence of singular events, regardless of whether those events involve the instantiation of a disposition of universal strength or of statistical strength, may be stated as follows:

(XI) A set of sentences S, known as the "explanans", provides *an adequate nomically significant non-causal explanation for the occurrence of a singular event described by another sentence E*, *known as its explanandum, relative to the language* $\mathfrak{L}zt$, if and only if:

(a) the explanandum is a deductive consequence of its explanans;
(b) the explanans contains at least one lawlike sentence of 'simple' form that is actually required for the derivation of the explanandum from its explanans;
(c) the explanans satisfies the requirement of strict maximal specificity ($RSMS$) with respect to its lawlike premise(s); and,
(d) the sentences constituting the explanans are true, relative to the language framework $\mathfrak{L}zt$.

The reference class description of any lawlike premise that occurs in such an explanation, therefore, must include *all and only* those predicates which are nomically relevant to the occurrence of that explanandum-phenomenon. A set of sentences satisfying conditions (a) and (b), from this point of view, may be regarded as a "potential" explanation, while those which also satisfy conditions (c) and (d) are "adequate", within the language framework $\mathfrak{L}zt$.

At least two aspects of these conditions deserve additional discussion. The first, of course, is that the requirements specified by conditions (XI) (a) through (d) for *non-causal* explanations are noticeably similar to those specified by conditions (V) (a) through (d) for *causal* explanations, except that condition (XI) (a) requires no reference to probabilistic consequences (since these arguments are of a 'simple' deductive form) and condition (XI) (d) requires no separate reference to the truth of the explanandum (which is logically guaranteed by the truth of its explanans). The second, moreover, is that the requirements specified by conditions (XI) (a), (b), and (d) are strikingly parallel to those originally proposed by Hempel and Oppenheim in their classic paper of [1948]; indeed, these conditions correspond to their requirements (R1), (R2), and (R4), respectively, and by implication satisfy their (dependent) requirement (R3). Insofar as axiom A9 of the probabilistic causal calculus $\mathscr{C}$ specifies that the *u-fork* entails the corresponding *fork* formulation, i.e., that a causal conditional of universal strength entails the corresponding subjunctive, the Hempel and Oppenheim requirements capture crucial conditions of adequacy for causal, as well as for non-causal, explanations involving strictly universal lawlike sentences, which certainly reflects the importance of their achievement.

The obstacle that Hempel and Oppenheim [1948] has ultimately been unable to overcome, therefore, is the successful formalization of their conditions of adequacy *within first-order extensional language*. Difficulties with securing an appropriate formal characterization, of course, have been the subject of a lengthy and growing number of articles and inquiries, including, especially, Eberle, Kaplan and Montague [1961], Kaplan [1961], and Kim [1963], but also Ackermann [1965], Ackermann and Stenner [1966], and, more recently, Nickles [1971], Morgan [1972], and Steinberg [1973]. None of these investigations, it seems, penetrates to the heart of the matter, from the dispositional point of view; indeed, all of them together should be regarded as supplying further evidence for the conclusion that extensional languages, in principle, afford an inadequate foundation for the formalization of any lawlike sentence. However, other studies, such as Tuomela [1976], [1976a] and especially Niiniluoto [1976], [1981], and [1981a], are evincing an increasing awareness of the necessity for intensional conceptions in relation to the problem of distinguishing between lawlike and accidental generalizations with respect to their support or non-support of subjunctive and counterfactual conditionals. The crucial point to appreciate within this context, therefore, is that the principal limitations of Hempel and Oppenheim's theory are rooted in their commitment to *deficient techniques of formalization* which depend upon an extensional language framework $\mathcal{L}$; for, from the perspective of an intensional language framework $\mathcal{L}^*$, it becomes apparent that these shortcomings should not be attributed to *inappropriate conditions of adequacy* instead, since their requirements, in general, are theoretically well-founded.

Perhaps the most interesting development in the theory of explanation in recent times has been the evolution of Salmon's explication of explanation on the basis of relations of *statistical* relevance to an alternative explication which assigns a central role to *causal* relevance relations; thus, for example,

Although this differs substantially from things I have said previously, I no longer believe that the assemblage of relevant factors provides a complete explanation — or much of anything in the way of an explanation. We do, I believe, have a bona fide explanation of an event if we have a complete set of statistically relevant factors, the pertinent probability values, *and* causal explanations of the relevance relations. (Salmon [1978], p. 699)

Although (what we might refer to as) Salmon's slogan, "Let's put the 'cause' back into 'because'", sounds like a tempting invitation, there appear to be at least three different grounds for doubting that his approach is likely to succeed, namely:

first, Salmon's objective, expressed in the formulation that, "to give scientific explanations is to show how events and statistical regularities fit into the causal network of the world" (Salmon [1977a], p. 162), runs the risk of *conceptual circularity*; for, while Salmon clearly intends to explain the occurrence of statistical relations on the basis of causal relations, he also contends that causal relations are, "a species of statistical relevance relations" (Salmon [1977a], p. 165, note 30); moreover, although he appeals to his analysis of theoretical explanation (Salmon [1975a]) for support for this conception, it is by no means obvious that he will be able to fulfill a pressing desideratum for this construction, namely: an independent analysis of causal relevance relations, a point which he pursues in Salmon [1978];

second, even if Salmon were successful in advancing an independent analysis of causal relevance relations, it is difficult to imagine that the notion he would thereby have articulated would be capable of successfully fulfilling its intended function; for, although the arguments which have been considered strongly support the theoretical necessity for an *intensional explication of causal connections*, Salmon himself remains a staunch defender of "The Humean Way": this view "gives rise to one of the most important problems in current philosophy of science, namely, to provide an explication of causality without violating Hume's strictures against hidden powers and necessary connections" (Salmon [1977a], p. 162); thus, while we will consider Hume's position in the chapter that follows, Salmon's approach does not appear especially appealing;

third, the principal thesis upon which Salmon's position essentially depends is that *adequate explanations are causal explanations*, i.e., that there are no non-causal explanations which qualify as adequate, in principle; consequently, unless the distinction between causal and non-causal adequate explanations we have uncovered above is either problematical or misconceived, it would appear as though there is no theoretical justification for Salmon's main contention, after all; moreover, that distinction itself receives informal support on the basis of the difference between explanations in which no change of properties occurs and those in which some change of properties occurs and formal support on the basis of the difference between lawlike sentences of 'simple' form and lawlike sentences of 'causal' form, which undermines his basic assumption.

Indeed, in several respects, the third of these issues is the most intriguing; for, in defense of his position, Salmon denies that "non-causal" laws such as *the ideal gas law*, i.e., $PV = nRT$, which specifies the relationship between the pressure, volume and temperature of n moles of an ideal gas relative to the universal constant R, are properly understood as affording a basis for "non-causal" explanations (as Hempel [1965], pp. 352–354, has supposed):

This attitude toward non-causal laws is surely too tolerant. If someone inflates an air-mattress of a given size to a certain pressure under conditions which determine the temperature, we can deduce the value of n — the amount of air blown into it. The *subsequent* values of pressure, temperature, and volume are thus taken to explain the quantity of air *previously* introduced. Failure to require covering laws to be causal laws leads to a violation of the temporal requirement on explanations. This is not surprising. The asymmetry of explanation is inherited from the asymmetry of causation — namely, that causes precede their effects. (Salmon [1979], p. 408)

Although Salmon maintains that this example provides support for the "causal" conception, others may suspect that his argument actually "begs the question"; for, surely intuitions conflict and the alleged *asymmetry of explanation* cannot be taken for granted. Even though some lawlike sentences do characterize changes in properties from time t to time t^* (where t^* actually is later than t), certainly other lawlike sentences involve no changes in properties from t to t^* (where t^* is actually identical to t). It would appear to be arbitrary to assume that some of these lawlike sentences can serve the role of covering laws in adequate explanations, while the rest cannot.

Hempel himself, of course, suggests a number of examples of lawlike sentences which are like the ideal gas law in exemplifying (what Hempel refers to as) *laws of co-existence* (or, *laws of co-variation*), including those that relate the length and period of swing of a pendulum, Ohm's law relating the resistance of a conductor to the potential difference between its ends, and Fermat's principle of least time in application to the propagation of light. Salmon's own selection of the ideal gas law to illustrate his claim, however, warrants further contemplation; for, surely this law, like the others Hempel has cited, might have application in *both "causal" and "non-causal" contexts, where the difference depends upon whether or not changes in properties occur*: the ideal gas law may be invoked to explain why the temperature at t^* is different than the temperature at t as a response to some *change* in volume at t, but it may *also* be invoked to explain why the pressure at t has a certain value as a function of volume and temperature at t, where no such changes occur; Ohm's law may be invoked to explain why the resistance of a conductor at t^* is high as a response to some *change* in the potential difference between its ends at t, but it may *also* be invoked to explain why the resistance at t has a certain value as a function of the potential difference between its ends at t, where no such changes occur; and so on.

Since the distinction between "causal" and "non-causal" explanations depends upon the difference between lawlike sentences of 'simple' and of 'causal' form, however, this analysis of the role of laws of co-existence may initially appear to pose an objection to the dispositional construction of explanations altogether; for, in order that laws of co-existence should be able to fulfill the role of covering laws within "non-causal" as well as "causal" explanation situations, it would be necessary for those laws to be expressible as 'simple' and as 'causal' lawlike sentences, respectively. Since no lawlike sentence is of *both* 'simple' *and* 'causal' form, however, evidently this account cannot possibly be correct — unless, of course, this dilemma is based upon an unwarranted premise. It is fascinating to observe, therefore, that axiom A9 provides the foundation for accounting for this otherwise anomalous situation; for, while it is indeed the case that no specific formulation of any lawlike sentence is *both* 'simple' *and* 'causal' in form, some 'causal' lawlike sentences are likewise expressible as 'simple' lawlike sentences. The ideal gas law, for example, not only entails that variations in volume may be brought about by variations in temperature (at a constant pressure), but also entails that constant volume invariably accompanies constant temperature (at constant pressure) for n moles of an ideal gas; thus, schematically, $Vxt/Vxt^* = Txt/Txt^*$ at pressure P (where P does not change from t to t^*), just as $Vxt \propto Txt$ at the pressure P. Without elaborating this point in formal detail, therefore, the ideal gas law may not only be expressed in 'causal' forms required for "causal" explanations, such as, say, '$(x)(t)[(nRTxt \cdot Vxt) \ni_u Pxt^*]$', but also in 'simple' forms appropriate for "non-causal" explanations, such as, '$(x)(t)[(nRTxt \cdot Vxt) \ni Pxt]$', a manifestation of the consideration that (having) a certain pressure is a permanent property *and* a causal consequence of the temperature and volume of a gas.

Consequently, there appears to be little evidence for Salmon's conception of adequate explanations as causal explanations; indeed, if the considerations adduced above are sound, Salmon's basic assumption would appear to be not only problematical but also misconceived. Salmon himself, however, has endeavored to sustain his position by suggesting that, "Non-causal regularities, instead of having explanatory force which enables them to provide understanding of events in the world, cry out to be explained" (Salmon [1979], p. 408); but the example he offers (of the correlation between the behavior of the tides relative to the moon) does not strengthen his conception so much as it discloses its ambiguity: "non-causal regularities", after all, could be *either* lawlike sentences (of 'simple' form) *or* accidental correlations, neither of which is likely to support Salmon's views or to weaken the dispositional construction. Perhaps the consideration of theories and of theoretical explanation will afford additional opportunities to evaluate Salmon's "causal" explication; but there appear to be ample grounds already for doubting the potential benefits of this novel approach.

THEORIES AND THEORETICAL EXPLANATIONS

Let us begin by considering the status of Salmon's example of the tides in relation to the moon as a correlation in need of explanation. Such a sentence, presumably, would characterize the behavior of the tides with respect to the relative positions of the Earth, the Moon, and the Sun. We may assume that sentences describing this relationship would take some such form as,

(1)   Whenever e, m, and s are in configuration C, the tides are high;

where these individual objects, of course could be described exclusively by means of permissible predicates, i.e., as objects of various volumes, masses, relative locations, and so forth, insuring that the individual constants 'e', 'm', and 's' do not occur as ineliminable constituents of (1). A similar example has been advanced by Achinstein [1971], who suggests as a possibility,

(2)   All of the men in this room are bald;

where, once again, "this room" could be replaced with a description expressed exclusively by means of permissible predicates, i.e., as a room of a certain size, decoration, furnishings and fixtures, insuring that this claim concerns every person of male sex who happens to enter any room of a certain kind. Another illustration has been proposed by Popper [1965], who views the sentence,

(3)   All moas die before the age of fifty;

as describing a species of snake which, under favorable conditions, might live well beyond fifty years of age but which, as it happens, is ravaged by disease with the consequence that no individual moa actually survives to enjoy a lifespan of fifty years or more.

Each of these examples shares a number of features. In the first place, all of them are expressible as general sentences which are not logically limited to any finite number of instances (during the course of the world's history) on syntactical or semantical grounds alone. In the second place, each of them might well be true as extensional generalizations which describe the actual course of the world's history, i.e., they could all happen to be true first-order extensional generalizations. In the third place, moreover, none of these sentences is capable of supporting either counterfactual or subjunctive conditionals, i.e., *each of them would be violated under uncontroversial circumstances*: the passage of a comet in the immediate vicinity, for example, could influence the behavior of the tides and thereby violate (1); Jones, who is very hairy, could enter some room where only the bald have gone before and thereby violate (2); a family of aborigines could have discovered a few young moas and raised them as pets, say, to the ripe old age of ninety-five, thereby violating (3) as well. Consequently, whether or not each of these sentences might happen to provide an accurate description of the course of the world's history in these particular respects, *none of them qualifies as lawlike, although all of them could qualify as logically unlimited true generalizations.* As a result, not one of these sentences could properly fulfill the role of a covering law within an adequate explanation.

On the dispositional construction, of course, this result should come as no surprise; for, although the attributes under consideration (of high tides, of baldness, and of fifty-year life-spans) all qualify as dispositional properties, they are, nevertheless, not properties no member of their respective reference classes (of instances of configuration C, of instances of men in a particular style of room, or of members of a certain species of snake) could lose without losing their membership within these classes, i.e., they are among the *transient properties* which things may gain or lose independently of their membership within these reference classes. These sentences are properly envisioned as accidental, rather than as lawlike, generalizations, therefore, precisely because they do not attribute permanent dispositional properties χ to every member of some reference class K (under some appropriate description 'K' within the language framework $\mathfrak{L}zt$). For the interpretation of χ as a *permanent property* of every member of reference class K (as described) requires not only (i) that the sentence function, '$Kxt \cdot -\chi xt$', is not satisfied during the course of history of the actual world W, but also (ii) that that same function is not satisfied in any world W that differs from W only with respect to its initial conditions, if at all, where (iii) the class of physically possible worlds is determined by the appropriate distribution of permanent dispositional properties to things of different kinds in relation to the actual world W. Sentences (1), (2), and (3), therefore, are accidental rather than lawlike because, although they satisfy the extensional condition (i), they do not also satisfy the weakly intensional condition (ii), which in turn, is determined by the strongly intensional condition (iii).

As a formal manifestation of the differences between accidental generalizations and (true) lawlike sentences, notice that, although the principle of *Universal Generalization* (UG) applies to nomological conditionals of the following form, '$(Kxt \cdot T^i xt) \ni_n O^i xt*$', it does not apply to "accidental conditionals" of the (superficially similar) form, '$Kxt \cdot (T^i xt \ni_n O^i xt*)$'. Consequently, the following reflects a valid principle of inference:

UNIVERSAL GENERALIZATION (UG): Where 'bt' represents either the concatenation of an individual constant and a determinate time or the concatenation of an ambiguous name and an ambiguous time, from a previously obtained singular nomological conditional of the form,

$$(Kbt \cdot T^i bt) \ni_n O^i bt *,$$

a universally quantified sentence of the following form may be inferred,

$$(x)(t)[(Kxt \cdot T^i xt) \ni_n O^i xt *],$$

provided that ... xt ... results from ... bt ... by replacing each occurrence of bt with xt (making no other changes);

However, the corresponding principle obviously does not reflect a valid principle of inference:

UNIVERSAL GENERALIZATION (UG)*: Where 'bt' represents either the concatenation of an individual constant and a determinate time or the concatenation of an ambiguous name and an ambiguous time, from a previously obtained conjunction of singular sentences of the form,

$$Kbt \cdot (T^i bt \ni_n O^i bt *),$$

a universally quantified sentence of the following form may be inferred,

$$(x)(t)[Kxt \cdot (T^i xt \ni_n O^i xt *)],$$

provided that ... xt ... results from ... bt ... by replacing each occurrence of bt with xt (making no other changes);

for just because one moa may happen to die before the age of fifty does not transform that transient property into a permanent property — even when every moa happens to possess that same transient property in common!

These considerations reflect the inability of (let us say) "extensional distributions", as either *constant conjunctions*, i.e., material conditionals, or *relative frequencies*, i.e., statistical correlations, to qualify as dispositional properties of the physical world. Moreover, these are not the only properties which do not properly qualify as dispositional. Consider, in particular, the distinctions between the spatial and the temporal properties of things, which appear to be as follows, namely: a spatial predicate designates a relation that things have to other things (such as x is higher than y and to the left of z relative to frame of reference R) as a feature of *atomic* events; while temporal predicates designate relations that things have to other things (such as x is earlier than y and lasts longer than z on the basis of standard T) as features of *molecular* events, i.e., relations that obtain between specific events (which may happen to be atomic or molecular), but are not properties of individual atomic events, *per se*. The sentence, "Books on physics are on the shelf above those on philosophy", for example, describes a feature of an atomic event, whereas the sentence, "I read the paper before I went outdoors", by contrast, orders a particular set of events and is therefore itself molecular. Indeed, *a maximal set of occasion sentences*, i.e., the totality of all occasion sentences (whether macro or micro) that are true together without any contradiction, provides a plausible conception of simultaneity between events (relativistic causal considerations notwithstanding), while recurring sequences of event instantiations (such as the periodic rotation of an electron around its orbit), of course, may furnish suitable standards for measuring their duration.

The differences between the spatial and temporal properties of things may also clarify the interpretation of dispositions as *actual physical states*; for particular dispositions are properties of objects and of arrangements of things that happen to be instantiated as features of atomic events, where *arrangements of things* are collections of things instantiating particular spatial relations, i.e., spatial predicates designate properties which are fundamental to such an arrangement's description. Since temporal predicates describe relations which atomic events only instantiate relative to other (atomic or molecular) events, therefore, the properties they designate are *not* actual physical states of any such object or arrangement; as a result, temporal predicates are not dispositional, i.e., *historical relations are not dispositions*. Since any event inherits an infinite number of temporal relations with other events merely by virtue of its occurrence (insofar as, for example, every event occurs *prior to*, *subsequent to*, or *simultaneous with* innumerable other events), these historical properties establish a convenient source for definite event descriptions, but nevertheless these predicates do not designate dispositions. Because Lincoln is born in 1809 and Vesuvius destroyed Pompeii in A.D. 79, Lincoln instantiates the property *born 1730 years after the destruction of Pompeii by Vesuvius*; but since this property is not instantiated as a feature of any atomic event, it could not be a disposition. Accordingly, from the dispositional perspective, there are (at least) three distinct kinds of properties and relations, namely: extensional distributions, historical relations, and dispositional tendencies.

Since these properties and relations are ordinarily associated with some reference class (or, specific event) description, Giere's ([1973a], p. 473) allegation that the propensities associated with an experimental arrangement, "are no more relative to any description than any other physical characteristics", should not be left unexamined; for Giere apparently subscribes to the mistaken conception that such properties as mass, specific gravity, and electrical conductivity are *not* relative to the appropriate specification of some experimental arrangement. Although we have already discovered that this position leads to triviality or to falsehood in the case of *statistical* dispositions, it should now be emphasized that it leads to identical results in the case of *universal* dispositions as well. Indeed, this claim is clearly refuted by even the simplest example of a bicycle and rider; for the mass of the bicycle-and-rider obviously differs from that of either the bicycle or the rider alone, which therefore requires some specification of the appropriate arrangement for an unambiguous determination. Indeed, from the perspective of explanatory adequacy, this point is crucial, since the features that distinguish the permanent from the transient properties of things depends upon some presupposed reference class description 'K', within the language framework $\mathcal{L}zt$. For properties such as a certain mass, a certain specific gravity or a certain propensity might be *either* permanent *or* transient properties of things, which makes an enormous difference with respect to explanation situations, since explanations of particular outcomes as the result of *transient* properties of things differs fundamentally from their explanation as the result of *permanent* properties of things, as we are about to ascertain.

Perhaps the most important point to appreciate with respect to examples such as (1), (2), and (3), let us note, is that the dispositional properties which they attribute to all of the members of their reference classes are not permanent properties of those classes (under corresponding descriptions) precisely because *there are processes or procedures − either natural or contrived − by means of which members of those classes could lose those attributes and still remain members of those classes* (the passage of a nearby comet, the entrance of a hairy-headed man, and the discovery and nurturing of a baby moa). The fundamental incapacity of extensional language to capture the lawlike properties of the actual world, therefore, reflects the intensional and subjunctive character of this conception, insofar as *these processes and procedures might or might not happen to be instantiated during the course of the world's actual history*. Indeed, it is for this reason that the further specification of these reference classes (through the conjunction of additional predicates) provides no guarantee, in principle, that the resulting generalizations will be any less immune to *potential violation*, whether or not they might be more accurate as historical descriptions. For the subjunctive and counterfactual character of possible instantiations of these processes and procedures (even when they never take place) substantially diminishes the plausibility of any purely extensional conception.

With respect to issues of explanation, therefore, Salmon's affirmation, that these "non-causal", i.e., accidental, regularities "cry out" for explanation, although well-founded, may or may not involve the mode of explanation Salmon himself considers; for, since each of these generalizations is susceptible to violation, the properties they ascribe to each member, say, $m_1, m_2, ..., m_n$, of their reference classes, as transient rather than as permanent properties, require independent explanation, i.e., in order to "explain why" these "non-causal" regularities obtain, it is necessary to explain why *each individual member* of those reference classes possesses the property in question. To "explain why" (1), (2), or (3) obtains, therefore, requires explaining why each of the *n* members of their respective reference classes happens to possess the relevant attribute: why each instance m_1 to m_n of configuration *C* has been an instance of high tides, why each instance m_1 to m_n of men entering such rooms has been an instance of baldness, why each instance m_1 to m_n of the members of the species of moas has been an instance of death before age fifty. And notice especially that the explanations in each such case are by no means required to be the same: some moas might have been killed by hunters for the value of their skins, while others succumbed to disease; some men in such rooms could be bald as a result of heredity, while others might have had their hair removed by electrolysis to please their wives; and so forth. For, so long as each of the members of these reference classes has the same transient attribute in common, those generalizations are accidental but explainable truths.

It is important to realize, of course, that, although one class of extensional generalizations will be accidental, as we have discovered, another class of extensional generalizations will be true as *deductive consequences of corresponding lawlike sentences which explain them*; indeed, it seems very likely that the members of this class are most readily mistaken for lawlike sentences themselves: if a melting point of 1063°C is among the permanent properties of gold, then the corresponding material generalization, namely:

(4)   Whenever a sample of gold is heated to 1063°C, it melts;

will be a logically unrestricted material conditional, which accurately describes the course of the world's history (since everything would have to be either not a sample of gold that is heated to 1063°C or a thing that melts); if a half-life of 3.05 minutes is among the permanent properties of polonium^{218}, then the corresponding material generalization, namely:

(5)   All polonium^{218} has a half-life of 3.05 minutes;

will be a logically unrestricted material conditional, which accurately describes the course of the world's history (since everything would have to be either not polonium^{218} or a thing with a half-life of 3.05 minutes); and if (water) solubility is among the permanent properties of (table) salt, then the corresponding material generalization, namely:

(6)   Table salt dissolves in water;

will be a logically unrestricted material conditional, which accurately describes the course of the world's history (since everything would have to be either not table salt or a thing which dissolves in water).

Unlike true extensional generalizations which attribute (what happen to be) transient properties (relative to their reference class descriptions), in other words, these true extensional generalizations attribute (what happen to be) permanent properties (relative to their reference class descriptions), although, of course, as logically unrestricted material conditionals, these sentences are not lawlike sentences, either. The differences between (purely) accidental and (let us say) *quasi-accidental* (or *quasi-lawlike*) generalizations, therefore, has at least three aspects, which could be exemplified as follows:

(a) for each (true) quasi-accidental (or quasi-lawlike) sentence of the form, '$(x)(t)(Kxt \supset \chi xt)$', there is a (true) non-accidental (or lawlike) sentence of the form, '$(x)(t)(Kxt \mathrel{\ni\mkern-5mu-} \chi xt)$'; but in the case of (true) accidental sentences of that form, there are no corresponding (true) lawlike sentences;

(b) for each (true) quasi-accidental (or quasi-lawlike) sentence of the form, '$(x)(t)(Kxt \supset \chi xt)$', interpreted as an explanandum, therefore, there is at least one adequate (nomically significant) *theoretical explanation*; but in the case of (true) accidental sentences of that form, there are none; finally,

(c) since every (true) quasi-accidental (or quasi-lawlike) sentence of the form, '$(x)(t)(Kxt \supset \chi xt)$', attributes (what happens to be) some permanent property to every member of its reference class (under the description 'K' in $\mathfrak{L}zt$), there exists no process or procedure — natural or contrived — by means of which a member of K could lose the property χ, without also losing membership in K; but this is not the case for (true) accidental sentences of that form.

In order to formulate conditions of adequacy for (nomically significant) theoretical explanations, however, several preliminary considerations require our attention. There are (at least) three kinds of sentences which may occur within the *explanans* of (well-formed) theoretical explanations, which we will differentiate as follows:

(i) one or more *lawlike sentence* (of 'simple' or of 'causal' form) may occur in the explanans of (well-formed) theoretical explanations;

(ii) one or more *property specification* (or partial definition) of the form, '$(x)(t)[\chi xt \mathrel{\ni\mkern-6mu-} (T^i xt \mathrel{\ni\mkern-6mu-}_n O^i xt^*)]$', for example, may occur in the explanans of (well-formed) theoretical explanations; and,

(iii) one or more *property identification* (or complete definition) of the form, '$(x)(t)[(Kxt \mathrel{\ni\mkern-6mu-} K'xt) \cdot (K'xt \mathrel{\ni\mkern-6mu-} Kxt)]$', for example, may also occur in the explanans of (well-formed) theoretical explanations.

Let us refer to sentences stipulating property specifications and property identifications indifferently as "meaning postulates"; then, *the explanans of (well-formed) theoretical explanations consist exclusively of at least one or more lawlike sentence and zero or more meaning postulates*; that is, the material premises of theoretical explanations consist of sentences of *only* these specific kinds (exclusive of logical truths).

Let us consider some examples of (intuitively satisfactory) theoretical explanations prior to formulating explicit conditions of adequacy, as follows. The logically unrestricted material conditional, "$(x)(t)(WPxt \supset Ixt = 30°C)$", i.e., white phosphorous ignites at 30°C, could be subsumed by an explanans consisting of the corresponding lawlike sentence, namely:

(XII) $(x)(t)(WPxt \mathrel{\ni\mkern-6mu-} Ixt = 30°C)$

$(x)(t)(WPxt \supset Ixt = 30°C)$

where the single line separating the explanans from its explanandum is intended to indicate that the logical properties of the relationship between this explanans and its explanandum are those of complete entailment, while the unrestrictedly general logical form of the sentences constituting this explanation — both the explanandum and its explanans — indicates that it is a theoretical explanation for a class of events, rather than a 'causal' or a 'non-causal' explanation for some singular event.

Analogously, the logically unrestricted material conditional, "$(x)(t)(ANxt = 15 \supset Ixt = 30°C)$", i.e., whatever is composed of the element with atomic number 15 ignites at 30°C, could be subsumed by an explanans consisting of an appropriate lawlike sentence and corresponding meaning postulate; for example,

(XIII) $(x)(t)(WPxt \mathrel{\ni\mkern-7mu-} Ixt = 30°C)$
$(x)(t)[(WPxt \mathrel{\ni\mkern-7mu-} ANxt = 15) \cdot (ANxt = 15 \mathrel{\ni\mkern-7mu-} WPxt)]$
——————————————————————
$(x)(t)(ANxt = 15 \supset Ixt = 30°C)$

where the explanandum sentence asserting that everything is either not composed of the element with atomic number 15 or ignites at 30°C is explained by deriving that sentence from an explanans consisting of a lawlike sentence asserting that if something were white phosphorous then it would have an ignition temperature of 30°C together with a property identification asserting that something would be white phosphorous if and only if it were composed of the element with atomic number 15.

Furthermore, the logically unrestricted material conditional, "$(x)(t)[ANxt = 15 \supset (Hxt = 30°C \supset Fxt*)]$", i.e., if a thing is composed of the element with atomic number 15, then if that thing is heated to 30°C, then it bursts into flame, could be subsumed by an explanans consisting of an appropriate lawlike sentence and corresponding meaning postulates; for example,

(XIV) $(x)(t)(WPxt \mathrel{\ni\mkern-7mu-} Ixt = 30°C)$
$(x)(t)[Ixt = 30°C \mathrel{\ni\mkern-7mu-} (Hxt = 30°C \mathrel{\ni\mkern-7mu-}_u Fxt*)]$
$(x)(t)[(WPxt \mathrel{\ni\mkern-7mu-} ANxt = 15) \cdot (ANxt = 15 \mathrel{\ni\mkern-7mu-} WPxt)]$
——————————————————————
$(x)(t)[ANxt = 15 \supset (Hxt = 30°C \supset Fxt*)]$

where the explanandum sentence asserting that everything is either not composed of the element with atomic number 15 or is not heated to 30°C or else bursts into flame is explained by deriving that sentence from the explanans consisting of a lawlike sentence asserting that if something were white phosphorous then it would have an ignition temperature of 30°C together with a property specification asserting that if anything were to have an ignition temperature of 30°C then heating it to 30°C would invariably bring about its bursting into flame and a property identification asserting that something would be white phosphorous if and only if it were composed of the element with atomic number 15.

This last example, moreover, illustrates the possibility that any particular logically unrestricted quasi-accidental (or quasi-lawlike) material conditional could be adequately explained by *alternative* (nomically significant) theoretical explanations; for surely the explanandum sentence of the preceding explanation would be similarly explainable by subsumption by an explanans consisting of the corresponding lawlike sentence, namely:

(XV) $\dfrac{(x)(t)[ANxt = 15 \ni\!\!\!- (Hxt = 30°C \ni\!\!\!-_u Fxt^*)]}{(x)(t)[ANxt = 15 \supset (Hxt = 30°C \supset Fxt^*)]}$;

consequently, it should not be assumed that there is any unique theoretical explanation for specified logically unrestricted explanandum sentences. It would also be a mistake to assume that *only* quasi-accidental (or quasi-lawlike) generalizations are amenable to theoretical explanation; for, as some of these explanans implicitly display, theoretical explanations may *also* be advanced for lawlike sentences themselves: the lawlike premise of (XV), for example, could be theoretically explained by means of its derivation from the explanans of (XIV), and so forth.

Indeed, an especially interesting illustration of the theoretical explanation for a lawlike explanandum sentence is that of a *statistical* lawlike sentence, such as that the strength of the tendency to obtain either side I or II as a result of a single toss with a homogeneous regular tetrahedron, i.e., "$(x)(t)[Rxt \ni\!\!\!- (Txt \ni\!\!\!-_{1/2} Sxt^* = I \vee Sxt^* = II)]$", which could be subsumed by an explanans consisting of other statistical lawlike sentences; for example,

(XVI) $\dfrac{\begin{array}{l}(x)(t)[Rxt \ni\!\!\!- (Txt \ni\!\!\!-_{1/4} Sxt^* = I)] \\ (x)(t)[Rxt \ni\!\!\!- (Txt \ni\!\!\!-_{1/4} Sxt^* = II)] \\ \Box(x)(t)(Sxt = I \supset -Sxt = II)\end{array}}{(x)(t)[Rxt \ni\!\!\!- (Txt \ni\!\!\!-_{1/2} Sxt^* = I \vee Sxt^* = II)]}$

where the explanandum sentence asserting that the single case propensity for obtaining either side I or side II as the result of a single toss with a homogeneous regular tetrahedron is explained by deriving that sentence from an explanans consisting of other lawlike sentences asserting that the single case propensity for obtaining side I under those conditions equals ¼ and for side II under those conditions equals ¼, where, so long as these outcomes are mutually exclusive, as asserted by the third premise, that conclusion follows. This example therefore also illustrates the consideration that, in order to deduce an explanandum from an explanans involving lawlike sentences, the necessary conditions specified by calculus $\mathscr{C}$ (such as axiom A26, in the case at hand) require explicit satisfaction.

Accordingly, the appropriate conditions of adequacy for (*nomically significant*) *theoretical explanations* for quasi-accidental (or, quasi-lawlike) or for lawlike sentences (of 'simple' or of 'causal' form) as themselves explanandum sentences may be specified as follows:

(XVII) A set of sentences S, known as the "explanans", provides *an adequate nomically significant theoretical explanation for a quasi-*

accidental *(or, quasi-lawlike) or lawlike sentence E, known as its explanandum, relative to the language* £zt, if and only if:

(a) the explanandum is a deductive consequence of its explanans;
(b) the explanans contains at least one lawlike sentence (of either 'simple' or 'causal' form) that is actually required for the derivation of the explanandum from its explanans;
(c) the explanans satisfies the requirement of strict maximal specificity (*RSMS*) with respect to its lawlike premise(s); and,
(d) the sentences constituting the explanans are true, relative to the language framework £zt.

The reference class description of each lawlike premise that occurs in such an explanation, therefore, must include *all and only* those predicates which are nomically relevant to the occurrence of their outcome attributes, respectively (which, of course, is intended to exclude nomically irrelevant predicates from the lawlike premise(s) of adequate theoretical explanations). A set of sentences satisfying conditions (a) and (b), from this point of view, may be regarded as a "potential" explanation, while those which also satisfy conditions (c) and (d) are "adequate", within the language framework £zt.

Although these conditions do not reflect such a conception, perhaps the strongest case for inclusion of a *requirement excluding irrelevant premises*, i.e., sentences not actually required for the derivation of the explanandum from its explanans, may be made in the case of explanations of this kind. I refrain from doing so, however, without attempting to establish a conclusive rationale for this omission, other than the ultimate concern of establishing a unified criterion for scientific explanations, regardless of their specific kind, which will be advanced in the section to follow. Nevertheless, since the principal function of *scientific theories* is to afford a foundation for theoretical explanations for classes of phenomena, such a requirement might serve a useful purpose by stipulating a necessary condition for a well-formed scientific theory, relative to an explanandum sentence E, as follows: *a set of sentences S, which qualifies as a "potential" (or "adequate") theoretical explanation for an explanandum sentence E, relative to the language* £zt, *qualifies as a "potential" (or, "adequate") scientific theory, respectively, provided that S consists exclusively of sentences that are actually required for the derivation of that explanandum from its explanans.* Consequently, the conception of a scientific theory which is offered here characterizes any set of sentences that satisfies conditions (a) and (b) as a "potential" scientific theory, while a set satisfying conditions (c) and (d) as well is an "adequate" scientific theory, *of the explanandum phenomenon E* (within a language framework $\mathfrak{L}zt$) – provided that this additional requirement is also satisfied.

The introduction of "meaning postulates" of forms (ii) and (iii) above, of course, represents one of the most important issues underlying the choice between an extensional and a non-extensional language framework, which emerges with particular clarity from reflection upon the differences that distinguish *properties* and *classes*. Quine ([1951], p. 120) has posed the problem and proposed a possible solution:

> classes are the same when their members are the same, whereas it is not universally conceded that properties are the same when possessed by the same objects. ... But classes may be thought of as properties if the latter notion is so qualified that properties become identical when their instances are identical.

In spite of its superficial plausibility, however, Quine's proposal appears to evade rather than to resolve this problem; for the principle that he recommends, namely:

(XVIII) $(F)(G)[(F = G) \equiv (x)(t)(Fxt \equiv Gxt)]$;

that is, "For all properties F and G, F is identical to G if and only if, for all x and all t, x is F at t if and only if x is G at t", fails in trial situations. For if all and only oval lockets happened to be made of gold, then it would be the case that for all x, x is an oval locket if and only if x is made of gold; but, surely the properties *gold* and *oval locket* are distinctly not the same, for the shape of something gold is not among the permanent properties of every member of that class, and being made of gold is likewise not a property that no oval locket could be without.

From the dispositional perspective, of course, this problem is perfectly intelligible; for the principle Quine adopts would be sufficient for its purpose only at the expense of its extensionality, i.e., by admitting 'possible worlds' into his sparse ontology or by embracing non-truth-functional logical connectives. For two properties would be identical if and only if all their instances in any possible world would be the same or if something would be an instance of one if and only if that something were also an instance of the other. The first of these conceptions thus requires quantification over every 'possible world' W as follows:

(XIX) $(F)(G)[(F = G) \equiv (W)(x)(t)(FxtW \equiv GxtW)$;

that is, "For all properties F and G, F is identical to G if and only if, for all W, x and t, x is an F at t in W if and only if x is a G at t in W", which dissolves one theoretical problem at the expense of creating another (in providing an analysis of the truth conditions for 'possible world' assertions). For surely the least that is required of any such ontology is an explanation of which worlds are possible and why.

The alternative provided by a dispositional conception, by comparison, affords both a principle for properties and a rationale for possible worlds as well. For the appropriate principle for the identity of properties is itself supplied by means of a subjunctive biconditional as follows:

(XX) $(F)(G)\{(F = G) \equiv (x)(t)[(Fxt \mathrel{\ni\!\!-} Gxt) \cdot (Gxt \mathrel{\ni\!\!-} Fxt)]\}$;

that is, "For all properties F and G, F is identical to G if and only if, for all x and all t, if x were an F at t then x would be a G at t and if x were a G at t then x would be an F at t", where sentences of the form, '$Kxt \mathrel{\ni\!\!-} \chi xt$', are true especially when either, (i) 'Kxt' entails 'χxt', within the language framework $\mathfrak{L}zt$, in which case the subjunctive is warranted on logical grounds; or, (ii) χ is a permanent property of every member of the reference class K (under the description 'K' in $\mathfrak{L}zt$), relative to the actual world W, in which case the subjunctive is warranted on ontological grounds. The satisfaction of the definiens of principle (XX), moreover, entails the satisfaction of the definiens of principle (XIX), which entails the satisfaction of the definiens of principle (XVIII), but not conversely, for those properties instantiated as features of atomic, rather than of molecular, events. The properties gold and oval locket, therefore, are identical according to principle (XX) if and only if either (i) the predicate "gold" entails the predicate "oval locket", and conversely; or, (ii) (being) *gold* is a permanent property of every member of the reference class *oval locket*, and conversely. Hence, if something could be gold without being an oval locket, or conversely, then those properties are not identical, while otherwise they are — which, of course, will be true or false independently of the historical contingency that, as a matter of fact, all and only oval lockets might be made of gold.

These considerations, moreover, suggest an alternative formulation of the dispositional conception of scientific theories; for a *theory* may also be characterized as *a set of sentences attributing permanent dispositional properties to the members of a reference class* (under an appropriate description within a language framework $\mathfrak{L}zt$), where, let us note, two theories, say, T_1 and T_2, are *identical*, i.e., two formulations of the same theory, relative to $\mathfrak{L}zt$, if and only if either (i) 'T_1' entails 'T_2' and conversely, relative to $\mathfrak{L}zt$, in which case the identification of these theories is warranted on logical grounds; or (ii) T_1 and T_2 attribute all and only the same permanent dispositional properties to things of all and only the same kinds, in which case the identification of these theories is warranted on ontological grounds. Condition (ii) could be satisfied even when condition (i) were not yet fulfilled, let us note, where an epistemic problem would arise in ascertaining that that is the case, especially in the absence of adequate translations into a common language ℒ. Even without fulfilling condition (i), however, there may be *empirical grounds* for assuming the identification of scientific theories; for the relation between these intensional formulations and their extensional implications is such that two theories will be identical *only if* their extensional implications are the same; that is, *T_1 and T_2 must attribute all and only the very same properties to all and only the very same things during the course of the world's actual history*. The discovery that predicate 'F' belonging to T_1 and predicate 'G' belonging to T_2 are or are not applicable to all and only the same things at all and only the same times, and so on, may afford significant and sometimes conclusive evidence of the identity or the non-identity of T_1 and T_2, which will partially, though not completely, overcome difficulties of this kind, as Carnap [1955] has explained.

Perhaps I should emphasize, by way of conclusion, that my employment of the phrase, "reference class under a description", is not supposed to entail that *permissible* predicates (in the sense of Chapter 4) be *non-dispositional* predicates (in the sense of Chapter 3). On the contrary, as Popper [1965], New Appendix *x, and Mellor [1974] have lucidly explained, *the existence of dispositional properties does not logically presuppose or nomically require the existence of corresponding (or underlying) non-dispositional properties*. Nevertheless, the employment of this phrase is intended to permit drawing an important distinction between *basic* and *derivative* natural laws along (more or less) the following lines, namely: whereas basic natural laws attribute dispositional properties to the members of *unrestrictedly general* reference classes (as instances of corresponding purely-dispositional – or universal – properties), derivative natural laws attribute dispositional properties to the members of *restrictedly general* reference classes (as instances of corresponding non-purely-dispositional – or non-universal – properties). Consequently, basic laws (such as those attributing gravitational attraction to everything having mass) may be characterized as specifying intensional relations between universals (as purely-dispositional properties), while derivative laws (such as those attributing particular configurations to the orbits of planetary bodies revolving about the Sun) may be characterized as specifying intensional relations between non-purely-dispositional reference properties, on the one hand, and purely-dispositional attribute properties, on the other. And while the defense of this conception depends upon the tenability of the identification of universals with pure dispositions (in the sense of Chapter 7), it appears to be thoroughly incompatible with the conception of laws as specifying extensional relations between properties as Dretske [1977] has proposed.

"INSTRUMENTALISM" AND THEORETICAL REALISM

The distinction between "realism" and "instrumentalism" is one of the most important and least understood in the history of philosophy. Perhaps the explanation for this phenomenon emanates from the discovery that a conception which *is* sufficient to distinguish "realism" from "idealism" is *not* sufficient to distinguish "realism" from "instrumentalism". In particular,

The Principle of Relativistic Realism, i.e., that the world exists as an entity apart from our beliefs about it, but the properties of the world are linguistically relativized, in the sense that there is more than one language, $\mathcal{L}^1$, $\mathcal{L}^2$, ... in which it may be described; hence, there is no unique descriptive language;

fulfills a sufficient condition of demarcation with respect to (let us say) *absolutistic idealism*, which must be "absolutistic", in the sense that, if any idealist were to concede that the world might be a construction, not of his own, but of someone else's, mind, his position would thereby have been demonstrated to be "self-refuting". Anyone who has ever stubbed their toe, hit their finger with a hammer, or cut themselves shaving, of course, will find that idealism is a difficult doctrine to sincerely maintain; thus, its appeal is largely confined to the infirm, the unstable, and the hypocritical. Whatever plausibility it may enjoy as a doctrine derives from the conflation of coherence as a *criterion* of truth with correspondence as its *definition*; but sadly there are no built-in guarantees that one or another view of things must certainly be correct if only we believe it enough.

The situation with "instrumentalism", however, is not at all comparable; for no instrumentalist needs to deny that the world exists apart from all of our beliefs about it. What the instrumentalist advocates instead, therefore, is that *only certain kinds of language* are properly regarded as descriptive of the physical world; in particular, *classical instrumentalism* (let us say) may be defined by the following theses: theoretical terms do not designate properties of the physical world; theoretical sentences, therefore, are not either true or false; and, the purpose of science is to describe and to predict, but not to explain, the phenomena of our experience (Morgenbesser [1969], pp. 200–218). Accordingly, the tenability of classical instrumentalism depends upon at least two sub-theses, namely: (i) that there is a defensible distinction between "observational" and "theoretical" language (for otherwise *empirical language* itself stands in jeopardy); and, (ii) that the thesis of the logical symmetry of explanation and prediction is untenable (for otherwise there are no predictions which are *not* potential explanations). To the extent to which the evidence thus far considered suggests that explanations and predictions do not invariably satisfy the same requirements of adequacy, therefore, classical instrumentalism finds support for the possibility of predicting and describing the phenomena of experience without also explaining it. To the extent to which considerations which are by now familiar suggest that a distinction between "observational" and "theoretical" language cannot be sustained (as summarized by Suppe [1972] and especially Suppe [1977], pp. 80–95, for example), however, classical instrumentalism requires additional defense, an ironic consequence of the apparent demise of the "standard conception" of scientific theories.

From an historical perspective, of course, there have been perhaps three major stages in the evolution and development of the conception of theories as an aspect of 20th century analyses of science as follows, namely:

stage one, i.e., the conception of theories as consisting essentially of an *abstract calculus* conjoined with an *empirical interpretation* [where both (a) the analytic/synthetic distinction and (b) the observational/theoretical distinction are taken for granted], advocated most forcefully by proponents of the movement known as *logical positivism* and represented by Carnap [1939], by Ayer [1946] and by Ayer [1959];

stage two, i.e., the conception of theories as consisting essentially of *internal principles* conjoined with *bridge principles* [where neither (a) an analytic/synthetic distinction nor (b) an observational/theoretical distinction fulfill a fundamental role], which has been recommended recently by a member of the movement known as *logical empiricism*, in particular, by Hempel [1966] and Hempel [1970]; and,

stage three, i.e., the conception of theories as consisting essentially of *theoretical structures* conjoined with *empirical claims* [an analysis where (a) the analytic/synthetic distinction itself might or might not be attended by (b) the observational/theoretical distinction], in which theories are represented by *predicates* rather than by *sentences*, which has been elaborated in formal detail by Sneed [1971] and by Stegmüller [1973a] and [1976].

Since relatively thorough presentations of these approaches are available elsewhere (especially in Suppe [1977] and Stegmüller [1976], for example), the objective of the present discussion is simply to isolate and to identify certain features or aspects of those accounts which are of special interest.

The crucial factor is that classical instrumentalism is in a bad way unless there is a defensible distinction, in principle, between "observational" and "theoretical" language; for if "observational language" is "theory-laden", while theoretical predicates do not designate properties of the physical world, then what is to inhibit the inference that "observational predicates" as well *do not designate properties of the physical world*? Moreover, if theoretical sentences, which employ a theoretical vocabulary, are neither true nor false, while the "observational vocabulary" is theoretically contaminated, what is to forestall the conclusion that "observational sentences" *are likewise neither true nor false*? Consequently, the collapse of the observational/theoretical distinction promotes the convergence of instrumentalism and idealism; for, if neither "observational" nor "theoretical" terms designate properties of the physical world, while all *empirical predicates* designate properties of the physical world, necessarily, then the conclusion appears inescapable that *there are, in principle, no empirical predicates*, i.e., that the world itself is nothing but a construction of the human mind. Conversely, moreover, if an "observational" predicate retains its empirical content, i.e., does designate some property of the physical word, in spite of its presumptive "theoretical contamination", then what grounds remain in support of the basic claims that theoretical terms do *not* designate properties of the physical world and that theoretical sentences are therefore *neither true nor false*? For without some hitherto unnoticed defense, classical instrumentalism appears to be a totally untenable and virtually incoherent philosophical position.

An imaginative and resourceful response to problems such as these has recently been advanced by Chalmers [1976], who introduces a position he refers to as "radical instrumentalism", in which the relationship between scientific theories and the external world is brought about by *community practice*, as an appropriate solution:

The radical instrumentalist extends an instrumentalist interpretation on the "theoretical" side of the dissolved boundary across to the "observational" side of the boundary. Neither our "observational" nor our "theoretical" language is correctly understood as a description of what the world is really like. (Chalmers [1976], p. 129)

Although there are fundamental differences between the conceptual systems of scientific theories and of ordinary language and the real world, nevertheless the real world and our conceptual systems are linked together by "scientific practice", which is a complex of experimental, mathematical, and theoretical techniques implemented by the members of the scientific community. Problems raised by this approach, however, are several and varied, including (a) ascertaining precisely *which complex* of experimental, mathematical, and theoretical techniques collectively constitutes *scientific* practice and (b) determining exactly *what relation* obtains between our conceptual systems and the real world which, in principle, they *cannot* describe. A tempting defense might be to adopt a Peircean posture that "conceptual systems" are behavioral dispositions which differ in the extent to which they promote successful interaction with the world itself. But unless the world is merely a construction of the human mind, once again, it is difficult to understand how to account for the difference between "successful" and "unsuccessful" interaction *without* availing ourselves of any distinction between true and false beliefs about a 'real' world.

One avenue of recourse from this dilemma, of course, might be to search for refuge among the tenets of stage two and stage three conceptions of theories, which, after all, must also cope with the demise of the classical conception of stage one. According to stage two conceptions, for example, there is a distinction between the language of internal principles and the language of bridge principles, namely: that the internal principles are characteristically expressed by means of a "theoretical vocabulary" which is related to the phenomena of experience by bridge principles expressed in terms of *antecedently understood* language, which is "pretheoretical" in a relative sense, i.e., in relation to that specific theory:

... the internal principles of a theory are concerned with the peculiar entities and processes assumed by the theory (such as the jumps of electrons from one atomic energy level to another in Bohr's theory) and they will therefore be expressed largely in terms of characteristic "theoretical concepts", which refer to those entities and processes. But the implications that permit a test of those theoretical principles will have to be expressed in terms of things and occurrences with which we are antecedently acquainted, which we already know how to observe, to measure, and to describe. (Hempel [1966], p. 74)

Thus, Hempel suggests, the crucial role to be fulfilled by "bridge principles" is that of relating theoretical language to its testable implications by means of *previously available* concepts and predicates which, however, may themselves have been acquired through a process involving the assimilation of other theories previously accepted, relative to the specific theory under consideration. Consequently, there appears to be no basis here for supporting the contentions that theoretical terms do not really designate properties of the physical world or that theoretical sentences are not really either true or false; for, while Hempel's conception of internal principles and bridge principles is pragmatical in application, i.e., relative to specific theories and "pretheoretical" conditions, the stage two construction of theories does not provide any foundation for the successful resolution of the instrumentalist's dilemma.

Perhaps stage three conceptions will prove to be more promising. According to stage three constructions, however, there may or may not be theory-free "observation language", i.e., the observational/theoretical distinction is not necessarily taken for granted, although its precise status is not therefore especially clear. As Stegmüller, for example, has explained, in relation to Lakatos' [1972] classification of methodologies of science,

... none of the titles he mentions fit the conception developed in this book. At best one could think of it as a variant of the *instrumentalism* he mentions on p. 95. But the objections raised there do not apply to this 'new' instrumentalism. It is not based on confusing truth with confirmation, nor does it overlook the possibility of something being both true and unprovable. Instead, it drops the fundamental presupposition that a theory *can* even be true or false. And this presupposition is *not* dropped because talk of truth is disqualified as 'dumb philosophical blather', but rather *because only propositions can be true or false and a theory is neither a proposition nor does it consist of propositions*. (Stegmüller [1976], p. 255, original italics)

Thus, Stegmüller here emphasizes that *theoretical structures* as definitions of set-theoretical predicates [such as, for example, "A physical system is a classical Newtonian particle system if and only if it is a system of objects which satisfies Newton's three laws of motion and the law of universal gravitation"] are, strictly speaking, either true, necessarily, as matters of definition, or else neither true nor false, as lacking empirical content. Nevertheless, the *empirical claims* that can be made by invoking these set-theoretical predicates [such as, for example, "The solar system is a classical Newtonian particle system" (lucidly discussed by Giere [1979a], especially pp. 63–73), however, *surely are either true or false*. Moreover, since Stegmüller acknowledges the need for evidential support for theoretical claims by observational data (Stegmüller [1976], p. 169), on the one hand, and the indispensable function of theoretical terms "in helping to preclude possible observable states of affairs" and in "sentences involving explanation, prediction, and the testing of hypotheses" (Stegmüller [1976], p. 66 and p. 70, respectively), on the other, there are ample grounds for hesitating to suppose that stage three constructions of theories are any more likely to support the resolution of the instrumentalist's dilemma than are those of stage two.

If these considerations are well-founded, therefore, then it appears as though instrumentalism may require *redefinition* to avoid the prospect of its *degeneration* into absolutistic idealism. What I want to suggest, in particular, is the importance of the third thesis of classical instrumentalism in relation to the structure of theories; for perhaps there is a suitable basis for distinguishing between *descriptions and predictions*, which are generally acceptable on instrumentalist principles, and *explanations*, which are not, by invoking the distinction between intensional and extensional language frameworks. The historical development of these alternative conceptions of scientific theories, after all, has largely proceeded within the context of a methodological commitment to exclusive reliance upon an *extensional language framework*, a commitment which, it appears, has been shared not only by Ayer, Carnap, Hempel, Stegmüller and Sneed, but also by the vast majority of those concerned to analyze the structure of theories. From this perspective, moreover, the crucial alteration involved in the shift from the logical positivist conception of theories as calculus-plus-interpretation to the logical empiricist conception of theories as internal principles-plus-bridge principles was presaged by the failure to provide complete explicit definitions for dispositional predicates, especially, within the context of a *first-order* extensional language framework, which not only led Carnap [1936–37] to develop the reduction sentence method for the partial specification of meaning, but also led Carnap and Hempel [1950], [1951], and [1965], in particular, to doubt the applicability of the analytic/synthetic distinction to the language of science as a whole. Logical empiricists were consequently motivated in the direction of liberalized standards of cognitive significance, such as those advanced by Hempel [1952] and by Carnap [1956], while remaining committed to an extensional language framework.

The continuing demise of the observational/theoretical language distinction, however, brought increased pressure for a reformulation of the "standard conception" of scientific theories, which Hempel [1966] and [1970], of course, took seriously enough to attempt to deal with explicitly. The development of sophisticated techniques of set theory, especially reflected by Suppes [1967] and [1967a] and by van Fraassen [1970], combined with the historical approach represented by Kuhn [1964], ultimately issued in the emergence of the set-theoretical conception, which is essentially equivalent to using a *second-order* extensional language framework. Although some, such as Kyburg [1979a], have expressed a degree of caution concerning the ultimate adequacy of extensional languages for their intended role with respect to the analysis of theories, a number of others, such as van Fraassen [1979a], especially, have declared the complete adequacy, in principle, of the extensional approach; thus, he claims,

1.1 *The language of science.* In some sense, surely this language can be taken as purely extensional! Namely, in the sense that it could be replaced by an extensional language, or could be made extensional by introducing quantification over possible physical states and possible physical systems. (van Fraassen [1979a], p. 282)

The position represented here, moreover, strikes a responsive chord relative to Kyburg's [1974] and [1978] attempts to formalize the single case propensity interpretation of probability by preserving extensionality across sets of possible worlds, a conception which van Fraassen is endorsing in general.

Considerations such as these, I believe, suggest the possibility of resolving the instrumentalist's dilemma by focusing upon the logical features of an acceptable language framework rather than upon the epistemic features of its vocabulary; for the contention that *descriptions and predictions* are acceptable, while *explanations* are not, may be infused with content by adopting the thesis that *scientific languages are extensional languages* as its obvious corollary. Consequently, the instrumentalist position may be redefined by embracing the principle that intensional languages, such as those we have relied upon in developing the conception of lawlike sentences as attributing permanent properties χ to every member of a reference class K (under the description 'K' within a language framework $\mathfrak{L}zt$), are unacceptable, in principle, for the rational reconstruction of the language of science. Indeed, from this point of view, it becomes clear that Kyburg [1976] and [1979], Levi [1977] and [1979], and van Fraassen [1970], [1977], [1979], and [1979a] are sophisticated representatives of the position thereby defined, which might be referred to as *neo-classical instrumentalism*, namely: *that extensional language frameworks, in principle, are adequate for the rational reconstruction of the language of science* — whether or not they happen to require quantification over "possible worlds", "possible physical states", "possible physical systems", and so on, as (let us say) 'weakly intensional' set-theoretical constructions. Thus, a 'strongly intensional' language framework, such as the probabilistic causal calculus $\mathscr{C}$, whose logical connectives, in principle, are not reducible to extensional operators across "possible worlds", is therefore unacceptable.

Neo-classical instrumentalism, thus defined, of course, is a very broad position, including, as it does, not only quantification within first-order extensional laguages but also quantification within second-order extensional languages over *possible*, as well as *actual*, objects and events. Perhaps I should emphasize, therefore, that neo-classical instrumentalism would not afford a plausible position were its commitments restricted to either first-order languages or actual objects and events; for the limitations that would be imposed by these classical constraints produce a conspicuously inadequate language framework. Indeed, *first-order* extensional languages which

forego quantification over possible objects and events are unable to cope with some of the most fundamental problems under consideration here, including these:
- (a) the analysis of counterfactual and of subjunctive conditionals;
- (b) the distinction between accidental and lawlike generalizations;
- (c) the discovery of adequate criteria for scientific explanations.

Moreover, even *second-order* extensional languages that provide for quantification over possible objects and events are not obviously satisfactory, insofar as they appear to encounter difficulties with problems such as these:
- (d) explicating the meaning of single case dispositional predicates;
- (e) providing an ontological foundation for nomological conditionals;
- (f) supplying non-teleological solutions to problems (a), (b), and (c).

The alternative advocated here (of "intensional realism", let us say), therefore, is rooted in its commitment to the existence of actual *single case dispositional tendencies* of objects and events within the physical world W, especially as those commitments are represented in Chapter 3 (a position which, however, appears to be compatible with "possible world's instrumentalism" as that conception has been advanced by Merrill [1978], especially pp. 320–321).

There are various direct and indirect indications that this interpretation provides a plausible conception of the realist/instrumentalist dispute from the perspective of logical theory; for the issues which we have considered throughout have themselves displayed different facets of this classical controversy in its contemporary guises: *the analysis of physical probability as a single case statistical disposition; the conception of lawlike sentences as attributing permanent properties to the members of reference classes; the distinction between lawlike sentences of 'simple' and of 'causal' form, with the conception of 'causal' and 'non-causal' explanations; the explication of dispositional definitions of nomic and of causal relevance relations; the differentiation between accidental, quasi-accidental and lawlike generalizations; the propagation of the appropriate demarcation between properties and classes* — all of these potential solutions to difficult problems presuppose reliance upon an intensional language framework of the 'strongly intensional' variety. Consequently, if the recommended reformulation of this debate is approximately sound, then we should expect that those who oppose *theoretical realism* (under this interpretation) will tend to support alternative approaches to problems of this kind; and, indeed, that does appear to be the case, not only relative to the positions we have already reviewed, but also with respect to issues we have not yet considered.

Perhaps the crucial test of this conception, moreover, occurs within the theory of explanation, since the instrumentalist position is thus understood as permitting *extensional formulations*, but not (strongly) intensional ones, for the purposes of *description and prediction*, while eschewing explanations altogether. An example introduced during the course of early deliberation on the relative merits of intensional and extensional languages for the analysis of lawlike sentences in Chapter 2 illustrates the importance of this issue, namely: "Whenever a bullet is fired into a piece of pine, it makes a hole", which is subject to formalization as a logically unrestricted material conditional, i.e., "$(x)(y)(t)[(Bxt \cdot Pyt \cdot Fxyt) \supset Hxyt^*]$", is *also* formalizable as a logically unrestricted nomological conditional, i.e., "$(x)(y)(t)[(Bxt \cdot Pyt \cdot Fxyt) \ni_u Hxyt^*]$"; yet the material conditional is logically equivalent with the corresponding material conditional, i.e., "$(x)(y)(t)[(Pyt \cdot -Hxyt^*) \supset -(Bxt \cdot Fxyt)]$", while the nomological conditional is *not* logically equivalent with the corresponding nomological conditional, i.e., "$(x)(y)(t)[(Pyt \cdot -Hxyt^*) \ni_u -(Bxt \cdot Fxyt)]$". As we observed during our previous discussion of this example, these material conditionals would appear to be capable of serving altogether useful and unobjectionable roles for the purposes of prediction and of retrodiction, i.e., for inferring *what* has happened rather than *why* it has happened, yet they would be both objectionable and useless as principles of explanation. It is interesting to discover, therefore, that the dispositional construction of lawlike sentences provides a theoretically elegant and intuitively satisfactory explanation for this difference, which an extensional approach could not possibly supply.

These considerations, therefore, reinforce the conception that the distinction between intensional and extensional language lies at the bottom of the realist/instrumentalist controversy. Since this is an issue with varied manifestations, moreover, it is intriguing to discover that philosophers who appear to represent the neo-classical instrumentalist position, such as van Fraassen, for example, also defend other theses which complement that basic orientation. In van Fraassen [1977] and [1979], he not only elaborates what he refers to as "the modal frequency interpretation of probability", which, of course, is an endeavor to provide an interpretation of probability as a physical magnitude while preserving extensionality across "possible worlds" — not quite succeeding by virtue of its 'weakly intensional' commitments in a situation where only 'strongly intensional' commitments will do — but also introduces a distinction between what a theory *says* and what we believe when we *accept* that theory: according to van Fraassen, to believe a theory is to believe that it is *true*, but to accept a theory involves only believing that it is *empirically adequate*, i.e., "that the actual phenomena are faithfully represented by the empirical substructures of one of its models" (van Fraassen [1979], p. 158). In a similar vein, van Fraassen [1979a] disparages the significance of physical laws:

> ... there seems to be no way to distinguish them from other general truths except by the fact that they apparently imply counterfactuals. I think there is probably some confusion in the very concept of law, and that this concept marks a mistaken reification of a pragmatical distinction. (van Fraassen [1979a], p. 283)

Indeed, these passages not only support the perception of van Fraassen as a *neo-classical* instrumentalist, but virtually represent the position defined by *classical* instrumentalism.

It seems quite clear, therefore, that an important distinction should be drawn between the neo-classical instrumentalistic position of van Fraassen, Kyburg, and Levi, for example, and the realistic position represented here. For no neo-classical instrumentalist can admit the *truth* of even one lawlike sentence, the *adequacy* of even one scientific explanation, the *necessity* to distinguish quasi-accidental from lawlike sentences, or the *suitability* of the definitions of nomic and of causal relevance relations under the dispositional construction; for each of these solutions involves going beyond restrictions imposed by adhering to a 'weakly intensional' language framework. Consequently, no neo-classical instrumentalist can embrace the unified criterion of adequacy for scientific explanations which emerges on that account, according to which (*nomically significant*) *scientific explanations* for either logically unrestricted or logically singular explanandum sentences are required to satisfy essentially the same conditions, namely:

> A set of sentences S, known as the "explanans", provides *an adequate nomically significant scientific explanation for the phenomenon — whether singular or general — described by another sentence E, known as its explanandum, relative to the language* $\mathfrak{L}zt$, if and only if:

(a) the explanandum is either a deductive or a probabilistic consequence of its explanans;

(b) the explanans contains at least one lawlike sentence (of either 'simple' or 'causal' form) that is actually required for the deductive or probabilistic derivation of the explanandum from its explanans;

(c) the explanans satisfies the requirement of strict maximal specificity (*RSMS*) with respect to its lawlike premise(s); and,

(d)   the sentences constituting the explanation — both the explanans and its explanandum — are true, relative to the language framework $\mathfrak{L}zt$.

Consequently, if the explanans is singular and the relevant lawlike sentence is of 'causal' form, then the explanation will satisfy the requirements for an adequate nomically significant *causal* explanation specified by (V) above; whereas if the explanandum is singular and the relevant lawlike sentence is of 'simple' form, then the explanation will satisfy the requirements for an adequate nomically significant *non-causal* explanation specified by (XI) above. And if the explanandum is unrestrictedly general, whether it happens to be quasi-lawlike or lawlike, then the explanation will satisfy the requirements for an adequate nomically significant *theoretical* explanation as specified by (XVII) above.

The conception of explanations as involving essential reference to permanent properties, of course, may be subject to philosophical appraisal on the basis of the relevant criteria of explicative adequacy. Among the most important considerations from this point of view, therefore, is the extent to which this conception accords with scientific practice or with ordinary language; in other words, the extent to which explanations that occur in technical scientific and in ordinary conversational contexts actually conform to this conception. And, indeed, it appears as though abundant support may be found in the form of such familiar "explanations" as follow: "The glass broke when it was dropped, because it was fragile"; "These iron filings were attracted to this metal bar, because it is magnetic"; "That sample of table salt dissolved when it was put in water, because it was soluble"; "This die comes up with an ace quite often when it is tossed, because it is loaded"; "That penny yields a head only about half the time, because it is fair"; "The lump of radon disintegrated by approximately half its mass in the past few days, because the half-life of radon is 3.82 days"; and so on. The evidence represented here, therefore, strengthens the impression that the dispositional interpretation of theoretical realism is a defensible position, where the most important questions which remain concern issues of testability within an epistemic context.

PART III

CORROBORATION

7. THE JUSTIFICATION OF INDUCTION

The "problem of induction" is perhaps as clear a paradigm of a purely philosophical difficulty as one is likely to find: it has been advanced in many different forms, for which a wide variety of disparate solutions have been proposed, about the correctness of which few philosophers have tended to agree. It has been variously characterized as subject to an inductive solution, a deductive solution, a linguistic resolution, or as incapable of solution; it has been diversely appraised as an important problem, as an unimportant problem, and as no problem at all. Perhaps even more remarkably, it appears to be a problem possessing no practical significance for anyone at all, since the particular form that its solution might take seems to be a matter of no more and no less practical consequence than whether it is amenable to any resolution after all. No doubt a problem whose "successful' resolution has no different practical implications than no solution at all has the *sine qua non* of every truly philosophical problem. It thus appears to be a paradigm, indeed.

One of the most interesting attempts to come to grips with the problem of induction is also one of the most controversial, namely: Popper's solution, which emphasizes the asymmetry between *verification* and *falsification* as (general) scientific procedures (especially Popper [1965], Part I; and Popper [1972], Chapter 1). The purpose of this chapter is to maintain that, although Popper's position may easily be misunderstood, when appropriate attention is given to,

(i) the fundamental distinction between lawlike and accidental generalizations;

(ii) the important differences between inductive confirmation and Popperian corroboration; and,

(iii) the theoretical significance of Popper's principle of empiricism as opposed to traditional principles of induction;

it becomes apparent that, whether or not Popper's arguments are mistaken in detail, his solution, in principle, is correct. The point, therefore, is to propose a Popperian defense of Popper's own position.

Not the least of the difficulties which this investigation confronts arises from a certain ambiguity afflicting our conception of induction itself, for the term, "induction", may be employed in at least two distinct senses. In its *broad* sense, an argument or an inference may be described as "inductive" whenever its conclusion, no matter how tentatively or provisionally entertained, contains more content than do its premises; while, in its *narrow* sense, an argument or inference is described as "inductive" whenever it conforms to certain specific "inductive rules" exemplified in particular by probabilistic conceptions of confirmation as well as by traditional principles of induction. Let us note, therefore, that, although procedures which are "inductive" in the narrow sense are also "inductive" in the broad sense, necessarily, those which are "inductive" in the broad sense are not necessarily also "inductive" in the narrow sense.

This distinction becomes especially important insofar as the position Popper has promoted fits within the category of those which are "inductive" in the broad sense but not in the narrow, a circumstance that has been obscured, at least in part, due to two quite different factors: first, the negative significance of *successful* attempts at falsification falls within the domain of deductive methodology; second, the very conception of "rules of induction" tends to be identified with precisely those procedures which Popper vehemently rejects. The complexity of the situation, however, calls for more subtle discriminations, insofar as Popper has sought to provide techniques and procedures, i.e., a set of rules, for assessing the acceptability of alternative hypotheses, where the positive significance of *unsuccessful* attempts at falsification clearly falls within the domain of inductive methodology, as the considerations which follow should explain.

Popper's analysis itself involves distinguishing between different versions of the *philosophical* problem (which he refers to as the "traditional" problem and its "methodological" variant, respectively) and what may be characterized as the *pragmatical* problem, which concerns decision-making and practical life. One significant feature of Popper's approach, therefore, is to seek to define the character of the problem at hand before undertaking its solution, which, in this case especially, should be viewed as an indispensable ingredient of rational procedure. In order to present Popper's solution within an appropriate context, however, I would like to begin by considering the distinction drawn between *validation* and *vindication* as modes of justification that apply to individual arguments and to inferential principles, respectively; for it will be my contention that preoccupation with vindication and validation has tended to obscure an important aspect of the problem of induction which, I believe, should be acknowledged as the foundation of any successful solution.

THE JUSTIFICATION OF INDUCTION

THE TRADITIONAL PROBLEM OF INDUCTION

Induction and deduction are generally supposed to be mutually exclusive and jointly exhaustive modes of propositional inference, i.e., which regulate inferences from propositional premises to propositional conclusions, as Chapter 1 has also explained. As modes of inference, of course, each may be envisioned as an accepted "set of rules" in accordance with which inferences must be drawn to qualify either as *deductively valid* or, let us say, as *inductively proper*. Analogously, proper arguments with true premises will be *correct*, just as valid arguments with true premises are *sound*. Provided we consider "arguments" as sets of propositions (or sentences) divided into two parts, i.e., "premises" and "conclusions", then whether or not a particular argument (or inference) belongs to the class of proper inductive arguments may only be determined relative to the set of accepted inductive rules; and whether or not a particular argument (or inference) is a member of the class of valid deductive arguments may only be determined relative to the set of accepted deductive rules. The establishment of a specific argument as in conformity with accepted rules of either set is thus sufficient to warrant the acceptance of that argument (or inference) as either inductively proper or deductively valid. The justification of an inference through such a procedure, of course, is known as *validation* (Feigl [1963]).

Since there are infinitely many possible rules of either kind, however, the selection of some finite set of rules of each type as acceptable rules has come to be regarded as a matter of considerable philosophical interest. The process of selection necessarily involves reference to certain characteristics which acceptable rules are supposed to possess. The establishment of a specific rule as possessing those characteristics thus constitutes a sufficient condition for the inclusion of that rule as a member of the set of acceptable rules. The justification of a rule by means of this procedure, moreover, is known as *vindication* (Salmon [1967]). Although validation and vindication are both familiar conceptions of justification, their application, in principle, depends upon a logically prior process of establishing those specific characteristics whose possession renders a rule acceptable. These characteristics, in other words, are ascertained in relation to the objective, purpose, or program of the enterprise itself, which, of course, in this case distinguishes a mode of inference, i.e., they are *pragmatically* determined.

The inductive and deductive modes of inference, therefore, may be analyzed into three constitutive components, namely: a class of inferences, a set of rules, and a program. The *class of inferences* which are acceptable within each mode of inference is determined by some set of accepted rules. The *set of rules* which are acceptable within each mode of inference, in turn, is determined by the accepted program. And the *program* which is acceptable for each mode of inference is determined by the purpose for which that mode of inference is intended. The residual problem, therefore, is one of ascertaining the purpose to be fulfilled by each of these modes of inference, a matter about which there appears to be considerable agreement with respect to the deductive mode. Thus, deductive inferences, in general, are supposed to be *truth-preserving*, in the sense that deductive rules of inference are acceptable if and only if, when applied to true premises, only true conclusions follow (Beth [1969]). The establishment of a set of rules and a class of inferences which fulfills this purpose thus defines the deductive program.

The selection of a set of accepted rules from within the set of acceptable rules normally involves both logical and non-logical considerations. The *logical* considerations, of course, include the soundness and completeness properties of a contemplated set of rules, since any such set which permitted some inferences to be drawn which are *not* valid would fail to satisfy the deductive program, just as such a set which does not permit other inferences to be drawn which *are* valid falls short of its complete satisfaction. Apart from these purely deductive conditions, however, *non-logical* considerations, such as balancing theoretical elegance against practical convenience, fulfill an important role; for, there are those who desire to derive a maximal set of theorems from a minimal set of axioms, even though the resulting proofs may be tedious and complex, while others wish to secure a modest set of axioms which, although not equally elegant, provides a more flexible foundation for constructing relatively simple and obvious proofs. Those who belong to the first group, of course, tend to take a dim view of those who belong to the second, while the latter, in turn, are themselves rather intimidated by the former — all of which tends to account for the enormous variety of different textbooks currently available within this field.

In the case of inductive inference, by contrast, there has been an absence of general agreement beyond the tendency to assume that induction is supposed to be *knowledge-extending*, insofar as inductive rules of inference are envisioned as regulating inferences whose conclusions contain more content than do their premises. This conception itself, however, provides (at best) a necessary but not sufficient condition for defining the program of the inductive mode of inference, since this requirement is satisfied by innumerable invalid deductive arguments that, nevertheless, would not qualify as inductively proper. One attempt to deal with this difficulty has been to seek to combine the *ampliative* (or knowledge-extending) function of the inductive mode with the *demonstrative* (or truth-preserving) function of the deductive mode to generate a class of ampliative-demonstrative inferences. Rules of inference that satisfy one of these conditions, however, do so at the expense of the other, which dictates the result that this contradictory conception of the inductive program cannot possibly succeed.

Another attempt to deal with this difficulty has been to recommend a rather specific conception of the purpose of induction, namely: to ascertain the limiting frequencies with which different attributes occur within different reference classes during the course of the world's history (Reichenbach [1949] and Salmon [1967], for example). Assuming this conception itself is logically consistent, it remains to be determined whether or not the inductive program is thereby adequately defined. Among the most important problems confronting this conception, for example, are (a) the extent to which scientific practice actually fulfills this interpretation and (b) the extent to which its objectives and procedures are illuminated by such a construction. The first problem might be described as one of *definitional relevance*, insofar as a conception that altogether ignores actual scientific practice is surely irrelevant as a definition of the inductive program; while the second problem might be described as one of *theoretical significance*, since a conception of induction that fails to clarify scientific procedure should certainly not define its foundations (conditions which parallel Hempel's [1952] requirements of adequacy for explications).

Not the least of the reasons for objecting to this conception, therefore, is that actual scientific practice appears to pursue the discovery of scientific laws rather than ascertaining limiting frequencies, to search for principles of prediction for the past and the present as well as for the distant future, and to value explanations for "the single case" in addition to those accessible for long and short runs. The point, let me emphasize, is not that the case for one conception rather than the other is clear-cut and beyond debate, but rather that there is an issue here which requires explicit attention and careful argumentation for its tentative resolution, namely: *the problem of defining the inductive program*. And insofar as there appear to be objective standards of logical consistency, of definitional relevance, and of theoretical significance which are applicable to such a dispute, the defense of some specific conception of the inductive program is an indispensable ingredient of attempts to provide an adequate justification for induction, an aspect we may refer to as *the process of exoneration*.

The three constitutive components of modes of inference, therefore, are attended by three distinct modes of justification, insofar as a class of inferences must be validated, a set of rules must be vindicated, and a program must be exonerated, for its comprehensive justification. [Distinctions such as these, incidentally, also apply to forming-beliefs, making-decisions, and so forth, which we will consider in chapters to come.] Thus, the foundation for Popper's analysis of induction consists in the conception of its program as one of *providing for the acceptance, rejection, or modification of hypotheses and theories of broad scope and systematic power which may be employed for the purposes of explanation and prediction, with special concern for the provisional and tentative quality of the results of scientific inquiry.* But the crucial difference between Popper's program and similar conceptions that others have endorsed is reflected by the difference between alternative formulations of "the problem of induction" itself (Popper [1972], p. 7), namely:

(1) as the *traditional* problem, which asks, "Can the claim that an explanatory universal theory is *true* be justified on empirical grounds, i.e., by assuming the truth of certain test statements or experiential findings?"; and,

(2) as the *methodological* problem, which asks, "Can the claim that an explanatory universal theory is *true* or that it is *false* be justified on empirical grounds, i.e., by assuming the truth of certain test statements or experiential findings?"

For Popper maintains that the answer to the traditional question is, "No", but that the answer to the methodological question is, "Yes", on the basis of certain asymmetrical characteristics of *verification* and *falsification* as (general) scientific procedures (Popper [1965], p. 265), which we shall now consider.

Not the least of Popper's contributions to philosophical discussion has been his unrelenting assault upon the *desideratum of certainty* as an indispensable foundation of scientific knowledge: in opposition to those who defend the conception of "incorrigible" sense data or of "irrevocable" observation reports, he has promoted the position that observation statements represent interpretations of perceptual experience by means of presupposed theories. Popper contends that, "since there can be no theory-free observation, and no theory-free language, there can of course be no theory-free rule or principle of induction" upon which all theories could be based (Popper [1976], p. 148). For the present, let us understand his position as undermining the conception of observation sentences as establishing an infallible evidential resource for subjecting scientific hypotheses and theories to empirical test, rather than as an objection to any principles of induction *per se*. Thus, to emphasize the fallibility of all experiential data, let us also refer to these reports as *basic statements* (Popper [1965], pp. 104–105).

These considerations make it clear that the differences between verification and falsification, whatever they may be, do not consist in the alleged incorrigibility of one of these procedures in comparison with the other: our supposition, therefore, denies the defensibility of "dogmatic" versions of either methodology, but instead only supports the possibility of relative, rather than absolute, verifiability and falsifiability, i.e., in application to basic statements, these procedures yield only *fallible* results (as Hempel [1965], pp. 39–46, has also observed). The principle of verifiability, accordingly, may be formulated as follows:

> An hypothesis h is *verifiable* within a language $\mathfrak{L}$ if and only if h is not analytic but follows from some logically consistent finite set of potential basic statements expressible in $\mathfrak{L}$;

where the cumbersome phrase, "potential basic statements expressible in $\mathfrak{L}$", is required to implement the semantical analogue of the pragmatical notion of basic statements as singular sentences whose truth or falsity is decided by tentative agreement among relevant members of the language-using community, as Popper [1965], pp. 100–106, suggests. Analogously, the principle of falsifiability may be formulated as follows:

> An hypothesis h is *falsifiable* within a language $\mathfrak{L}$ if and only if the negation of h is not analytic but follows from some logically consistent finite set of potential basic statements expressible in $\mathfrak{L}$;

where, as before, the results of neither process are regarded as infallible.

The principal attraction of these principles, of course, is that they offer the prospect of characterizing scientific procedure entirely by means of the application of deductive principles to basic statements as empirical premises; as a consequence, however, they fulfill the ampliative function of inductive inference in a psychological, but not in a logical, sense, insofar as the results of these procedures may amount to the *transformation* of what we may consider as "existing knowledge", but not in its addition, rejection, or modification — at least to the extent to which that "knowledge" itself is consistent. The psychological benefits that accrue from reliance upon these demonstrative procedures, of course, are not therefore without value; on the contrary, the deductive transformation of tentatively agreed-upon conclusions concerning the data of more-or-less immediate experience may not only expose unexpected contradictions in the state of "existing knowledge" but also dictate further observation and experimentation to re-establish consistency, a circumstance which may become acute with the sudden acquisition of enormous quantities of unfamiliar evidence through the introduction of technological innovations such as the microscope, the telescope, x-ray diagnosis, and so on. [Thus, Ackermann [1976] refers to the sets of data available by means of fixed sets of technical instruments as "data domains".]

Of course, if all of the objectives of scientific inquiry could be attained by means of deductive procedures alone, there would be nothing more to be desired; it seems clear, however, that verification and falsification, individually or in combination, do not suffice for such a purpose, since in particular, as is indeed well-known, different classes of sentences are amenable to each of these procedures. Thus, for example,

Sentence Type	Verifiable	Falsifiable
Existential Generalizations	Yes	No
Finite Frequency Statements	Yes	Yes
Limiting Frequency Statements	No	No
Mixed Quantification Statements	No	No
Statistical Lawlike Statements	No	No
Universal Lawlike Statements	No	Yes

which reflects the incapacity of deductive methodology alone to justify the claim that some specific limiting frequency statement, statement of mixed quantification, or statistical lawlike statement either *is true* or *is false* on the basis of any logically consistent and finite set of basic statements (an issue Hempel [1965], especially pp. 102–107, has also discussed). A number of explanatory comments could be made relative to these considerations; for example, if statistical lawlike sentences were to be *identified* with limiting frequency statements, then only five, rather than six, distinct sentence types would actually be required. Other considerations, however, suggest that a *seventh* kind is called for instead.

The distinction between a criterion of cognitive significance (which distinguishes between meaningful and meaningless sentences) and a criterion of demarcation (for distinguishing between "scientific" and "non-scientific" sentences), I take it, is sufficiently familiar to us all as to warrant no detailed elaboration. Popper's employment of falsifiability as a criterion of demarcation appears to have the uningratiating consequence of rendering existential generalizations, limiting frequency statements, statements of mixed quantification, and statistical lawlike statements, not meaningless, but *non-scientific*. This, of course, creates an immediate quandary, since such sentences as,

"There are black holes", "In arbitrarily large populations of natural childbirths, the limiting frequency for male children is .514", and, "Every metal has a melting point", seem to represent at least *some* of the kinds of knowledge which scientific inquiry should be able to supply. [Popper himself, of course, has discussed most of these problems; see, for example, Popper [1965], pp. 100–103, and especially Schilpp, ed., [1974], pp. 1038–1039.] Rather than evaluate Popper's criterion of demarcation, however, I would like to consider the problems implicitly created for his program.

These difficulties, moreover, are less severe with respect to limiting frequency statements and statements of mixed quantification, for example, provided they do not also qualify as *lawlike* generalizations; for the principal objective of Popper's program is to provide for the acceptance, rejection, and modification of hypotheses and theories of broad scope and systematic power which are applicable for the purposes of explanation and prediction, i.e., *the members of the class of scientific laws*. Thus, if statistical lawlike statements, for example, are not amenable to empirical test on the basis of Popper's principle of falsifiability, then either (a) Popper's set of rules of induction lacks an adequate vindication, or (b) Popper's program lacks a satisfactory exoneration, or, perhaps, (c) both. If statistical lawlike statements are identifiable with limiting frequency statements, however, then a solution may present itself, insofar as limiting frequency statements in turn may be identified with non-decreasing sequences of finite sets of elements, each of which might be described as a finite frequency and therefore be not only falsifiable but also verifiable.

This problem, of course, was very much on Popper's mind while writing *The Logic of Scientific Discovery*, which was in large part devoted to this problem's resolution. According to the methodological principles developed there, important aspects of testing and evaluating hypotheses and theories are determined by conventional decisions and procedures of several different kinds, namely: (1) an hypothesis or theory under test must be delineated from (provisionally unproblematic) background knowledge and auxiliary hypotheses; (2) decision procedures must be agreed upon for the acceptance and rejection of basic statements describing experimental outcomes as well as initial conditions; (3) the decision must be made to regard particular experiments (or sets of experiments) as tests of the relevant kinds (usually, although not always, by means of experimental repetitions); and, (4) in the case of statistical (or probabilistic) hypotheses, agreements must also be reached rendering specific patterns of outcome distributions "compatible" or "incompatible" with the hypothesis under investigation. [Concerning (4), see especially Popper [1965], pp. 198–205, and Lakatos [1970], especially pp. 106–111.]

The peculiar difficulty with statistical hypotheses, of course, stems from the fact that any limiting frequency over an infinite sequence is logically compatible with any relative frequency within a finite segment; but Popper recognized that, although arbitrary deviations from limiting values within (normal) sequences were *logically possible*, they were not therefore *equally probable*. Thus, by exploiting the results of Bernoulli's theorem, among others, Popper extended the principle of falsifiability for statistical statements by adopting the *supplementary principle* that, since almost all possible finite segments of large size will very probably exhibit relative frequencies that are very close to the limiting frequencies of these infinite sequences, *highly improbable outcomes deserve systematic neglect*. [See especially Popper [1965], pp. 410–419.] Popper has thus sought to defend the position that the testing of statistical hypotheses, like that of all other scientific hypotheses, is deductive: for, the probabilities of various frequency distributions over large (but finite) samples are mathematically derived employing the principles of the calculus of probability, highly improbable results are supposed not to occur, and the hypotheses under consideration are confronted by experience. [See Gillies [1973] as well.]

Now while these considerations tend to justify Popper's program and to salvage his criterion of demarcation, there appear to be further grounds in support of the principle of falsifiability as a fundamental scientific rule, namely: the theoretical distinction between *lawlike sentences* and (merely) *accidental generalizations*. For, as we have discovered, by introducing the conception of a single case statistical predicate as a refinement of Popper's propensity interpretation of probability, it has proven possible to develop an account of statistical laws as universal generalizations attributing statistical properties, i.e., probabilistic dispositions, to every member of appropriately specified reference classes, which in turn has led to the emergence of a unified theory of the character of universal and statistical laws on the basis of a dispositional analysis. The pursuit of this objective has also brought about the elaboration of a single case dispositional predicate possible-world formal semantics in order to formalize the logical relations that obtain between logical, subjunctive, and nomological conditionals within scientific language, which was presented in Chapter 3.

The account of *scientific conditionals* that has resulted from these investigations reflects the following suppositions:

(1) that subjunctive conditionals may be justified on the basis of either

logical or nomological considerations, but not on other grounds;

(2) that nomological conditionals are themselves of two different kinds: 'simple' nomic conditionals and 'causal' nomic conditionals;

(3) that 'causal' nomic conditionals are likewise of two kinds, i.e., 'universal' and 'statistical' (or 'probabilistic') causal conditionals;

(4) that true scientific conditionals satisfy a requirement of maximal specificity, which is not generally applicable in ordinary discourse; and,

(5) that lawlike sentences are logically general nomological conditionals, i.e., that nomological conditionals are instantiated lawlike sentences.

The most important feature of this analysis for Popper's inductive program, moreover, is that it reflects three desiderata concerning the character of scientific laws which Popper has sought to emphasize, namely: (a) that lawlike sentences are unrestrictedly general; (b) that all universals are dispositional; and, (c) that lawlike sentences are not extensional (or, truth-functional) generalizations (especially Popper [1965], pp. 420–441).

Indeed, it is the characteristic of lawlike sentences as *intensional generalizations* that provides the ultimate grounds for preferring Popper's conception of the importance of falsifiability over alternative principles. Whereas the truth-values of extensional generalizations are determined by the history of the actual world — they are true if they describe that history and otherwise they are false — the crucial ingredient for an adequate solution to the problem of induction consists in the recognition that lawlike sentences as intensional generalizations entail, but are not entailed by, corresponding extensional generalizations. For, the truth conditions for generalizations that describe classes of possible worlds are strictly determined by the truth conditions for generalizations describing the history of the actual world *only when those extensional generalizations happen to be false*. Since the actual world is the only world in which generalizations may be subjected to (more or less) direct empirical test, however, it should be apparent that Popper's solution, in principle, is correct.

If the objects of scientific inquiry are intensional generalizations, then it is as important to distinguish between universal lawlike sentences and universal material conditionals as it is to differentiate statistical lawlike sentences from limiting frequency statements. For every universal lawlike sentence semantically entails a corresponding set of universal material conditionals, and every statistical lawlike sentence probabilistically implies innumerable possible frequency distributions for runs short and long; yet intensional statements of neither kind are logically equivalent to their extensional counterparts. As a result, therefore, one more class of sentences is required for a more adequate classification, namely: *universal material generalizations*, which, of course, are falsifiable but not verifiable — unless the universe itself is finite. These observations, we now know, suggest an appropriate criterion for separating lawlike sentences from accidental generalizations and display the inadequacy of any scientific program focused exclusively upon extensional distributions.

The falsifiability of universal material generalizations, moreover, strongly suggests that Popper's criterion of demarcation may be necessary but not sufficient to fulfill its intended objective, so long as (merely) accidental generalizations should not qualify as "scientific" statements. This difficulty, of course, remains whether sentences of this kind happen to be true or happen to be false, since *falsifiability* does not depend upon a sentence's truth-value; however, it does support the view that, as a criterion of demarcation, falsifiability may be too weak instead of being too strong, depending upon matters of interpretation. The most important difference distinguishing accidental generalizations and lawlike sentences, therefore, is not that sentences of one of these kinds are falsifiable and the other not, but that lawlike sentences possess nomological significance, i.e., they reflect non-logical relations of necessary connection.

The differences involved here, although enormously important, are not infrequently obscured because the "deep structure" of a lawlike sentence is not always represented by its "surface grammar"; thus, the statement, "Fat men are jolly", might be interpreted *either* as a material conditional, say, (A) "For all x and all t, if x is a fat man at t, then x is jolly at t", which does not pretend to be lawlike; *or* as a subjunctive conditional, say, (B) "For all x and all t, if x were a fat man at t, then x would be jolly at t", which does. The truth of an accidental generalization, such as (A), however, is compatible with the existence of processes or procedures which could produce or provide a fat man who is not jolly, while the truth of a lawlike statement, such as (B), precludes that possibility. Falsification practices provide crucial opportunities for discriminating between them in principle, therefore, since statements which are not both *lawlike* and *true* are theoretically vulnerable to the eventual discovery of such refutations — where the class of quasi-accidental (or, quasi-lawlike) sentences qualifies as the intriguing case that "proves the rule": for, *if it were physically impossible to falsify* (A), *then the corresponding lawlike sentence*, (B), *would have to be true*!

Popper ([1968], pp. 54–55) has portrayed the problem of induction as arising from an apparent contradiction between Hume's realization that ampliative arguments are deductively invalid, necessarily, and *the principle of empiricism*, i.e., that, in science, only observation and experiment may decide the acceptance or the rejection of hypotheses and theories. Most prior attempts to resolve this dilemma have tended to flounder upon a certain unwarranted assumption, namely: that all scientific statements must be *completely* (if fallibly) *decidable*, in principle, i.e., both verifiable and falsifiable. But once this (misleading) assumption has been abandoned, it becomes possible to resolve this apparent contradiction by interpreting scientific laws and theories as only *partially* (and fallibly) *decidable*, in principle, i.e., as statements that may be tested by systematic attempts at their refutation (as Popper [1965], especially pp. 312–313, has emphasized). Moreover, there is a telling precedent for Popper's asymmetrical conception, since whether or not a specific conclusion *follows from* specified premises, within the context of deduction, may only be established, in the absence of a valid demonstration, in general, by systematic attempts to provide one.

THE "PARADOXES" OF CONFIRMATION

One measure of the theoretical significance of the Popperian program, of course, is the extent to which it succeeds in clarifying various difficulties that have beset alternative conceptions. Given the character of the enterprise before us, it would appear appropriate to consider whether or not this approach is capable of illuminating such perplexing aspects of the underlying problems as Goodman's *riddle of induction*, Hempel's *paradox of confirmation*, and Hume's *critique of causation*. In order to undertake this arduous endeavor, however, it is necessary to review at least some of the details of the dispositional construction of lawlike sentences; thus, as an alternative definition, *S is a lawlike sentence of 'simple' form in a language* $\mathfrak{L}zt$ if and only if (a) S is logically general, (b) S is not logically true, and (c) S is of the form, ⌜For all x and all t, if x were K at t, then x would be χ at $\bar{t}$⌝, where 'χ' is a (purely) dispositional predicate of universal or statistical strength in $\mathfrak{L}zt$ (a formulation suggested to the author by Evan K. Jobe). On this analysis, let us note, a lawlike sentence S, as before, is an unrestrictedly general subjunctive conditional attributing a (universal or statistical) dispositional property χ to every member of a reference class K under an appropriate description.

Goodman's *riddle of induction* may be viewed as a special case of the general difficulty which Hempel has referred to as *inductive inconsistency*, i.e., the problem of ascertaining which members, if any, of a set of inconsistent hypotheses are inductively confirmed by a logically consistent but finite set of basic statements (Hempel [1965], pp. 53–73). The essential elements of this predicament, if such it be, are illustrated by curve-fitting problems, in which, no matter how many points appear upon a graph, so long as they are finite, an infinite number of mutually incompatible lines may be drawn to connect those points, where all these lines differ with respect to other locations. The statistical version of this difficulty, of course, arises due to the circumstance that any relative frequency within a finite segment of an infinite sequence is logically compatible with any limiting frequency, which suggests that an infinite number of mutually inconsistent hypotheses might be confirmed by that same data. Goodman presented a meta-linguistic version of this difficulty, while seeking to define its ramifications and significance (Goodman [1965], especially pp. 72–92).

A simple example of Goodman's riddle is represented by a finite set of logically consistent basic statements reporting observations of green emeralds. All of these reports, of course, have been made prior to some time t, which, let us assume, might be midnight tonight. Goodman notes, however, that, although this evidence does not contradict the hypothesis H_1: "All emeralds are green", it also does not contradict the alternate hypothesis H_2: "All emeralds are grue", where something is "grue" just in case it is green before midnight tonight and blue thereafter; nor the hypothesis H_3: "All emeralds are gruple", where something is "gruple" just in case it is green before midnight tonight and purple thereafter; and so forth. But if all of these hypotheses are regarded as confirmed by this same data, then should one predict that an emerald that might be observed tomorrow will be green? Or blue? Or purple? And so on. The solution Goodman was to recommend consisted in the suggestion that the attribute predicates, "green", "grue", and so on differed significantly in their linguistic histories within the language-using community, insofar as some of these, such as "green", have frequently occurred in successful predictions in the past, while certain others, such as "grue", have not. On the basis of considerations such as these, therefore, Goodman developed his *theory of projectibility*, according to which the reliability-rating (as it were) of different predicates in the formulation of predictions is taken to be a function of their past successful employment, i.e., of their *degree of entrenchment*, which might be described as the result of a process of linguistic evolution.

The general conclusions that Goodman derives from his investigations, however, are especially noteworthy, since Goodman infers that *only a statement that is lawlike is capable of being confirmed by its instances*: the residual problem, therefore, is one of ascertaining, "what distinguishes lawlike or confirmable hypotheses from accidental or non-confirmable ones" (Goodman [1965], p. 80). From the perspective of the present program, of course,

Goodman's analysis is confused and his solution is mistaken. For, in spite of its apparent plausibility, the thesis that accidental generalizations are incapable of being confirmed by their instances cannot be sustained: recall the examples of Chapter 6, including, "All moas die before the age of fifty". Indeed, Goodman's contention strongly suggests that the truth of accidental generalizations is beyond the scope of human knowledge. If that were the case, then the problem of separating lawlike from accidental generalizations would have a trivial, but effective solution, namely: if the truth of a generalization is determinable on the basis of its instances, then that generalization must be lawlike. The very opposite, however, appears to be far closer to the truth.

Consider, for example, *the satisfaction criterion of confirmation* that Hempel has constructed in application to Goodman's problem. The basic idea behind this conception is that *an hypothesis h is confirmed by a logically consistent finite set of basic statements* for example if that hypothesis is satisfied by the finite class of individuals mentioned in those statements (Hempel [1965], especially pp. 35–39). Let the evidence class consist of the sentences, "John Jones is a professor and John Jones is bald", "Bill Smith is a professor and Bill Smith is bald", and, "Mary Clark is a professor and Mary Clark is bald"; and let the hypothesis under consideration be H_4: "All professors are bald". Then, according to Hempel's satisfaction criterion of confirmation, H_4 is confirmed by the specified class of evidence; yet surely no one doubts that the hypothesis, even if it were true, is not a lawlike statement, since (to cite at least one desideratum with which Goodman has been concerned) it does not support the subjunctive conditional, "For all x and all t, if x were a professor at t, then x would be bald at t". But this hypothesis *is* capable of being confirmed by its instances, nevertheless, as we have ascertained.

It might be objected, of course, that Goodman himself was attempting to overcome, if not outright rejecting, Hempel's satisfaction criterion, which should therefore not be relied upon as a source of counterexamples, even if the hypotheses under consideration contain only well-entrenched constituent predicates, such as "professor" and "bald". Indeed, the force of the objection to Goodman's conception does not depend upon the source of our examples at all and *a fortiori* does not depend upon specific accounts of confirmation. Consider, for example, any circumstance of custom, practice, or tradition in which it is plausible to suppose that a certain pattern of behavior might be conformed to for an indefinite temporal interval. Imagine that within many agrarian communities, say, it is a tradition to clip the ears of every

190   CORROBORATION

dog at thee years of age; then, when the only dogs of a species D happen to belong to those communities, it might easily turn out to be the case that the hypothesis, "All dogs of species D over three years old have clipped ears", is true. But surely the only way to ascertain whether or not this claim *is* true would be to examine its instances, for customs may change, traditions may wane, and practices may be violated. This phenomenon, moreover, may be characterized quite generally; for with respect to any arbitrarily selected reference class, R, and any arbitrarily selected attribute, A, either,

>All R's are A's ; or,
>.99 R's are A's; or,
>............... ; or,
>.01 R's are A's; or
>All R's are non-A's.

So long as the presence or the absence of the attribute, A, is not entailed by the description of the reference class, R, the truth of sentences such as these will be determinable, in general, only on the basis of their instances, even though, as extensional distributions, none of them happens to be lawlike.

The difficulties with Goodman's analysis in attempting to differentiate lawlike from accidental generalizations on the basis of their capacity to be confirmed by their instances arises from implicitly mistaking a semantic and ontological problem for an epistemic and pragmatic one. The crucial difference to be drawn is that lawlike sentences are subjunctive generalizations, which characterize classes of possible worlds by means of their ontological structure, while accidental generalizations are extensional sentences which characterize the actual world by describing (some segment of) its history. Thus, the relevant issue concerns the kind of connection claimed to obtain between the reference class property, K, and the attribute class property, χ, where *there is a nomological relationship between K and χ*, for example, when χ is a permanent property of every member of the reference class, K, under the appropriate description within the language $\mathcal{L}zt$. Consequently, it is not the case that the hypothesis, "All professors are bald", would be a lawlike statement, provided it were true, because there *are* processes and procedures, including hair transplants, by means of which individuals could lose their baldness without also losing their professorships.

The situation is quite different, of course, with such hypotheses as, "Gold is malleable", "Sugar is soluble", "The half-life of polonium^{218} is 3.05 minutes", and so on, since there are no processes or procedures that would render gold non-malleable without altering its atomic structure, or make sugar non-soluble without changing its chemical composition, and so forth. Indeed, it seems quite plausible to assume that Goodman's account of the character of protectible predicates is similarly beside the point; for the explanation is not merely that predicates such as "grue" lack the history of past successful projection that typify such predicates as, say, "green", but that predicates such as "grue" are the wrong kind of predicate, namely: "green" is a *purely dispositional predicate*, whereas "grue" is not. "Grue" is a dispositional predicate to the extent to which the property it designates (i) is a (complex) tendency to display appropriate outcome responses when subjected to appropriate singular trials, and (ii) is an actual (complex) physical state of some object individually or of some arrangement of things collectively (when it happens to be satisfied at all). By virtue of its explicit definition by means of a temporal reference uniquely denoting a particular moment in the history of the actual world, however, "grue" is not a property that is predicable, in principle, of any member of the history of any possible world and therefore fails to designate a *pure* disposition: *for pure dispositions are universals*, and conversely.

Goodman could respond, of course, with the contention that this claim cannot possibly be correct, insofar as any "purely dispositional predicate" may be explicitly defined by a suitable combination of unprojectible predicates. Assume, for example, the following:

x is grue at $t =_{df}$ x is green and t is before midnight tonight or
x is blue and t is after midnight tonight; and,
x is bleen at $t =_{df}$ x is blue and t is before midnight tonight or
x is green and t is after midnight tonight.

Then, although "grue" and "bleen" are both clearly unprojectible, nevertheless they provide a sufficient basis for defining predicates that are; for,

x is blue at $t =_{df}$ x is bleen and t is before midnight tonight or
x is grue and t is after midnight tonight; and,
x is green at $t =_{df}$ x is grue and t is before midnight tonight or
x is bleen and t is after midnight tonight.

Definitions such as these, of course, are proposed by Goodman himself. To the extent to which projectible predicates and unprojectible predicates are mutually definable, therefore, the implied distinction between "purely" and "partly" dispositional properties upon which the above analysis depends, it seems, should be discarded as specious and without theoretical support.

The relevant considerations, however, concern the role of indexical expressions whose denotations are implicitly determined by specific sequences characterizing the histories of different possible worlds; for the sense of the phrase, "midnight tonight", not only varies with the contexts of its utterance, but also denotes some particular moment during the history of some specific world on each of those occasions (such as 24 December 1979 in the actual world W). The definitions Goodman recommends, therefore, are inadequate for purely dispositional predicates 'F', including "blue" and "green", whose intensions do not depend upon the history of any particular world at all. Thus, purely dispositional predicates are *rigid property designators*, whose instances (denoted by "this" if any) satisfy the following condition:

$$(W)(x)(x \text{ is } F \text{ in } W \equiv x \text{ bears the relation } same_k \text{ to "this" in } W);$$

that is, ⌜For all worlds W and all objects x, x has the property F in W if and only if x bears the relation $same_k$, i.e., of being a member of the same kind as, to any object denoted by "this" in the actual world W⌝ [where kinds are understood as defined in Chapter 2, which entails that some presupposed reference class requires specification within an appropriate framework $\mathfrak{L}zt$], while partly dispositional predicates are not [as that concept is defined by Putnam [1973], especially pp. 707–709]. Thus, the explicit introduction of uniquely denoting temporal referents has the logical consequence of converting dispositional properties into (at least "quasi-") historical relations.

Hempel's *paradox of confirmation*, by contrast poses different kinds of difficulties, which the Popperian program also clarifies. The paradox itself, of course, arises from an intuitively plausible "condition of adequacy" for any acceptable conception of confirmation, namely: *the equivalence condition*, according to which, if a logically consistent finite set of basic statements confirms an hypothesis h, then it also confirms every hypothesis which is logically equivalent with h (Hempel [1965], especially pp. 14–20). Although this requirement appears to be logically impeccable, it nevertheless creates a paradox to the extent to which evidence that intuitively confirms one formulation of an hypothesis h does not invariably intuitively confirm another: for example, the hypothesis H_5: "All non-black things are non-ravens", is extensionally equivalent with the hypothesis H_6: "All ravens are black"; yet certain evidence, such as the observation of red pencils, yellow cows, or white handkerchiefs, which would intuitively confirm H_5 does not intuitively confirm H_6, which instead appears to concern only ravens and their color.

Hempel's own explanation of this paradox relies upon the recognition that the specific phrasing by which an hypothesis is formulated does not dictate its logical content, since, in particular, within an extensional language framework, hypotheses H_5 and H_6 are both logically equivalent to hypothesis H_7: "All things are either non-ravens or black", which surely receives intuitive confirmation from observations of red pencils or yellow cows as well as those of black ravens. Hempel ([1965], pp. 14–20 and pp. 47–48) concludes, therefore, that the paradoxical cases have to be counted as confirmatory, while their counter-intuitive character may be accounted for by virtue of *pragmatical circumstances*, such as background knowledge. The explanation that emanates from our preliminary deliberations, however, suggests that these paradoxes are not merely psychological, but are rooted in an inadequate conception of the logical structure of lawlike sentences, which cannot be formulated by means of extensional language alone; hence, properly understood, these paradoxes invite attention to one of the most important philosophical problems confronting the theory of science.

In order to perceive this connection, remember that the hypothesis, "All ravens are black", interpreted as a lawlike sentence, represents a *subjunctive generalization*, namely: "For all x and all t, if x were a raven at t, then x would be black at t", which will turn out to be true if and only if, although "(being) black" does not follow from the reference class description, "(being a) raven", nevertheless, *being black* is a permanent property of everything that happens to be a *raven*, which, of course, will be the case only if there is no process or procedure whereby something could lose the color *black*, while remaining throughout a *raven*. But the most important syntactical feature that distinguishes these lawlike sentences from material conditionals, on the one hand, and logical necessities, on the other, is that these intensional generalizations are not subject to principles of transposition, which would affect a trivialization of the significance of the conception of permanent properties as the ontological foundation for an adequate explication of scientific laws. For with transposition, the distinction between subjunctive conditionals which are warranted on logical grounds as opposed to those which are warranted on nomological grounds cannot be sustained, while the invocation of permanent properties becomes theoretically superfluous (for the recognition of which the author is indebted to Charles E. M. Dunlop).

In order to appreciate the theoretical necessity to exclude transposition for nomological conditionals, in general, let us consider a proof formulated by Nute [1976], which warrants publication. Nute's proof of the identity of subjunctive conditionals, generally, with strict implications *if* the logical equivalence of '$p \mathrel{\supset\mkern-5mu\relbar} q$' with '$-q \mathrel{\supset\mkern-5mu\relbar} -p$' is assumed, i.e., if principle YZ is accepted, namely:

YZ. $(p \mathbin{\ni} q) \equiv (-q \mathbin{\ni} -p)$,

depends upon the following rules and definitions:

(1)   if 'p' is a tautology, then 'p' is a theorem [tautology];
(2)   if n is a positive integer and '$(p^1 \cdot \ldots \cdot p^n) \supset q$' is a theorem, then '$[(r \mathbin{\ni} p^1) \cdot \ldots \cdot (r \mathbin{\ni} p^n)] \supset (r \mathbin{\ni} q)$' is also a theorem [the principle of strengthening of subjunctive antecedents (SSA)];
(3)   '$\Box p$' if and only if '$-p \mathbin{\ni} p$' [the definition of '$\Box$' (df. $\Box$)]; and,
(4)   the principle of hypothetical syllogism as a derived rule of inference [H.S.].

The proof itself then proceeds as follows:

1.	$q \supset (p \supset q)$	tautology
2.	$(p \mathbin{\ni} q) \supset [p \mathbin{\ni} (p \supset q)]$	ln. 1, SSA
3.	$[p \mathbin{\ni} (p \supset q)] \supset [-(p \supset q) \mathbin{\ni} -p]$	YZ
4.	$(p \mathbin{\ni} q) \supset [-(p \supset q) \mathbin{\ni} -p]$	lns. 2, 3 H.S.
5.	$-p \supset (p \supset q)$	tautology
6.	$[-(p \supset q) \mathbin{\ni} -p] \supset [-(p \supset q) \mathbin{\ni} (p \supset q)]$	ln. 5, SSA
7.	$(p \mathbin{\ni} q) \supset [-(p \supset q) \mathbin{\ni} (p \supset q)]$	lns. 4, 6 H.S.
8.	$(p \mathbin{\ni} q) \supset \Box(p \supset q)$	ln. 7, df. $\Box$

Consequently, unless one or another of these assumed rules and definitions ought to be rejected, the principle of transposition YZ itself cannot be accepted.

The ontological explanation for this logical distinction, moreover, seems to be as follows. A logical necessity of the form, '$\Box(p \supset q)$', for example, is a sentence whose truth follows from the syntactical and semantical rules of the language framework $\mathfrak{L}$ alone; in particular, in the case of the definitions of *logical* "possibility", "necessity", and "impossibility" advanced in Chapter 3, its truth must follow from the syntactical rules of that framework $\mathfrak{L}$ alone. Otherwise, the linguistic status of a conditional true only of all physically possible worlds could be converted into that of a conditional also true of all logically possible worlds as well by *definitional stipulation*. [Any statement which is not syntactically contradictory, of course, may be "held to be true come what may", by making sufficiently drastic alterations of our framework $\mathfrak{L}$, as Quine [1953], pp. 20–46, especially has emphasized. But only true lawlike sentences and true nomological conditionals may be converted to logical truths without thereby altering the class of true *subjunctive conditionals* within the language framework $\mathfrak{L}$.] When a scientific conditional of the form '$p \mathrel{\exists\mkern-6mu{-}} q$', is *true-in* $\mathcal{L}$, therefore, the corresponding logical necessity, '$\Box(p \supset q)$', cannot be *true-in* $\mathcal{L}$, since the truth of the former is logically contingent, i.e., it depends upon the permanent properties of the actual world W ; similarly, when a scientific conditional of the form, '$-q \mathrel{\exists\mkern-6mu{-}} -p$', is *true-in*-$\mathcal{L}$, the corresponding logical necessity, '$\Box(-q \supset -p)$', once again cannot be *true-in*-$\mathcal{L}$, since it too is logically contingent, i.e., it depends upon the permanent properties of the actual world W . Consequently, even when all p-worlds happen to be q-worlds *as a matter of physical necessity* (in relation to the actual world W), there must still remain $-q$-worlds that are *not* $-p$-worlds *as a matter of logical necessity* (in relation to other possible actual worlds W *)! The imposition of transposition upon scientific conditionals thus appears to be theoretically unwarranted.

The deductive principle, *modus tollens*, of course, continues to serve as a valid rule of inference relative to arguments with 'simple' lawlike premises in general. Among other authors who have also rejected transposition for subjunctive or for lawlike conditionals, moreover, are Stalnaker [1968], especially p. 110, and Pollock [1972], especially p. 307. Indeed, when considered along with certain other semantical properties distinguishing the logical from the physical modalities, the non-transposability of scientific conditionals does not seem to be especially surprising. In particular, when an 'if ... then ———' statement, S, is interpreted as *nomological*, then:

(i) S is not always true when its antecedent is false, in contradistinction to corresponding material conditionals;

(ii) S is never true when its antecedent is logically impossible (or necessarily false), in contradistinction to corresponding logical truths;

(iii) S is not always true when its consequent is true, in contradistinction to corresponding material conditionals; and,

(iv) S is never true when its consequent is logically necessary (or necessarily true), in contradistinction to corresponding logical truths;

which underscores the theoretical necessity for distinguishing between them (as Popper [1965], pp. 432–441 and especially p. 434, has also emphasized).

It is extremely interesting to observe, moreover, that, even though principles of transposition for lawlike sentences would reinstate the paradoxes of *confirmation* within our intensional language, there are no corresponding paradoxes of *falsification* (intensional or otherwise). Thus, notice that the evidence which would falsify the hypothesis, "All ravens are black", and the evidence which would falsify its extensional equivalent, "All non-black things are non-ravens", is *identical*, i.e., ravens that are non-black or non-black things that are non-non-ravens. Consequently, considered purely as a *methodological* problem, for example, concerning the techniques of hypothesis testing, rather than as a *semantical* one, i.e., concerning the logical structure of lawlike sentences, this asymmetry itself might reasonably be regarded as the foundation for a resolution of Hempel's paradoxes in favor of falsification; indeed, this is essentially the position adopted by Popper himself (Schilpp, ed. [1974], especially pp. 990–991) and by Watkins [1964].

As a result, Hempel's paradoxes not only implicitly display the essential inadequacy of extensional language for the formalization of lawlike sentences but also serve to emphasize the fundamental role of falsification procedures; for, as Popper, especially, has maintained, all laws have *the force of prohibitions*, i.e., they forbid the occurrence of any event of a certain kind, not only during the history of the actual world but also during the history of any other possible world. An "inventory" of the history of the actual world, therefore, does not provide a basis for evaluating the truth of every lawlike statement, for the particular sequence of events which constitutes that history does not necessarily reflect all of the world's physical possibilities: the problem, therefore, is *to attempt to arrange this world's history* that it should include sets of events evidentially relevant to testing alternative hypotheses and theories, which requires, in general, controlled experiments and systematic observations. We thus have to decide, in part, which hypotheses to test. [The view that deliberate experimentation affords an avenue for testing subjunctives which might otherwise remain counterfactual during the history of the actual world is also advanced by von Wright [1971].]

Of course, it would be wrong to conclude that no hypothesis is subjected to test during the course of the world's "natural" history, i.e., without intervention by means of deliberate experimentation. Numberless hypotheses are tacitly tested by the observation of a single red pencil, for example, which could provide sufficient evidence to *falsify* enormous numbers of lawlike hypotheses, such as, "All pencils are purple", "All pencils are elastic", "All pencils are soluble", and so on, while simultaneously affording appropriate evidence to *confirm* enormous numbers of extensional statements, such as, "All pencils are red", "All pencils are wooden", "All pencils are broken", and so forth. Although the equivalence class of extensional generalizations is closed under transposition, moreover, the equivalence class of intensional generalizations is not. The problem, therefore, is not to distinguish between *confirmable* and *accidental* generalizations, as Goodman misleadingly suggests, but to distinguish between *testing lawlike hypotheses* and *confirming extensional generalizations*; for the evidence that confirms a generalization does not necessarily support a law.

A CRITIQUE OF HUME'S CRITIQUE

Hume's *critique of causation*, finally, is important not only because permanent properties as intensional relations are a species of necessary connection but also because every lawlike sentence, in principle, may be expressed in the form of a causal conditional as well. This result is a consequence of the conception of dispositional properties as *single case causal tendencies*, which is reflected by their mode of definition, since a dispositional predicate may be (partially) defined by a conjunction of relevant-test/outcome-response conditions by invoking the "probabilistic causal conditional" as follows: "x is malleable at t", means, by definition, "hammering x into shape S^1 at t would invariably (or, probably) bring about x's having shape S^1 at t^*; and, pressing x into shape S^2 at t would invariably (or, probably) bring about x's having shape S^2 at t^*; and, ...", where t^* is either simultaneous with or subsequent to t. The 'causal' form of a lawlike sentence is thus obtained by substituting one of its defining conjuncts for a dispositional predicate, as follows: "For all x and all t, if x were gold at t, then hammering x into shape S^1 at t would invariably bring about x's having shape S^1 at t^*", which is a partial reflection of the causal significance of the lawlike statement, "Gold is malleable", on this intensional program. The question therefore arises of the extent to which conceptions such as these actually satisfy Hume's strictures.

Without providing a detailed reconstruction of the specifics of Hume's position, let us assume that Hume's motivation was not merely to establish the meaning of a certain term within a community of language users, but to ascertain the extent to which claims of a certain kind might be justified on the basis of experiential considerations alone, i.e., on the basis of (more or less) direct observation and relatively simple measurement. The ordinary usage of causal language thus suggests that the term "cause" is characteristically employed to assert that one event (called "the cause") bears a certain relationship to another event (called "its effect") where this relation of causation embraces the following components, namely: (a) *resemblance*, i.e., similar causes are attended by similar effects (which, we may assume, may be either their invariable or their probable outcomes); (b) *regular association*, i.e., causes are temporally simultaneous or prior to and spatially contiguous with their effects, conforming to a repeatable pattern which, for probabilistic causes, is that of a relative frequency; and, (c) *necessary connection*, i.e., effects are necessarily, rather than accidentally, related to their causes, even though this relation is not a matter of logical necessity.

The classical result of this inquiry was Hume's determination that, although the relations of resemblance and of regular association were experientially ascertainable, relations of necessary connection could not be detected on the basis of experiential considerations alone. Consequently, the attribution of relations of necessary connection between events, Hume reasoned, must be a result of human frailty, i.e., an inevitable habit of the mind, which, however, is epistemically unwarranted and therefore philosophically inexcusable. The benefit of this analysis was a clarification of the concept of causation that it should no longer embrace any notion of necessary connection between events but encompass only those aspects whose presence or absence could be experientially established: resemblance and regular association. And as a philosophical legacy, Hume (implicitly) bequeathed the following argument:

Necessary connections between events are either observable and objective or psychological and subjective; but they cannot be ascertained on the basis of experiential considerations alone; consequently, they must be merely psychological and subjective.

Indeed, this argument has been among the most pervasive and influential in the history of Western philosophy.

It is my contention, however, that Hume has presented a misleading dilemma in the form of the premise that necessary conditions must be *either* observable and objective *or* psychological and subjective. This crucial assumption precludes the possibility that necessary connections, as theoretical properties of the physical world, might be unobservable and *nevertheless* objective. If the logical foundation supporting Hume's argument should prove to be insecure, therefore, then the critique of the concept of causation based upon it should pose no insurmountable obstacle to adopting an intensional conception of lawlike statements. Moreover, once these logical foundations are clearly displayed, it becomes apparent that there are important reasons for abandoning them; indeed, my argument will be that rigid adherence to the epistemological principles implicit in the Humean approach would itself pose a virtually insurmountable obstacle to the progress of empirical science.

The logical foundation of Hume's argument, therefore, is the thesis that *properties of the physical world are objective if and only if they are observable*. Notice that, although this thesis is directly applicable to terms that designate properties of the physical world, it is indirectly applicable to statements that describe the world as well. In its direct application, in other words, it serves as a criterion for distinguishing "objective" from "subjective" *predicates*; in its indirect application, by contrast, it functions as a criterion for differentiating between "cognitively significant" and "cognitively insignificant" *sentences*. Thus, the obvious complement to Hume's position is the thesis that an hypothesis or a theory is empirically meaningful (or scientifically significant) if and only if it is *verifiable* on the basis of a logically consistent finite set of observation sentences.

The first difficulty encountered by this thesis is that it imposes an excessive restriction upon the language of empirical science; for the Humean position entails that scientific language, if it is to be "objective", must be constructed exclusively from a non-logical vocabulary that is limited to observational predicates alone. The only permissible predicates, presumably, would be those designating (more or less) directly observable properties of (more or less) directly observable entities, perhaps, e.g., the color of the sky or the shape of a die. Those that designate (more or less) unobservable properties of observable entities, e.g., the fragility of a vase or the flexibility of a rubber band, or designating unobservable properties of unobservable entities, such as the strength of gravitational fields or the transmission of a recessive trait, would not be permissible predicates. All dispositional and theoretical predicates would accordingly be excluded from the language of empirical science. [Compare Hempel's [1965] discussion, especially pp. 107–113.]

The second difficulty is that scientific laws, whether of universal or statistical form, would have to be given an entirely extensional analysis in order to guarantee their complete verifiability. But this would entail abandoning established criteria for lawlike statements, since their "objectivity" would require (i) that they be logically equivalent with a finite conjunction of singular sentences and (ii) that they be incapable of providing support for counterfactual and for subjunctive conditionals. Indeed, under such a strict conception, subjunctives and counterfactuals would have no scientific significance at all. Any true generalizations describing (segments of) the world's history accessible to observations and measurement might have to be accorded the status of a scientific law; for there would be no syntactical or semantical basis for distinguishing between genuinely lawlike and merely accidental generalizations. [Again compare Hempel's [1965] discussion, especially pp. 338–343.]

The third difficulty is that empirical science would have to dispense with scientific theories, at least insofar as such theories are envisioned as encompassing "objective" properties of the physical world. But scientific theories have proven to be exceptionally beneficial to the progress of science, for they provide the opportunity to systematically incorporate broad ranges of experiential data by means of general principles in terms of which past events may be explained and future events predicted. A further advantage of scientific theories is their heuristic fertility and conceptual economy, in the sense that their pursuit has stimulated the discovery of unsuspected connections between natural occurrences in an attempt to account for the largest variety of phenomena while relying upon the smallest set of assumptions. Indeed, the development of scientific theories has been a distinguishing characteristic of the history of science itself. [Compare Hempel's [1965] discussion once again, especially pp. 210–222.]

Not the least of the difficulties confronting this Humean position, I might add, is the defense of the very distinction upon which its fundamental principles depend, for the difference between observational and non-observational language has proven notoriously difficult to successfully define; indeed, if the Popperian conception of universals as dispositional is correct, there is, in principle, no theoretical warrant for that distinction at all, since observational, dispositional, and theoretical language represent only differing degrees of epistemic accessibility, from the ontological point of view. But even if the distinction at issue were to be assumed, the underlying inadequacy of Hume's analysis should by now have been made apparent; for Hume relies upon an excessively narrow conception of objectivity. Surely a scientific assertion need not be couched exclusively in observation language in order to be cognitively significant in relation to the data of experience; for an empirical hypothesis is epistemically permissible when it can be subjected to objective tests by means of systematic observation and controlled experimentation in accordance with Popper's principle of empiricism.

The deficiencies of Hume's critique of causation, however, should not be supposed to affect the results of his critique of induction; for whether or not there are any non-logical necessary connections, there are certainly no ampliative-demonstrative arguments. Thus, to perceive that this is the case, simply notice that no logical inconsistency arises from admitting:

(a) the kinds of tests that have verified in the past might not verify in the future;

(b) the kinds of tests that have falsified in the past might not falsify in the future;

(c) the predicates that have been projectible in the past might not be projectible in the future; and,

(d) the necessary connections that have obtained in the past might not obtain in the future.

Perhaps more importantly, however, if we acknowledge that our conception of scientific laws entails that *when they are true, they do not change*, then a conditional argument that relates the past and the future is possible, after all.

with respect to the members of the class of presumptive scientific laws as follows, namely:

> Even if the world *is* as we believe it to be in these specific respects, it remains *logically possible* that they might still change tomorrow; but *if* the world is as we believe it to be in these specific respects, then it is not *physically possible* that they actually will change tomorrow.

And this, perhaps, is as much as we should ask of any "inductive" procedure.

The general conception of the Popperian program that emerges from these considerations at last, therefore, suggests that empirical science endeavors to fulfill (at least) two complementary objectives: *first*, to ascertain the historical patterns displayed by the world's history, which are expressible in extensional language and are capable of confirmation by their instances; and, *second*, to discover the lawlike relations constitutive of the world's structure, which are expressible in intensional language and, in principle, are subject to procedures of falsification but not to those of verification. These historical patterns, moreover, may be described by means of frequency distributions and by material conditionals to which traditional conceptions of induction apparently apply; their principal importance, however, resides in their usefulness as *evidence* for discovering the lawlike relations which they covertly display and which in turn implicitly explain them. Although the confirmation of extensional generalizations thus fulfills a necessary condition for the conduct of scientific inquiry, therefore, the principles of fallibilistic falsification are indispensable for its ultimate success.

8. CONFIRMATION AND CORROBORATION

The question that remains, of course, is whether science can ultimately succeed if its methods are confined to fallibilistic falsification, that is: surely an adequate conception of scientific knowledge should provide for the *acceptance* as well as the *rejection* of some hypotheses and theories, perhaps in the form of merely *probable knowledge*. Popper ([1965], p. 368, note 3), however, denies even this modest conception of inductive capabilities:

> Empirical knowledge *in some sense* of the word 'knowledge' exists. But in other senses — for example in the sense of *certain* knowledge, or of *demonstrable* knowledge — it does not. And we must not assume, uncritically, that we have 'probable' knowledge — knowledge that is probable in the sense of the calculus of probability. It is indeed my contention that we do not have probable knowledge in this sense. For I believe that what we call 'empirical' knowledge, including 'scientific knowledge', consists of guesses, and that many of these guesses are not probable (or have a probability zero) even though they may be very well corroborated.

In order to evaluate this controversial claim, therefore, it is important to ascertain the characteristics that distinguish probabilistic conceptions of confirmation, traditional principles of induction, and Popperian procedures of corroboration, with particular concern for the sense in which these alternatives provide for the acceptance as well as the rejection of their objects of inquiry, if, indeed, they do so at all. Before turning to an examination of their more detailed features, however, some of their more general aspects deserve preliminary consideration.

Current research on probability and induction typically operates within the context of a pair of interlocked and reinforcing presuppositions, namely:

first, that adequate measures of evidential support must satisfy certain mathematical relationships specified by the calculus of probability, which we shall refer to as *the thesis of the probabilistic framework*;

second, that the cumulative influence of acquired evidence may properly be understood as a process of conditionalization by means of Bayes' theorem, which we shall refer to as *the thesis of conditionalization*.

The broad range and wide appeal of these theses receives ample testimony from their diversified and influentieal adherents; for the Bayesian tradition, thus defined, includes among its members not only *logical* Bayesians, such as

Carnap [1962] and Hintikka [1966], *subjective* Bayesians, such as Savage [1954] and de Finetti [1964], and *empirical* Bayesians, such as Salmon [1967] and Rosenkrantz [1977], but also a number of *critical* Bayesians (as they might be described), such as Suppes [1966] and Jeffrey [1965], who hold to Bayesian principles with critical reservations, and *quasi* Bayesians, including Levi [1967] and Shimony [1970], whose differences from the rest appear to be even more important than do their similarities. Nevertheless, in spite of significant divergence concerning the specifics of their formulations, (virtually) all of the positions represented here (and those of others unnamed) converge in their agreement upon the dual claims cited above. [For example, Lindley [1971], Ramsey [1931], Good [1976], Reichenbach [1949], Hilpinen [1968], Niiniluoto and Tuomela [1973], Burks [1977], and others still unnamed. For excellent reviews, see both Giere [1979b] and Niiniluoto [   ], which include abundant references.]

The principal proponents of alternative conceptions, let us note, adhere primarily to the principles of *orthodox statistical hypothesis testing*, as it has evolved from the efforts of Fisher [1956], Neyman and Pearson [1966], and Wald [1950], whose primary advocate among philosophers has been Giere [1975], [1976], and [1977], on the one hand, or to the *Popperian program of corroboration*, which I identify with the work of Peirce [1955], Popper [1965], Braithwaite [1953], and Hacking [1965] (although it too has been heavily influence by Fisher), a rather important defense of which has recently been advanced by Gillies [1973], on the other. Perhaps their most important principles, moreover, are those they share in common; for, the statistical tradition and the Popperian program, which differ in significant respects, reject alike the underlying principles upon which Bayesian analyses ultimately depend, namely: the probabilistic framework and the conditionalization theorem — not as mathematical machinery *per se*, but as essential to the character of confirmation. Since this point could easily be misunderstood, let us consider it in somewhat greater detail.

The term "probability" itself, we are in the process of discovering, is highly ambiguous, occurring in at least three distinct contexts with various shades of meaning; in particular, "probabilities" may be envisioned: (i) as physical properties of physical systems *within an ontic context*, in which a "probability statement" attributes a statistical property to (some specific feature of) the physical world, represented by the frequency and the propensity interpretations in their varied formulations; (ii) as logical relations of certain sets of sentences *within an explanation context*, in which a "probability statement' acribes a degree of nomic expectability to (a particular explanandum of) an explanatory argument, represented by the "partial entailment" interpretation as a purely deductive conception; and, (iii) as logical or *quasi-logical* properties of certain sets of sentences *within an inductive context*, in which a "probability statement" attributes some measure of evidential support to an hypothesis (relative to an inductive argument), reflected, in particular, by the Bayesian conception of confirmation. The crucial point to appreciate, therefore, is that the Popperian program and the statistical tradition do *not* require "probabilities of hypotheses" in sense (iii), even though they do depend upon "probabilities of events" in senses (i) and (ii). Consequently, although "probabilities" as they occur in senses (i) or (ii) are expected to satisfy the mathematical properties of the probability calculus, in general, neither the Popperian program nor the statistical tradition are thereby committed to *the thesis of a probabilistic framework* for inductive relations of evidential support. The purpose of this chapter (and of the one to follow), therefore, includes exploring the differences between the Bayesian and the non-Bayesian traditions by an investigation of several problems encountered on indispensable Bayesian principles, together with an exploration of their resolution within the context of the Popperian program.

BAYESIAN CONCEPTIONS OF CONFIRMATION

The central core of the Bayesian conception has been lucidly summarized by Rosenkrantz as follows:

There is no sound principle of induction as traditionally understood. Induction, or learning from experience, is just the process of revising probability assessments in the light of additional information. This is done by Bayes' rule:

$$P(h/e) = P(e/h) \cdot P(h)/P(e),$$

where h is an hypothesis and e an experimental outcome or datum. The problem of justifying induction is thus, ..., the problem of justifying conditionalization or Bayes' rule. (Rosenkrantz [1977], p. 48)

According to Bayes' theorem, in other words, the probability of an hypothesis h, given evidence e, is equal to the *likelihood* of h, given e, i.e., the probability of e if h is true, multiplied by the ratio of the *prior probability* of h to the *prior probability* of e. These "prior probabilities", moreover, are envisioned as the probabilities of h and of e "prior to" the acquisition of e, where their precise interpretation provides the central grounds differentiating between the various alternative conceptions of Bayesian confirmation. Although these differences are enormously important, in principle, when any

CONFIRMATION AND CORROBORATION

specific version is under consideration, I shall attempt to focus upon problems which concern Bayesians of every persuasion (because they are rooted in the probabilistic framework, in general, and in the conditionalization theorem, in particular), in what follows.

Whether or not Bayesians ever attribute "probabilities" to events within an ontic context, i.e., as physical properties of physical systems, their principal preoccupation is with the *probabilities of hypotheses* in relation to specified evidence; consequently, let us assume that the argument places h and e represent sentences p and q within a language $\mathfrak{L}$ (or, instead, stand for *propositions* as equivalence classes of sentences within that framework). Hacking has emphasized the necessity, on Bayesian principles, to distinguish between (what he refers to as) "probabilities given facts", say, as

(I) $\quad P_f(h)$,

"a number representing the person's personal probability for h, when he knows f; for short, his probability given f" (Hacking [1967], p. 313), by contrast to *conditional probabilities*, which assume the corresponding form,

(II) $\quad P_f(h/e) = P_f(h \& e)/P_f(e)$,

for positive values of e, where these probabilities "indicate how confident a person knowing only f judges that he would be if he knew e as well" (Hacking [1967], p. 314). For, in spite of the familiarity of formula (II) — absent the subscript "f" — formula (I) is fundamental in representing (what he calls) "the static assumption" of Bayesian confirmation, namely: *that for all persons z, there exists a set of propositions A (forming a Boolean algebra) such that, for any hypothesis h belonging to A and at least any fact f known to z included in A, $P_f(h)$ is defined and satisfies the probability axioms* (Hacking [1967], p. 313). While Hacking's use of the term, "fact", may be questionable on the grounds that these *accepted sentences* as sincere beliefs are not invariably true, the point which he makes is not; for Bayesians invariably assume that a person's beliefs should conform to these probabilistic constraints, a condition whose fulfillment qualifies such a person (or his beliefs) as *coherent*.

In relation to the discussion of Chapter 1, the condition of coherence may very plausibly be regarded as the *probabilistic application* of the requirements of rational belief (CR-1) and (CR-2); for, the conception of this set of belief-propositions as forming a Boolean algebra implements the condition of *deductive closure* (CR-1) for probabilistic beliefs, while the conception of this set of belief-propositions as conforming to the probability axioms likewise enforces the condition of *logical consistency* (CR-2). As a result, the condition of coherence represents a standard for a set of probabilistic beliefs to qualify as a *(minimally) rational set of beliefs*, which, once again, is not strictly satisfied by many of us during the history of the world. Consequently, like the conditions of deductive closure and of logical consistency themselves, the requirement of coherence reflects a "normative", rather than a "descriptive", conception of (minimal) rationality; in other words, it represents (what Hacking thinks of as) an idealized and approximate "model" of rational beliefs, as that conception was elaborated during our discussion in Chapter 1.

The rationality requirement thereby imposed, of course, is rather liberal, insofar as any person's sincere beliefs will qualify as "coherent" so long as they continue to satisfy this conception, even though (as Kyburg especially has emphasized) that person could be "mad as a hatter": however diverse the specific contents of their accepted beliefs, for example, n different persons are entitled to n different distributions of prior probabilities at any one time with equal rationality, provided only that their beliefs are "coherent" (Kyburg [1970], pp. 69–70). Bayesians of various persuasions nevertheless entertain different solutions to this problem, insofar as they may disagree upon the *source* of these prior probabilities themselves: subjective Bayesians, for example, may adopt a "betting quotient" analysis according to which "coherence" is both necessary and sufficient for rationality of belief, but logical and empirical Bayesians impose other constraints with respect to language frameworks and to specific information, respectively, which render this condition necessary but not sufficient for a person's beliefs to qualify as "rational". Carnap [1962], Hintikka [1966], Salmon [1967], and Rosenkrantz [1977], therefore, all embrace (what come to) much stronger conceptions of rationality of belief, which, however, nevertheless entail "coherence" as an indispensable requirement.

In order to appreciate the role of *conditionalization* within the Bayesian scheme, of course, (what Hacking refers to as) "the dynamic assumption" assumes a central role, since "old" posterior and "new" prior probabilities are supposed to be related as follows (Hacking [1967], p. 314):

(III) $\quad P_{f \cup \{e\}}(h) = P_f(h/e)$,

which asserts that a person's personal probability for h upon the acquisition of the belief e as an addition to his previously accepted beliefs f should be equal to the conditional probability of h given e when that person accepts f. A complete characterization of the acquisition of knowledge on Bayesian principles, therefore, requires specification of a prior belief function satisfying condition (I) together with the assumption that *changes in beliefs* occur in accordance with condition (III). However appealing this conception might initially appear, however, it suffers from serious drawbacks; for, as Putnam [1963], Suppes [1966], and Levi [1974], among others, have observed, the belief that any person should at all times conform to the requirements of conditionalization on the basis of some single distribution of prior probabilities cannot be sustained: adherence to such a strict construction, normatively or descriptively, although receiving support from Carnap [1962], de Finetti [1964], and Savage [1954], does not afford a reasonable analysis of rationality. Confronted by this paradox, Levi recommends a generalization of Bayesian procedures by dispensing with a *single* such function in favor of a single *set* of functions, while Suppes has gone so far as to doubt the applicability of conditionalization altogether (Suppes [1966], p. 64):

It is even possible to question whether any changes [in beliefs] can be so expressed. The measure P effectively expressing my beliefs at time t cannot be used to express what I actually observe immediately after time t, for P is already "used up" so to speak in expressing the *a priori* probability of each possible event that might occur, and cannot be used to express the unconditional occurrence of that which in fact did happen at t.

If the Bayesian "problem of induction" is that of justifying conditionalization, as Rosenkrantz suggests, therefore, it would appear to pose a formidable task, indeed. [See especially Teller (1973).]

This difficulty is crucial for Bayesian confirmation, since without some justification for (*continuous*) *confirmational consistency* by restricting the range of acceptable prior belief functions persons may employ at any time t, a Bayesian may rationally accept or reject the conclusion of any argument by the expedient of adopting a "more congenial" prior probability distribution: as Kyburg has observed, "If one coherent credence function is as good as another, a single person at two different times has as much right to two credence functions as two people at the same time" (Kyburg [1970], p. 71). The only condition beyond coherence itself that Bayesians, in general, tend to endorse, however, seems to be that of *strict coherence,* namely: *that an hypothesis should be assigned a prior probability of one (or of zero) given f if and only if that hypothesis is true, necessarily (or false, necessarily) given f* (compare Lehmann [1955], Kemeny [1955], and Shimony [1955]). Thus, when f consists of a language $\mathfrak{L}$ alone, an hypothesis should be assigned the prior probability of one (or zero) if and only if it is logically true (or logically false). Since the "if" clause already follows from the condition of coherence itself, only the "only if" clause is subject to debate, where, in spite of plausible rationales for its retention (such as that in matters of empirical belief, persons should "keep an open mind" and similar "betting behavior" analogues), this condition has been rejected by Suppes and by Levi alike on identical grounds, namely: *that "accepting an hypothesis" (on whatever grounds) requires assigning that hypothesis a probability of one*, which will often involve changing its probability value from *less than one* to *one* (compare Suppes [1966] and Levi [1970], [1971], and [1974], for example). The difficulty here, of course, is to provide for the possibility of a subsequent revision among accepted beliefs on the basis of subsequent evidence, a desideratum which Levi especially has sought to accommodate through a process of contraction, deliberation, and expansion to resolve conflicts in beliefs − a problem avoided by those who eschew acceptance and rejection altogether. Although we shall return to the comparison of Bayesian and Popperian approaches in sections to follow, let us first consider whether logical Bayesians (such as Carnap and Hintikka) or empirical Bayesians (such as Reichenbach and Salmon) contribute potential solutions to some of these difficult problems.

TRADITIONAL PRINCIPLES OF INDUCTION

Reichenbach [1949] and Salmon [1967], of course, attempt to establish a foundation for ascertaining "prior probabilities" without assuming the availability of background knowledge, in general, by employing (what we shall refer to as) traditional principles of induction, which may be represented for our purposes by at least the following rule, namely:

(IV)   From "m/n observed A's are B's", infer to "m/n A's are B's";

with the understanding that (i) a large number of instances of A's have been observed (ii) under a wide variety of conditions, where (iii) all such inferences must be consistent with the observational data available at subsequent times, i.e., previously accepted generalizations must be rejected or revised on the basis of further inquiry, whenever their results no longer correspond (compare Chalmers [1976], pp. 3−5). [This conception, incidentally, follows the conditions of rationality (CR-3) and (CR-4) introduced in Chapter 1.] A special case of this rule, moreover, occurs when every instance of an A is an instance of a B or, in other words, when m equals n, as follows:

(V)   From "All observed A's are B's", infer to "All A's are B's".

These principles, of course, are especially applicable within an inductivist program which envisions the aim of science as that of determining the limiting frequencies and constant conjunctions with which particular attributes occur within various reference classes during the history of the world; indeed, in relation to that goal, these rules are subject to a *pragmatical vindication*, insofar as, if an attribute B occurs within a reference class A with a limiting frequency m/n (where m may equal n), then the (perpetual) application of principles (IV) and (V) will eventually ascertain that limit, necessarily (a rationale recently subjected to scrutiny by Salmon [1968] and Hacking [1968]).

In order to illustrate (what I take to be) the function of these principles within a Bayesian framework, therefore, let us consider the interpretation of Bayesian formulations such as (I), (II), and (III), from a frequency perspective, which requires that the probabilities occurring therein must be *conditional probabilities*, necessarily, since, under the frequency construction, probabilities only exist relative to specified reference classes, $R^1, R^2, \ldots$, for any attribute, $A^1, A^2, \ldots$. The frequency interpretation of formula (I), therefore, might reasonably be expressed as follows,

(VI) $P_R(A)$,

where this "prior probability" for the attribute A, relative to the reference class R, is determined by the application of principles (IV) and (V). Hence, the frequency analogue of formula (II) might be expressed as follows,

(VII) $P_R(B/A) = P_R(B\&A)/P_R(A)$,

where the "posterior probability" for the attribute B, given A, relative to the reference class R, is equal to the "prior probability" for both A and B, relative to the reference class R, divided by the "prior probability" for A, relative to the refrence class R. Consequently, the frequency analogue of formula (III) might be expressed as follows,

(VIII) $P_{R \cap A}(B) = P_R(B/A)$,

where the "new" prior probability for the attribute B, relative to the reference class R and A, is equal to the "old" posterior probability for that attribute, given A, relative to the reference class R (Teller [1973]).

The significance of these formulations of Bayes' theorem, however, is crucially dependent upon the precise interpretation of the variables A, B, and R, where there appear to be (at least) three possibilities, as follows:

first, these variables may be understood as assuming the values of *specific properties* (or, predicate constants), such as, 'A', 'B', and 'R', in which case their class-designating character reflects their role for the description of *kinds of events*; thus, within this context, the probability operator '*P*' receives the familiar (empirical) limiting frequency interpretation, with which we have been generally concerned, especially in Chapter 4;

second, these variables may also be understood as assuming the values of *singular sentences* (or, sentence functions), such as 'Ax', 'Bx', and 'Rx', in which case their occasion-sentence character reflects their role for the description of *particular events*; thus, within this context, the probability operator '*P*' receives the less familiar (logical) truth frequency interpretation, which was elaborated (to some extent, at least) in Chapter 5; and,

third, these variables might perhaps be understood as assuming the values of *general hypotheses* (or, scientific theories), such as, '$(x)(t)(Gxt \supset Mxt = 1063\,°C)$', i.e., 'Gold melts at 1063 °C', in which case their logically unrestricted character reflects their role for the description of *classes of phenomena*; thus, within this context, the probability operator '*P*' would (presumably) receive an unfamiliar (logical) truth frequency interpretation *relative to classes of hypotheses and theories*.

The first and second interpretations, of course, differ insofar as the second, in effect, provides the meta-language analogue of the first, since, as we have already ascertained, the (logical) truth frequency construction depends, in principle, upon the (empirical) limiting frequency construction, where the best estimate of the value of the logical probability will equal the value of the best estimate of the value of the corresponding empirical probability, necessarily, *within an epistemic context*, provided (CR-3) and (CR-4) are satisfied, just as the value of the logical probability will equal the value of the corresponding empirical probablity, *within an ontic context* (since those sentences will be true if and only if the events which they describe occur). Perhaps the most important feature of these interpretations, in general, however, is that they are both appropriately envisioned as attributing "statistical probabilities" to events, rather than as attributing "inductive probabilities" to hypotheses.

It would not be unreasonable to suppose that the conception underlying the third interpretation of Bayes' theorem arises from reflection upon the circumstance that, for any finite class of evidence sentences, there will be an infinite class of possible hypotheses or theories (consisting of general sentences and auxiliary hypotheses) which could be invoked to explain them. This suggests the possibility that perhaps '*R*' should be entertained as a description of the relevant empirical explanandum, i.e., as the evidence to be explained, while '*A*' consists of specific conjunctions of auxiliary hypotheses and '*B*' represents some class of alternative hypotheses or theories. The "inductive probability" of B, given A, relative to R, would then presumably represent the truth-frequency for an hypothesis or theory of kind B, relative to a class of auxiliary hypotheses of kind A and a class of explanandum phenomena of kind R. Apart from difficulties in establishing suitable "prior probabilities" (which Salmon [1967] and [1970] has acknowledged) and in selecting appropriate classes of hypotheses and theories (which Nagel [1936] and Rogers [1977] have discussed), the principal objection to this approach would seem to be that, within such classes of mutually exclusive theoretical alternatives, A and B, at most *one* hypothesis or theory could possibly be true, in which case the required "probabilities" are equal to zero or not well-defined.

Furthermore, although frequency conceptions of inductive procedure provide some foundation for the acceptance, rejection, and modification of universal generalizations and statistical hypotheses, there are (at least) three general grounds upon which their adequacy may be subject to question, namely:

(a) since the evidence available at any particular time is necessarily restricted to relative frequencies within finite classes, while the "conclusions" to be drawn concern limiting frequencies over infinite sequences, Reichenbach's pragmatical vindication may be important in principle, but not in practice; for, frequentist principles of induction afford no solution to problems of inductive inconsistency, whether with regard to the infinite class of limiting frequencies that are logically compatible with any finite frequency or with regard to corresponding classes of possible Goodmanian alternatives, as Hempel [1965], pp. 57–73, has observed;

(b) to the extent to which the relevant evidence for the application of principles (IV) and (V) is envisioned as characterizing objects, properties, and relations that are (more or less) directly accessible to intersubjective observation, the adoption of such a conception of inductive procedure surely would have the consequence, whether intentional or otherwise, of eliminating hypotheses and theories couched in dispositional, theoretical, or continuous mathematical concepts from the language of empirical science, at least to the extent to which those predicates and relations were not translatable into an observational vocabulary; and,

(c) insofar as these procedures are restricted, in principle, solely to the actual occurrence of different outcomes in various reference classes during the history of the world, they afford no basis for distinguishing between genuinely lawlike and merely accidental generalizations; as a result, even if these principles are potentially beneficial for the inductive confirmation of extensional generalizations, they are incapable, in principle, of sustaining inferences to the truth of any corresponding lawlike statements (even though, to be sure, their inductive "conclusions" themselves may happen to be true), as we have previously ascertained.

Upon initial consideration, of course, the situation might appear to be even worse, since, on frequency criteria of statistical relevance, the only ontically homogeneous reference classes are those for which either all A's are B's or those for which no A's are B's, as we discovered in Chapter 4. I should therefore like to emphasize that, insofar as *predictions*, in general, occur within an epistemic, rather than an ontic, context, the features that qualify as inadequacies with respect to *explanations* are not necessarily also inadequacies with respect to predictions; in particular, it should not be thought to be a defect that there might be no *non-epistemic* adequate predictions for singular events for which $r \neq 0$ and $\neq 1$, i.e., that there might be no ontic (or true) statistical *predictions* for singular events. For if the purpose of a prediction is to establish grounds for believing that a certain sentence (describing an event) is true, rather than to establish grounds for explaining why that event (described by a certain sentence) occurs, i.e., if predictions are appropriately interpreted as "reason-seeking" why-questions, rather than as "explanation-seeking" why-questions, then it is clearly plausible to suppose that the conditions of adequacy for predictions and the conditions of adequacy for explanations need not necessarily coincide; indeed, within the context of prediction, it would appear to be a virtue rather than a vice that, if only we could know enough about the event under our consideration, it would be possible, in principle, to determine with probability $r = 0$ or $= 1$ whether or not that event occurs.

There appear to be distinct past-oriented and future-oriented varieties of explanation-seeking and of reason seeking-why questions. (i) Both varieties of *explanation-seeking* why-questions are asked regarding statements that are known to be true: (a) the past-oriented species asks, "Why did the event described by this statement occur?"; (b) the future-oriented species asks, "Why will the event described by this statement occur?" (ii) Both varieties of *reason-seeking* why-questions are asked regarding statements that are *not* known to be true: (a) the past-oriented species asks, "What grounds are there for believing that this statement (describing an event belonging to the past) is true?"; (b) the future-oriented species asks, "What grounds are there for believing that this statement (describing an event belonging to the future) is true?" From this point of view, moreover, Hempel's [1965], [1968] positions and Salmon's [1975a], [1978] positions appear to be the same, namely: that every adequate answer to a (i)(a) question is *necessarily* a potential answer to a (ii)(a) question, but not conversely; and, every adequate answer to a (i)(b) question is *necessarily* a potential answer to a (ii)(b) question, but not conversely.

This explication of the relationships obtaining between the two kinds of why-questions, however, should not be confused with the original thesis of the symmetry of explanations and predictions initially proposed by Hempel and Oppenheim [1948], which, it may reasonably be supposed, was intended to pertain exclusively to questions to type (i) and should have been expressed as follows: every adequate answer to a type (i)(a) why-question is *necessarily* a potential answer to a type (i)(b) why-question, and conversely. Thus understood, the symmetry thesis would appear to be philosophically tenable: in this case, adequate explanations (explaining why some *explanandum-event* occurs) would necessarily be potential predictions (providing grounds for anticipating the occurrence of the corresponding *predicandum-event*, let us say), and conversely. Consequently, the appropriate conclusion to draw is that *every adequate explanation is potentially a prediction, but not conversely*. Indeed, as we observed in Chapter 5, Hempel's own revised requirement of maximal specificity (RMS^*), when employed on the basis of frequency criteria of statistical relevance, affords a perfectly appropriate condition of adequacy for predictions which are not potential explanations.

If the preceding analysis is correct, therefore, then the conception of scientific knowledge supported by traditional principles of induction is not theoretically appealing; for these principles are not only restricted to the confirmation of extensional generalizations that are expressed in purely observational predicates, but they supply no assurance that their "conclusions" are lawlike rather than merely accidental generalizations. Although each of these principles allows for the acceptance of various statistical hypotheses and universal generalizations, moreover, their inductive incapacities appear to be directly proportional to the extent to which science aims at establishing sets of laws and theories of broad scope and systematic power. Furthermore, since there apparently are *no* theory-free observations to which these principles might be applied, the underlying conception of theory-free rules of induction upon which all scientific theories should be based affords an untenable foundation for the explication of induction procedures, as Popper has frequently emphasized. [For further discussion, see Lakatos [1968], especially.] For reasons such as these, therefore, traditional principles of induction, such as (IV) and (V), are unlikely to provide the essential elements for an adequate account of scientific knowledge, even when their use occurs in conjunction with Bayes' theorem.

Carnap [1962] and Hintikka [1966], by contrast, pursue an approach in which "prior probabilities" are ascertained relative to appropriate language frameworks, where the inductive relation is envisioned as one of *logical probability* in a sense quite different than that of Reichenbach. Probabilistic conceptions of confirmation, of course, may or may not incorporate acceptance and rejection rules; but those which do encompass rules of this kind seek to provide a quantitative characterization of the degree of support which an hypothesis h receives from evidence e (within a language framework $\mathfrak{L}$) more or less reflecting the following conception:

(IX) An hypothesis h is acceptable on evidence e at time t when: (i) e supports h to degree r (where r equals or exceeds r^*); (ii) r^* is the level of acceptance (usually equal to .5); and, (iii) e embodies the total relevant evidence available at t.

[Compare Radnitzky [1973], p. 100, who, however, misconstrues condition (iii).] The particular feature that renders this conception "probabilistic", however, is that the quantitative measure, say 'm', which is introduced as the measure of the degree of support conferred upon hypothesis h by evidence e at time t is supposed to conform to certain mathematical properties that are characteristic of the calculus of probability, namely:

(X) '$m(h/e) = r$' is a measure of the degree of support which the evidence e confers upon the hypothesis h (at time t) only if: (i) '$m(h/e)$' assumes a value in the interval from zero to one; (ii) '$m(h \vee h^*/e) = m(h/e) + m(h^*/e)$' when h and h^* are incompatible; (iii) '$m(h \vee \ldots \vee h^n/e) = 1$' if $h, \ldots, h^n$ are exclusive and exhaustive; and, (iv) '$m(h \& h^*/e) = m(h/e) \cdot m(h^*e \& h)$' is the rule for conjunctions.

The degree of support afforded h by e is then generally interpreted as follows:

(XI) $m(h/e) = m(h \& e/t')/m(e/t')$,

where t' represents logically true (or "tautological") evidence (or, let us say, no empirical evidence at all); and a formulation equivalent to (XI) is,

(XII) $m(h/e) = m(h \& e)/m(e)$,

in which both $m(h \& e)$ and $m(e)$ represent so-called "absolute" (or "*a priori*")

CONFIRMATION AND CORROBORATION 215

probabilities, i.e., probabilities based upon no empirical evidence at all. [Since $m(h\&e) = m(h) \cdot m(e/h)$, $m(h)$ and $m(e/h)$ are equally the values desired.] Thus, the systematic development of a probabilistic account of confirmation might yield a probabilistic acceptance and rejection rule such as follows:

(XIII) *Tentative Rule for Inductive Acceptance:* when $m(h/e)$ is greater than .5, accept h; when $m(h/e)$ is less than .5, reject h; and when $m(h/e)$ equals .5, h may be accepted, rejected, or left in suspense, i.e., neither accepted nor rejected, given e.

[Difficulties with this particular rule, which was proposed by Hempel [1962], p. 155, however, shall be reserved for discussion in the chapter to follow.] This conception of inductive procedures not only provides for the acceptance, rejection, and modification of alternative hypotheses and theories, therefore, but also promises to satisfy a further desideratum of considerable intuitive appeal, namely: our tendency to assume that, as the relevant evidence available increases with the accumulation of observational and experimental data, the inductive probabilities of hypotheses and theories which are true may be expected to increase (while those of hypotheses and theories which are false will decrease correspondingly) with maximal values of one (and of zero) were all possible relevant evidence also available. Nevertheless, an approach of this kind encounters certain distinctive difficulties, as follows:

(a) the application of principle (XI), which is a simple form of Bayes' theorem, depends upon some method of assigning values to the *a priori* probabilities without which no *a posteriori* probabilities of the form, '$m(h/e)$', may be obtained; but, although Carnap, especially, displayed great ingenuity in attempting to deal with this problem, his solutions, in principle, depend upon (at least) two parameters presenting considerable latitude of variation, namely: (i) the determination of an *a priori* probability distribution to each of the world's logically possible descriptions; and (ii) the determination of some particular value (known as "lambda") to regulate the relative weight of acquired empirical evidence in relation to *a priori* probabilities for degrees of confirmation; with the result that the justification of particular values for these parameters seems to be logically arbitrary and empirically vacuous (see especially Carnap [1962], pp. 74–76 and pp. xxi–xxii; Nagel [1963], pp. 790–794; Lakatos [1968], pp. 324–325; and Schilpp, ed. [1963], pp. 966–979);

(b) to the extent to which this approach envisions a continuous progression toward hypotheses and theories that are *true* with the continuous accumulation of observational and experimental evidence, it appears to rest upon an indefensible presumption; for, (i) the evidence available at any specific time depends in part upon then existing technology, changes in which appear to be extremely difficult, if not altogether impossible, to systematically predict (where the discovery of techniques for microscopic examination, say, might provide evidence completely undermining hypotheses and theories which were previously well-supported by macroscopic observations); moreover, (ii) the very conception of probabilistic appraisals of hypotheses and theories in terms of varying numbers of tests of various different kinds presupposes that it is theoretically possible, in principle, to systematically arrange evidence available at any given time with respect to the number of tests of a specified kind and with respect to the variety of kinds of tests by means of a linear ordering; but this conception faces problems of its own (Nagel [1939], pp. 68–70, Carnap [1962], pp. 230–231, and, Popper [1965], p. 408, discuss this problem, which Niiniluoto and Tuomela [1973] seek to resolve);

(c) even assuming that these difficulties could be overcome, however, it remains important to realize that, insofar as the evidence available is always restricted, necessarily, to observations and experiments during the history of the actual world, probabilistic conceptions of confirmation are likewise confronted with the extraordinary difficulty of distinguishing between genuinely lawlike and merely accidental (true) generalizations, which arises not only because (i) hypotheses and theories which are confirmed are not necessarily true, and conversely; but also because (ii) lawlike hypotheses and theories concern classes of possible worlds rather than the history of the actual world; as a result, even if an hypothesis or a theory were to receive maximal confirmation, say, $m(h/e) = 1$, on all possible relevant evidence, so long as that evidence describes (even all of) the world's history, it remains the case that the corresponding lawlike statement might be false; for those statements that are true of the history of the actual world may or may not be true of other possible words as well.

Upon initial consideration, at least, there would appear to be several potentially promising avenues of escape from problems such as these. Thus, the formal analysis of causal necessity advanced by Burks [1977] and Uchii [1973], especially, presented within the framework of an "inductive logic", might be thought to solve objection (c). Yet this approach defines "causal necessity" so broadly that a sentence S is *causally necessary* if and only if (i) S is true, (ii) S is purely universal, i.e., unrestrictedly general, and (iii) S is formulated without the occurrence of indexical predicates — or S is a logical consequence of other sentences which satisfy (i), (ii), and (iii) (Uchii

[1973], pp. 277–278). This conception not only fails to distinguish between genuinely lawlike and merely accidental true generalizations, therefore, but does not even differentiate between *logical truth* and *nomological truth:* indeed, it is a theorem of this system that, if 'p' is a logical necessity, then 'p'is a causal necessity (cf. Uchii [1973], p. 282). The objective Bayesian approach represented by Rosenkrantz [1977] and Jaynes [1967], moreover, attempts to exploit certain relations between "information" and "energy", especially in the form of the second law of thermodynamics, as the solution to problem (a). Jaynes' *maximum entropy rule* for example seems to provide a rational warrant for preferring one particular distribution of prior probabilities to all others, namely: that which maximizes entropy as a measure of uncertainty (cf. Rosenkrantz [1977], pp. 52–62). The application of this principle, however, depends upon certain constraints, i.e., averages and symmetries, which are not invariably satisfied; moreover, when these are fulfilled, this method yields results exemplifying a special case of Carnap's continuum (as Dias and Shimony [unpublished] have shown). The comparative merits of this approach thus remain difficult to discern (see also Seidenfeld [1979a]).

Not the least of the objections that have been raised to probabilistic conceptions of confirmation, however, concerns a different kind of problem; for, since universal generalizations within extensional languages are logically equivalent with infinite conjunctions of singular sentences within an infinite domain, the degree of confirmation of any universal generalization, say, g, will be measured by the probability of an (infinite) conjunction of singular sentences, $h^1, h^2, \ldots$, where,

(XIV) $\quad m(g/e) = m(h^1/e) \cdot m(h^2/e \& h^1) \cdot \ldots$;

which, given the multiplication axiom of the calculus of probability, seems to entail the result that the degree of confirmation of any universal generalization will always equal *zero* for any consistent, finite set of evidence. Thus, confronted with this consequence, the principal options appear to be: (a) to abandon the probabilistic conception of degree of confirmation given this counter-intuitive implication; or, (ii) to accept this implication but deny its counter-intuitive character to retain the probabilistic conception. Popper chose option (i), while Carnap pursued option (ii). Hintikka [1966], however, persevered in developing a probabilistic metric for which universal generalizations may receive non-zero degrees of confirmation by adopting instead "slightly more optimistic" parameters of inference than did Carnap himself. [For an extension of Hintikka's approach, see Niiniluoto [1976].]

In defense of this position, Carnap advanced the contention that the concept of "degree of confirmation" under consideration is, after all, a rather vague and ambiguous notion; thus, in its place, Carnap recommended the more precise and less ambiguous concept of *rational betting quotients* supplemented by the introduction of the "qualified-instance confirmation" of an hypothesis or theory. Carnap suggested that the principal desideratum in appraising hypotheses and theories is not their truth-values but rather their *reliability* for the purpose of prediction; hence, their importance depends, not upon the infinite class of singular instances which constitutes their content, but upon their reliability for each successive individual instance, i.e., their reliability for the *next* instance *per se*. Consequently, as Popper pointed out, an hypothesis or theory might retain a high qualified-instance confirmation for its next instance, even though that hypothesis or theory were actually *refuted* by the available evidence: the hypothesis, "All tossed pennies always show heads", for example, would have the qualified-instance confirmation, say, $m^*(h/e) = ½$ rather than 0. Popper concludes that, regardless, of its potential value for other areas of inquiry, the implied notion could not possibly serve as an appropriate measure of the degree of evidential support for an hypothesis or a theory. [See, for example, Carnap [1962], pp. 571–575; Popper [1968], pp. 282–283; and Lakatos [1968], pp. 333–335.]

Popper persisted in his criticism of the probabilistic conception of confirmation and offered several arguments intended to demonstrate its indefensibility; thus, for example, he constructed the following critique: (a) since science aims at hypotheses and theories of broad scope and systematic power (which is inversely related to their probability, according to the axioms of that calculus), science does not aim at high probability; (b) since the degree of testability of an hypothesis or a theory increases with its content (which facilitates obtaining a high degree of evidential support), science needs high content (and thus low *a priori* probability); (c) those who identify confirmation with probability, therefore, must believe that the most probable hypothesis is the most desirable; but those hypotheses and theories that are the most probable have the least content (Popper [1968], pp. 285–287; Carnap, however, emphatically rejected (c)). Thus, Popper argued, aiming at high probability entails the adoption of a policy favoring *ad hoc* hypotheses and theories, which go as little beyond the evidence as possible. As Lakatos was subsequently to observe, Popper occasionally conflated *corroboration* with *corroborability* (or the degree to which an hypothesis or a theory *is* supported by the available evidence with its *testability*, which varies directly with its content) in at least some of his arguments, with invalid and counter-intuitive results; still, while the (strong) thesis that corroboration and content are directly related should be abandoned, the (weaker) thesis that corrobor-*ability* and content are directly related should be retained; for, when h^1 entails h^2 and the empirical content of h^1 exceeds that of h^2, then, for some e, it should be possible for the corroboration of h^1 to exceed that of h^2 (see Lakatos [1968], pp. 353–355, for example; but keep in mind that *corroboration* proceeds by attempting to *refute* hypotheses and theories).

Carnap, however, remained undaunted by Popper's criticism, going so far as to maintain that Popper's criticisms were irrelevant misunderstandings of Carnap's own position, which, he said, concerns the concept of a rational betting quotient and *not* that of a degree of evidential support (Schilpp, ed. [1963], pp. 995–998). And, indeed, the intuitive properties of these concepts may well be quite distinct and require clarification, as Lakatos ([1968], p. 352), for example, has claimed:

According to our *betting intuition*, any conjunction of hypotheses, whatever the evidence, is at least as risky as any of the conjuncts. (That is, $(h)(h')(e)[p(h/e) \geq p(h\&h'/e)]$.) According to our intuition of evidential support, this cannot be the case: it would be absurd to maintain that the evidential support for a more powerful theory (which is, in the Carnapian projection of the language of science onto the distorting mirror of particular hypotheses, a conjunction of hypotheses) *must not* be more than the evidential support for a weaker consequence of it (in the case of the Carnapian projection, for any of its conjuncts). Indeed, intuition of evidential support says that the more a proposition says, the more evidential support it may acquire.

The question to be raised, therefore, is not whether Carnap's construction is immune from Popper's criticism, but why it should be thought worthwhile.

Nowadays those who hold with Carnap that betting quotients are central to inductive procedures have become a large and varied group, including, in particular, the so-called *subjective* Bayesians, who endorse the personalist interpretation of probability developed in detail by de Finetti [1964] and by Savage [1954]. The most important elements of this position may be described as follows:

(a) individuals ought to distribute their beliefs in accordance with the principles of the calculus of probability, since otherwise a *book* (or set of bets) may be made against that individual which he is sure to lose; thus, according to the Ramsey-de Finetti theorem, a betting system is *fair* (or, as Carnap puts it, a set of beliefs is *coherent*) if and only if it is consistent with the axioms of probability (see Kyburg [1970], pp. 69–70, or Lakatos [1968], pp. 336–337, for example); furthermore,

(b) individuals should depend upon Bayes' theorem in developing their beliefs, but in doing so they should "keep an open mind" in the sense that no statement should be assigned an *a priori* probability of zero, unless it happens to be impossible; thus, through the cumulative impact of evidence acquired *a posteriori*, a convergence of opinion should occur as an expectable outcome of reliance upon Bayes' theorem, even between those who have begun with initially divergent *a priori* probability distributions (for example, see Levi [1970], [1971], [1974], and especially Shimony [1970]).

In spite of its widespread appeal, however, it is not easy to imagine that an adequate theory of scientific procedure could be developed on the basis of the personalistic interpretation; for, (i) as we observed before, if an individual's conclusions are unacceptable (for any reason whatever), one always has the option of simply changing one's premises (by adopting a more agreeable distribution of *a priori* probabilities), since, on personal principles, there is as much justification for the same individual having n different *a priori* probability distributions at n different times as for n different individuals to have n different distributions at the same time — the issue of "(continuous) confirmational consistency" referred to above; moreover, (ii) although reliance upon Bayes' theorem for regulating changes between states of belief as a result of the acquisition of further evidence promises to promote a convergence of opinion between different individuals and to guarantee their asymptotic independence from the effects of specific assignments of *a priori* probabilities, this "long run" justification serves its intended purpose with one important qualification; for, as Burks has remarked, at any given stage of inquiry, the available evidence will always be finite: those who rely upon Bayes' theorem, therefore, will find themselves in arbitrarily close agreement on posterior probabilities, in general, only when they also agree upon their priors, which underscores the importance of this problem (Burks [1977], especially pp. 90–92; see also Shimony [1967]).

The ultimate irrelevance of the subjective approach to the investigation of scientific knowledge, however, involves more subtle issues; for, as even de Finetti sought to emphasize, there is no necessary correlation, in principle, between subjective probabilities and relative frequencies; that is,

No relation between probabilities and frequencies has an empirical character, for the observed frequency, whatever it may be, is always compatible with all opinions concerning the respective probabilities; these opinions, in consequence, can be neither confirmed nor refuted, once it is admitted that they contain no categorical assertions such as: [that] such and such an event *must* occur or *can not* occur. (de Finetti [1964], pp. 117–118)

As a result, the occurrence of certain phenomena with specified frequencies may or may not exercise any influence upon the formation of an individual's opinion concerning their probabilities, for the connection between them is one of a *causal*, rather than of a *logical*, kind: "it can influence in some way the judgment of an individual on the probabilities in question" (de Finetti [1964], p. 120). The problem, in other words, is the lack of any determinate systematic connection between the principles of Bayesian personalism, relevant evidence, and the occurrence of physical phenomena, which promotes the convergence of opinion between different investigators as a matter of theoretical necessity, rather than as the occasional outcome of fortuitous conditions.

The ultimate import of this circumstance, I surmise, is such as to deprive the personalist conception of all claims to epistemic significance in general; for, apart from its intrinsic interest as a mathematical analysis concerning the behavior of a certain class of abstract objects ("exchangeable events"), the absence of a specifiable inductive relation between the occurrence of physical phenomena, on the one hand, and the formation of subjective opinion, on the other, disqualifies such an approach as an analysis of scientific inference. [I therefore agree with Levi [1970], p. 141, that this interpretation is ultimately incompatible with empiricist epistemology.] Note, in particular, that the subjective conception not only *violates the principle of empiricism* (by permitting the acceptance and the rejection of hypotheses and theories without reliance upon observation and experimentation), but also *disregards the objective of inquiry* (namely, the discovery of hypotheses and theories of broad scope and systematic power that are applicable for the purposes of explanation and prediction, i.e., *laws as properties of the physical world*). I am inclined to believe, on grounds such as these, therefore, that the differences between Popperian and Bayesian methodologies are both genuine and important. [Difficulties encountered in bridging the gulf between subjective opinion and objective phenomena on personalist principles are discussed, for example, by Braithwaite [1957], Hintikka [1971], and Suppes [1966]. See also Suppes [1973], especially pp. 515–516. For a discussion of methodological issues, see in particular Jeffrey [1975] and Giere [1975a]. For a general critique, see especially Kyburg [1978a].]

The shift in Carnap's approach from the confirmation of hypotheses and theories to rational betting quotients for singular propositions, therefore, not only signaled a difference in emphasis but also entailed an exchange of inductive programs; for, to the extent to which science aims at the acceptance, rejection, and modification of hypotheses and theories of broad scope and systematic power, it may be said that, "together with the *term* 'confirmation', went his *theory* of confirmation, that is, his theory of evidential support" (Lakatos [1968], p. 356). And, while the consistency of a set of beliefs may be supposed to require conformity to the principles of the calculus of probability as a necessary condition, it by no means follows that the principles in terms of which the members are accepted into or rejected from such a set must therefore be *probabilistic principles* themselves; for (as we shall ascertain in detail in the very last chapter) the consistency of a set of beliefs concerns the contents of the accepted sentences of some *object-language*, while the principles of acceptance and of rejection belong to the *meta-language* of inductive procedure: it is reasonable to conclude, therefore, that "The Bayesian Way" should not be entertained as an inviting path toward the goal of understanding the character of scientific knowledge.

POPPERIAN PROCEDURES OF CORROBORATION

Popperian procedures of corroboration, as Jeffrey [1975] has suggested, are easy to characterize but difficult to formalize. Insofar as science progresses through a process of bold theoretical conjectures and ruthless experiential refutations, an hypothesis h should warrant serious consideration as a possible candidate for scientific investigation, in general, if and only if (i) h is falsifiable, (ii) h is high in content, and (iii) h has not yet been falsified (see, for example, Popper [1968], pp. 240–244). Existential generalizations, finite disjunctions, singular sentences, finite conjunctions, and universal generalizations represent successively increasing content on the basis of their logical forms; but comparisons of content are not thereby effectively determined, since all universal generalizations, for example, would appear to possess equally (maximal) high content (see Lakatos [1968], p. 344, note 1; and especially Popper [1965], New Appendix *vii). The Popperian imperative to, "Always choose the most improbable hypothesis", therefore, is perhaps best understood in relation to a definite *domain of phenomena* to be entertained as the object of inquiry — which, of course, has to be pragmatically determined — where lawlike hypotheses relevant to that domain may be partially ordered on the basis of their potential systematic power, i.e., their explanatory and predictive significance (for those phenomena), where *lawlike hypotheses of different strength* are assigned a measure that conforms to the following principle: *the content of a lawlike hypothesis is directly proportional to its systematic power for a particular domain*, i.e., relative to the occurrence of outcomes of kind O^i, given conditions of kind T^i, for example, under a description 'd', within the language $\mathfrak{L}zt$.

Whenever a lawlike hypothesis, h, attributes a dispositional property, χ, of strength r to every member of a reference class, K, then (let us say) the *a priori content measure of h* for such a singular outcome is equal to r, i.e., $E(d/h)$, say (or the explanatory power of h for the domain d), is thus equal to *the degree of nomic expectability of d as an explanandum, given h as its explanans*. Since every lawlike hypothesis of universal strength has a content measure of one (relative to the classes of phenomena which they would potentially explain if they were true), therefore, these hypotheses may be further ordered on the basis of entailment relations that may obtain between them, since, if an hypothesis h^i entails an hypothesis h^j, but not conversely, then the content measure of h^j cannot exceed that of h^i. If an hypothesis H_8: "All planets move in circles", for example, entails an hypothesis H_9: "All planets move in ellipses", but not conversely, insofar as circles are ellipses of zero eccentricity, then $E(d/H_8)$ exceeds $E(d/H_9)$ with respect to their common domain, i.e., the planetary orbits, even though both hypotheses are of universal strength (Popper [1965], pp. 373–374). Moreover, this is a matter of considerable significance, since there will then be tests of H_8 which are not tests of H_9, but not conversely; for example, H_8, but not H_9, may be refuted by discovering some eccentricity greater than zero in any planet's orbit. Consequently, the comparative *testability* of different hypotheses does indeed appear to be a direct function of their *content*, where, in principle, hypotheses of greater explanatory power, in relation to a particular domain d, are more severely testable, in general, and therefore are capable of greater corroboration (Popper [1965], pp. 395–402, especially p. 401).

The degree of nomic expectability of d as an explanandum, relative to h as its explanans, of course, is not only equal to the strength of the dispositional tendency for outcomes of that kind under those conditions, but also *equals the likelihood of that hypothesis as an explanans, given the datum d as its a posteriori evidence e*; that is, the likelihood of h, given e, say, $L(h/e)$, is determined by the nomic expectablity of e, if h is true, namely:

(XV) $L(h/e) = E(e/h)$;

where, as before, e describes m O^i-outcomes over n T^i-trials, as appropriate; thus, as a simple example, if a marksman has fired at a target with his rifle twice and hit the target both times, then the likelihood that the shots were independent, with the propensity for a hit of .9, is equal to .81; and, with the propensity for a hit of .8, is equal to .64; and so forth (Hacking [1965], especially pp. 56–57). Typically, of course, such outcomes may be derived from such hypotheses only in conjunction with appropriate auxiliary hypotheses concerning, for example, the quality of the ammunition which is being used, the condition of the rifle being fired, and so on; however, under the dispositional interpretation of lawlike sentences, all of the nomically (or, causally) relevant factors that contribute to bringing those outcomes about must be included within the reference class description of that potential explanans – *provided*, of course, *it is true*. In order to subject an hypothesis of this kind to a systematic test, therefore, it is necessary to entertain the *supplementary hypothesis*, relative to that test situation, that the evidence *e* itself has been obtained under the relevant circumstances, a difficulty which we shall consider in detail with respect to the problems of acceptance and of rejection in the chapter to follow.

Since the likelihood of an hypothesis will tend to be high when either the available evidence is scanty or the hypothesis itself is vague, likelihoods, in general, should only be interpreted as partial measures of degree of corroboration when their values represent "a good fit in a large sample", i.e., when they reflect the results of severe tests (Popper [1965], p. 411). Perhaps the most important element of Popper's conception of corroboration, therefore, is his insistence that a quantitative measure of the evidential support that evidence *e* affords an hypothesis *h* may qualify as a *degree of corroboration* only if *e* is *a report upon the most severe test that we have been able to devise*, i.e., when *e* does not report the results of a sincere attempt to falsify *h*, then, $C(h/e)$, say, does not represent *any* degree of corroboration (Popper [1965], p. 418). Popper has proposed that, where *e* is the test under consideration and *b* is our background knowledge prior to that test, *the severity of e as a test of h, relative to b*, could be measured by,

(a) $P(e/h \& b) - P(e/b)$;

i.e., as the "probability" of *e*, given *h* and *b*, minus the "probability" of *e*, given *b* alone (Popper [1968], pp. 390–391, especially formula (6) on p. 391). Strictly speaking, however, the intent of (a) is more suitably captured by a formulation in which it becomes explicit that the difference involved here is one of *differences in expectations* with respect to outcome distributions over sets of appropriate trials. A more precise formulation of this measure could therefore be advanced by employing the appropriate difference with respect to *the nomic expectability of e, given h, relative to b*, as follows,

(b) $E(e/h \& b) - E(e/b)$;

i.e., as the degree of nomic expectability of *e*, given *h* and *b*, minus the degree of nomic expectability of *e*, given *b* alone.

In contrast to Bayesian procedures, therefore, these "probabilities" are properly envisioned as attributing probabilities to *events* as the outcomes of singular and multiple trials within the physical world, i.e., as representing logical relations between the members of a certain set of sentences *within an explanation context*, in which a probability statement ascribes some degree of nomic expectability to a particular explanandum 'e', relative to a certain explanans, where 'b' includes a description of those trials that are assumed to have occurred. Thus, when 'b' entails 'e', then $P(e/b) = 1$ and, for all 'h', $P(e/h \& b) = 1$ as well; consequently, the severity of such a test, for any 'e', is minimal, i.e., zero. Conversely, when 'h' entails 'e', while 'b' entails '$-e$', then the severity of 'e' for 'h' relative to 'b' is maximal, i.e., one. Since 'h & b' would, then be inconsistent, necessarily, which *could* also happen otherwise, however, it appears appropriate to interpret 'e' as a description of m outcome of kind O^i, for example, while 'h' and 'b' are interpreted, respectively, as the conjunction of a "new" hypothesis, say, 'h^2', with a description of n trials of kind T^i and as the conjunction of an "old" hypothesis, 'h^1', say, with a similar trial description, in which case the desired result is obtained. For simplicity, however, we may still employ,

(c) $\quad P(e/h^2) - P(e/h^1)$,

as our formal measure, keeping in mind that, strictly speaking, the "absolute difference" is the quantity required [since, for some 'e', $P(e/h^1)$, of course, may exceed $P(e/h^2)$].

Whether or not such an 'h^1' actually belongs to 'b', therefore, *severe tests promote competition among competing hypotheses*. One might maintain instead, as does Lakatos, however, that severe tests *require*, rather than merely *promote*, competition among competing hypotheses (Lakatos [1968], especially pp. 414–415). However, $P(e/b)$ will have a useful value even when 'b' does not actually entail an hypothesis, 'h^1', which should be regarded as in competition with the hypothesis, 'h', i.e., 'h^2', under test, namely: (i) when 'b' entails '$-e$', apart from any such 'h^1', $P(e/b)$ will equal zero, necessarily; and, (ii) when 'b' entails neither 'e' nor '$-e$', $P(e/b)$ should equal zero, by convention; then, $P(e/h) - P(e/b)$, will reflect the extent to which 'h' either contravenes or exceeds accepted background knowledge, even though 'b' does not entail any such 'h^1'. Perhaps it should be observed at last, moreover, that the measure of severity, with respect to competing hypotheses 'h^1' and 'h^2', is a *symmetrical relation*; that is, if 'e' is a severe test of 'h^2' (as a "new" hypothesis, relative to 'h^1'), then 'e' is also a severe test of 'h^1' (as an "old" hypothesis, relative to 'h^2'), necessarily. Perhaps the importance of so-called "crucial experiments", such as Eddington's 1919 expedition to measure the deflection of a ray of light during an eclipse of the Sun, therefore, derives from the discovery of some test sufficiently severe to virtually eliminate *one or another* of two or more competing hypotheses or theories.

Based upon these preliminary considerations, therefore, a rather plausible measure of the *degree of corroboration of an hypothesis h, given evidence e, relative to background knowledge b*, would appear to be as follows:

(XVI) $C(h/e \& b) = L(h/e) \cdot [|P(e/h \& b) - P(e/b)|]$;

that is, as the product of the likelihood of h, given e, multiplied by the severity of e as a test of h, relative to b. (XVI) is a Popperian measure of corroboration, let us note, but it is not necessarily Popper's; indeed, Popper suggests (as one possibility) a measure that, for lawlike hypotheses, may be represented as follows:

(XVI*) $C(h/e \& b) = \dfrac{P(e/h) - P(e/b)}{P(e/h) + P(e/b)}$;

which Popper himself does not find entirely satisfactory (see Popper [1965], p. 400) and which Lakatos severely criticizes (see Lakatos [1968], especially pp. 408–416). Our previous consideration of (a) as a measure of the severity of evidence e for a test of hypothesis h, relative to background knowledge b, however, suggested that (c) is preferable for its intended purpose; in which case the appropriate measures of degrees of corroboration are *corroboration ratios*, i.e., relative degrees of evidential support:

(XVII) $\dfrac{C(h^2/e)}{C(h^1/e)} = \dfrac{L(h^2/e) \cdot [|P(e/h^2) - P(e/h^1)|]}{L(h^1/e) \cdot [|P(e/h^1) - P(e/h^2)|]}$;

which, of course, reduce to the corresponding likelihood ratios of $L(h^2/e)$ divided by $L(h^1/e)$ as a measure of comparative corroboration for competing hypotheses (of particular importance when the absolute value of (c) *is* high).

At least three features of this conception of corroboration warrant special mention. The first is that this standard is intended as a measure of *preferability* between competing hypotheses, relative to a specified domain, rather than as a measure of *acceptability* with respect to particular hypotheses, a problem to be considered in Chapter 9. Thus, in effect, the adoption of (XVII) reflects Hacking's so-called "law of likelihood", i.e., (i) that data d supports hypothesis h^i better than hypothesis h^j when the likelihood of h^i, given d, exceeds that of h^j, given d; and, (ii) that data d supports hypothesis h^i better than it supports h^j if the likelihood ratio of h^i to h^j exceeds one (Hacking [1965], pp. 70–71). The second is that in cases in which the

CONFIRMATION AND CORROBORATION 227

evidence e has not actually occurred, but is only being contemplated hypothetically, as it were, the standard proposed serves as a measure of the degree of comparative corroboration which would obtain, were that evidence to actually occur, i.e., as an *a priori* measure of comparative *corroborability*, relative to revidence e, rather than as a measure of the degree of comparative corroboration which does obtain, where those results actually are available, i.e., as an *a posteriori* measure of comparative *corroboration*, relative to (CR-3) and (CR-4).

The third, moreover, is that these methods, alone among those which we have considered here, provide a method for testing *lawlike sentences*, which are properly interpreted as logically unrestricted nomological conditionals, on the basis of evidential resources which are restricted, necessarily, to the occurrence of events during the course of the world's actual history; i.e., Popperian procedures of corroboration *do* provide a method for testing lawlike sentences as *intensional generalizations* on the basis of evidence consisting of *extensional distributions*, specifically, the outcomes that occur as results of singular and multiple trials of appropriate arrangements during the course of the world's actual history. Consequently, of all of the conceptions of inductive principles under consideration here, only the Popperian construction affords a suitable basis for distinguishing between genuinely lawlike and merely accidental generalizations; for, as we ascertained in Chapter 7, *only statements that are both lawlike and true are theoretically invulnerable to the eventual discovery of refuting evidence* – where the role of background knowledge and auxiliary hypotheses reflects the consideration that even the results of apparent refutations must be envisioned as tentative and fallible, relative to conditions (CR-3) and (CR-4), which serves to emphasize the fallibilistic character of a conception of scientific knowledge as the outcome of a process consisting essentially of "conjectures and refutations".

Not the least of the benefits of this Popperian conception is that it generates agreeable philosophical consequences; thus, for example, provided the severity of evidence e as a test of an hypothesis h is low, the degree of corroboration of h given e will also be low; and when the severity of e is high, then the degree of corroboration of h given e reflects the likelihood of h given e (rather than its probability). Insofar as probabilities and likelihoods are directly related, however, this interpretation of the degree of evidential support afforded an hypothesis h by evidence e partially conforms with *probabilistic intuitions* regarding evidential support, even though these measures are not additive and do not satisfy the axioms. Moreover, when the severity of a test is high, the application of the procedures of corroboration will tend to provide the strongest support to all those hypotheses in which the value of r is close to the frequency m/n, as the evidence reflects, which thus corresponds with *traditional principles of induction*. The results of particular applications of Popperian procedures, however, are perhaps even more significant, where the competition between classical mechanics and general relativity provides an illustration; for, although Newton's and Einstein's theories of gravitation are strictly inconsistent, they yield the same results for weak gravitational fields and for slowly moving objects (in relation to the speed of light). Thus, tests within their overlapping domain of application could not possibly serve to discriminate between them, since both theories entail similar outcomes for similar tests, which therefore support them both; but beyond this range of overlap, they offer incompatible outcomes, which provide opportunities to differentiate between them by severe observational and experimental tests, which have corroborated Einstein while refuting Newton, suggesting further that *scientific practice* may also accord well with a Popperian conception.

The comparison of Popperian corroboration with Bayesian confirmation, of course, may be facilitated by making their respective measures as nearly similar as possible; in particular, the Bayesian formula (III) requires expansion in order to explicitly display the implicit role of likelihoods as follows:

(XVIII) $P_{f \cup \{e\}}(h) = P_f(e/h) \cdot P_f(h)/P_f(e)$,

in which the posterior probability of h, given the "new" facts e together with the "old" facts f, equals the likelihood of h, if that person were to know e as well as f (which is already known), multiplied by the ratio of his prior probability for h, given f, and his prior probability for e, given f. Let us assume (1) that the Bayesian conception under consideration is a *normative construction*, in which logical relations and mathematical computations are taken for granted, rather than varying with each person's intellectual capabilities; (2) that any problems involving the *acquisition of evidence e* by our observers and experimenters are likewise susceptible to uniform resolution, such that Bayesians and Popperians alike have equal access to the results of tests and trials; and, (3) that the "facts" f accepted by an imaginary Bayesian *convey the same content* as the "background" b considered by an hypothetical Popperian.

Substituting "b" for "f" in formula (XVIII), therefore, and leaving the likelihoods they share aside, the Bayesian conception makes measures of posterior probabilities functions of *our ratio of prior probabilities*, namely:

(XIX) $P_{b \cup \{e\}}(h) \propto P_b(h)/P_b(e)$;

which, of course, will naturally vary from person to person with variations in their priors; while, by also deleting likelihoods from formula (XVI), the Popperian construction renders degrees of corroboration functions of *the severity of our tests*, namely:

(XX) $C(h/e \& b) \propto |P(e/h \& b) - P(e/b)|$;

which, by contrast, do not vary from person to person — as objective features of test situations. Perhaps the most important difference between them becomes quite obvious, namely: *that, on Popperian principles, there is a univocal solution to questions of confirmation*, i.e., concerning the degree of support afforded an hypothesis h by evidence e, given background b, *while on Bayesian principles, in general, there is not only no univocal solution, but not even a univocal class of solutions* (other than the trivial class of all possible available values). And this, in turn, reflects their most important theoretical difference; for, the Popperian construction automatically resolves the problem of (continuous) confirmational consistency, which is the major difficulty that Bayesian conceptions appear to be unable to overcome, by establishing completely objective inductive standards.

The rationale for normative epistemology, broadly construed, could be characterized as that of undertaking the discovery of principles and procedures whose systematic application, in principle, would yield the following result: for any individual z, if z were to apply those principles and procedures to any logically consistent finite class of accepted beliefs or evidence sentences e, then z would be rationally warranted in accepting *all and only* those members $h^1, h^2, \ldots$, belonging to some suitably related class of sentences or beliefs. A necessary condition for the fulfillment of this conception, of course, is establishing appropriate standards for appraising the degree of evidential support afforded a particular hypothesis, h, by an appropriately specified set of evidence, e, where an interpretation intended to establish the character and structure of the relations suitable for this purpose remains *subjective* to the extent to which it fails to define a univocal class (or a univocal set of classes) whose members are rationally warranted for z, relative to e, provided only the members of those classes are specified within the language framework $\mathfrak{L}zt$. From this point of view, therefore, it would appear to be a significant advantage of the Popperian account that, unlike its Bayesian alternative, it provides the foundation for a completely *objective* measure of degrees of evidential support, thereby contributing to the development of a completely objective conception of the nature of scientific knowledge.

9. ACCEPTANCE AND REJECTION RULES

Although the "law of likelihood" (which Hacking himself [1973], p. 495, no longer endorses without qualification) has been advanced as (part of) the solution to the *problem of preference* with respect to alternative hypotheses, it should be conceded that that principle does not appear promising as a solution to the *problem of acceptance* within the context of normative epistemology. According to its precepts, the data d supports an hypothesis h^i better than it supports an hypothesis h^j so long as the likelihood of h^i, given d, exceeds that of h^j, given d; thus, the best supported members of a specified set of incompatible hypotheses for their common domain will always be those with *maximum likelihood* in relation to the available evidence d, i.e., those whose likelihood on that evidence is not exceeded by that of any other member of that set. It should not be difficult to discern, however, that *more than one* hypothetical alternative may possess maximum likelihood relative to d at any particular time: if innumerable hypotheses, $h^1, h^2, \ldots$, for example, all entail that data, then they will all be supported by that evidence with likelihoods of one, which, of course, confer upon each of them not only maximum likelihoods as *the best supported hypotheses* (in the comparative sense), but maximal likelihood as receiving *the best possible support* (in the non-deductive sense), in relation to the law of likelihood itself. Since not more than one of these hypotheses could possibly be true, therefore, it should be obvious that likelihood relations, at best, establish necessary but not sufficient conditions for resolving the problem of acceptance.

Another difficulty which arises with likelihood principles as such might be referred to as, say, "the paradox of minimal evidence", namely: that the best supported hypotheses, given data d, will tend to be generalizations that restrict propensity ascriptions *only to outcomes which have actually occurred*. Suppose, for example, that the hypotheses under consideration, $h^1, h^2, \ldots$, were probabilistic claims concerning the propensity distribution for some coin (or for coins of some kind) to turn up *heads* and to turn up *tails*, under well-defined test conditions. Then, if the only available evidence happened to be the result of a single toss with that coin, say, *heads*, the best supported of these alternative hypotheses would ascribe a propensity of one to heads and a propensity of zero to tails! This consequence, moreover, is entirely general:

similar results would arise in the case of dice, spheres, and so on. Thus, if a six-sided die with sides marked "one" through "six" were given a single toss and came up "two", for example, the best supported hypothesis would ascribe a propensity of one to coming up "two" and a propensity of zero to any other outcome; if a sphere were rolled across an irregular surface just once and came to rest upon a specific point on its surface, the best supported hypothesis would ascribe a propensity of one to stopping at just that point and a propensity of zero to stopping at any other; and so forth. Although these ascriptions would appear perplexing were they misconstrued as "conclusions" rather than regarded as "preferences", however, surely hypotheses ascribing (higher) propensities to outcomes which *never* occurred should receive less support than hypotheses ascribing (higher) propensities to outcomes which *have* occurred.

A parallel phenomenon, moreover, would affect traditional principles of induction were their application not constrained by the requirements that the relevant evidence should consist of a large number of trials conducted under a wide variety of conditions; otherwise, those principles would both warrant the inference *from* "All observed flips of this coin come up heads" *to* "All flips of this coin come up heads", *from* "All observed tosses with this die turn up 'two' " *to* "All tosses with die turn up 'two' ", and so forth — even if the evidence were to consist of the result of merely one trial! The source of "paradox", however, appears to be explicable in relation to *background knowledge*; for experience and reasoning with other coins, other dice, and other spheres has established the presumption that usually, if not always, the probabilities of outcomes under conditions of these kinds tend to be distributed more or less evenly over the class of possible outcomes (as Reichenbach [1949] explained). Mutiple outcome possibilities clearly require multiple actual trials to afford any reasonable expectation of securing frequency displays approximating their generating propensities. These considerations thus imply that the application of principles of likelihood ought to be similarly constrained with respect to numbers of trials and variety of conditions; but it would be valuable to discover some objective foundation for ascertaining the "appropriate" number of trials and variety of conditions in any individual case.

A reasonable inference, therefore, is that the "best supported" members of a set of alternative hypotheses will warrant inductive acceptance into the knowledge context $\mathscr{K}zt$ *only if* the available evidence consists of a large number of trials over a wide variety of conditions. The "best possible" support an hypothesis may receive, it appears, is not simply a likelihood of one,

ACCEPTANCE AND REJECTION RULES 233

per se, but instead a likelihood of one relative to a large number of trials over a wide variety of conditions. This desideratum is at least partially fulfilled by the measures of degree of corroboration introduced in Chapter 8; for the severity of e as a test of h^2, relative to h^1, i.e., $|P(e/h^2) - P(e/h^1)|$, is a function determined by the difference between the strengths of the propensities ascribed to each single trial *and* by the number of trials that constitute that test: if h^1 asserts that the propensity for $K \cdot T^i$ to bring about O^i equals .9, while h^2 asserts that the propensity for $K \cdot T^i$ to bring about O^i is actually .7, for example, then the degree of nomic expectability for O^i on a single trial is .9, given h^1, and .7, given h^2, where the difference equals .2; for O^i twice on two trials is .81, given h^1, and .49, given h^2, where the difference equals .32; and so forth. Indeed, the successive occurrence of O^i ten times over ten trials has a nomic expectation of .348, given h^1, and .028, given h^2, for a difference of .32; but, even though h^1 would be better supported than h^2, given evidence of this kind,, it would not *therefore* deserve acceptance: this result would have a nomic expectability of .904, if the propensity were .99, for example, and of 1.0 if it were 1.0 instead. Although these measures *are* sensitive to the large number desideratum, therefore, the conditions under which an hypothesis *should be accepted* or *should be rejected* require further investigation.

Since the principles advanced by Fisher [1956], by Neyman and Pearson [1966], and by Wald [1950] bear upon this problem, perhaps their consideration may contribute toward its solution. Indeed, the technique of "significance tests", which Fisher [1956] recommends, appears to formalize the Popperian strategy for dealing with statistical hypotheses, namely: the *supplementary principle* that, since almost all possible finite segments of large size will very probably exhibit relative frequencies that are very close to their generating propensities, *highly improbable outcomes deserve systematic neglect*. Thus, the logical properties of "significance tests" themselves are not difficult to define relative to a designated hypothesis, h, specified number of trials, n, and a "rejection region", say, r, whose particular values are determined by the objectives of a specific research project. The *rejection region* represents a class of outcomes whose occurrence would qualify as "incompatible evidence" in the specific sense that *either* the hypothesis under consideration, h, is false *or* the outcome that has been obtained as the result of n trials occurred (as they say) "by chance", i.e., with only some probability, when that hypothesis is actually true. The difficulty, of course, is that for statistical hypotheses every combination of possible outcomes of n membered trials will occur *only with some probability*, even when hypothesis h *is* true; consequently, the specification of an appropriate rejection region presupposes the selection of *how high* or *how low* the probability of possible outcomes over n trials must be for the evidence it yields to be "strong enough" to warrant the rejection of hypothesis h, where the value of this variable reflects the "significance level" (or, the "size") of that test.

The *significance level* of an empirical test of a statistical hypothesis h, typically denoted by 'α' (alpha), therefore, represents the degree of *improbability* of acquired evidence that is required for the rejection of hypothesis h; but it should not be overlooked that alpha likewise designates the *probability* for the (mistaken) rejection of hypothesis h when h itself happens to be true! An alpha of .05, for example, not only indicates that the rejection region is defined by a class of outcomes having a probability of occurrence of (at most) .05, *if* the hypothesis under investigation is true, but also discloses that an outcome within this rejection region should be expected to occur approximately once-in-twenty tests, when that particular hypothesis *is* true. Although Hacking ([1973], p. 499) has raised the issue of finding a rationale for so-called "tail area" rejection regions, i.e., those distant from the most probable outcomes, relative to any such hypothesis h, a far more serious problem for normative epistemology arises from the most crucial feature of significance tests, namely: for any specific hypothesis h and number of trials n, whether or not h is rejected will vary with the specific significance level that is employed. Unless there is an objective foundation for selecting one value of alpha from the continuum of possible values, therefore, the objectivity of scientific inquiry would appear to be undermined by this conception. Moreover, insofar as these procedures facilitate the *rejection* of hypotheses but not their *acceptance*, perhaps more adequate principles may be found in the work of Neyman and Pearson [1966] and of Wald [1950], which we shall now consider.

"ORTHODOX" HYPOTHESIS TESTING

Although orthodox statistical hypothesis testing constitutes the core of contemporary techniques of statistical analysis, the philosophical justification for reliance upon Neyman-Pearson procedures of inference and of decision-making remains a subject of controversy among theoreticians and statisticians; indeed, there appear to be grounds for supposing that Neyman and Pearson themselves were not altogether unequivocal on the fundamental question of whether their procedures are primarily applicable to *decision-making* or to *inference*. Neyman himself, of course, has advocated the concept of "inductive behavior", an approach that emphasizes the function fulfilled by hypothesis testing within the context of decision stategies formulated by Wald; but it should be observed that the interpretation of Neyman-Pearson procedures as "principles of decision" under probabilistic conditions leaves as its residue the definition of "principles of inference" concerning statistical conclusions: for, unless these methods are equally suitable for *both* purposes, the eventual prevalence of either "solution" would resolve one of these problems without affecting the other. [Compare Neyman [1977], for example.]

The classic situation required for orthodox statistical hypothesis tests consists of the specification of some *pair* of alternative hypotheses h and h^*, where these are envisioned as exclusive and as exhaustive alternatives, i.e., h will be accepted if and only if h^* is rejected, given the relevant evidence. Since the significance levels of these alternatives cannot be simultaneously maximized, in general, one has to be chosen as the hypothesis of *primary* interest (for which the consequences of mistaken rejection are regarded as more serious), while the other remains as the hypothesis of *secondary* interest (for which the consequences of mistaken rejection are supposed to be less serious). Then mistakes that arise as a result of *rejecting* the primary hypothesis, h, when it is true due to the role of chance are referred to as *errors of Type I*. Failure to reject an hypothesis at a specified significance level as the result of an empirical test, of course, affords no guarantee that that hypothesis is true. Mistakes that arise as a result of *accepting* the primary hypothesis when it is false, therefore, are referred to as *errors of Type II*. The probability of mistakenly accepting the primary hypothesis h when it is true, moreover, depends upon the specific alternatives to h under consideration h^*.

Where h specifies some particular value v of some parameter (such as the mean, for example) of a theoretical population R, alternatives to h may specify that the value of that same parameter is different from, greater than, or less than v, such that h: '$f(R) = v$', h^*: '$f(R) \neq v$', and so forth (where 'f' designates the property of the population that is represented by the value v). In cases such as these, of course, the h^* hypotheses that are alternatives to h represent classes of alternative values for the parameter v instead of a single specific value v^*; for this reason, they are called "composite" as opposed to "simple" alternatives. Once the selection of the hypothesis of primary interest, h, and its alternative, h^*, has taken place and the significance level α of the test has been determined, the number of trials required may be calculated by specifying the desired probability for errors of Type II, typically denoted by 'β' (beta), where the numerical value of 1 minus beta $(1-\beta)$ represents the "power" of that test and the potential benefits of hypothesis tests are supposed to be maximized by locating tests of *small size and large power*. As Seidenfeld ([1979], p. 40) has observed,

> The connection between size, power and empirical frequencies, rests on the ability to actualize the hypothetical experiment used to justify the application of the mathematical model to the problem at hand. That is, using the law of big numbers, it is 'practically certain' that the frequencies of the two errors resulting from applying the statistical test to a sequence of 'random' experiments will agree (in the long run) with the values α and β. In the long run, the proportion of 'false rejections' of h will be α (at most) and the proportion of 'false acceptances' of h will be $1 - \beta$ (at most).

Among the most important of Neyman and Pearson's mathematical contributions to statistical methodology, therefore, is the "Neyman-Pearson Lemma", which specifies the principles for determining the most powerful tests for h^* alternatives to designated hypotheses h and significance levels α for continuous distributions (where, for discrete distributions, "randomizers" may be required; see, for example, Seidenfeld [1979], pp. 41–47).

The *prima facie* appeal of Neyman-Pearson hypothesis tests largely stems from the apparent foundation they provide for selecting between alternative hypotheses on the basis of objective statistical procedures, a circumstance which has undoubtedly contributed to their widespread utilization. Whether or not these practices are appropriate for their intended purpose, of course, depends upon what that purpose is, where Birnbaum in particular has suggested the necessity to distinguish between *evidential* and *behavioral* conceptions of Neyman-Pearson principles: "accepting an hypothesis" could be taken to mean *either* deciding upon a course of action (purchasing produce, marketing equipment, vaccinating children, etc.) *or* drawing a conclusion concerning what is true (that less than 5% of those tomatoes are rotten, that more than 95% of these wrenches will sell, that this vaccine tends to prevent smallpox, etc.). The underlying distinction between *making-a-decision* and *drawing-an-inference*, however, is by no means cut and dry; for decision-making typically entails acting *as though* an hypothesis were true (under suitable conditions), while inferring that an hypothesis is true likewise typically entails *acting* as though it were true (under similar circumstances). The theoretical warrant for this distinction, therefore, requires further contemplation (cf. Birnbaum [1977], pp. 28–34, and Smith [1977], pp. 81–82).

In both cases, of course, "inferences" or "decisions" are being drawn or made on the basis of information concerning 'random' samples. Perhaps it may be useful, therefore, to schematize the difference involved here for the sake of abstract comparison. In the case of *drawing-an-inference*, for example, an inference is drawn by accepting or rejecting an hypothesis h, as follows:

Inference based on sample:	Truth about the Population:	
	h is true	*h* is false
Reject *h* (as false)	Type I Error	Correct Inference
Accept *h* (as true)	Correct Inference	Type II Error

This diagram thus represents the possible outcomes of *a two-hypothesis inference problem* (let us say), in which inferences are drawn on the basis of Neyman-Pearson principles. Analogously, in the case of *making-a-decision*, a decision is made by accepting or rejecting a course of action A, as follows:

Decision based on sample:	Truth about the Population:	
	Does meet standards	Does not meet standards
Avoid Act A	Type I Error	Correct Decision
Perform Act A	Correct Decision	Type II Error

This diagram thus represents the possible outcomes of *a two-alternative decision problem* (let us say), in which decisions are made on the basis of Neyman-Pearson principles (cf. Moore [1979], p. 297). Not the least of the features distinguishing "decisions" from "inferences", of course, is that "decisions" are *usually irrevocable*, often involving choices between *restricted alternatives*; while "inferences" are ordinarily *subject to revision*, where the possible alternatives are *virtually unlimited*. Nevertheless, in spite of these apparent differences of degree, it would be valuable to discover some theoretical foundation for their distinction.

Perhaps the best approach to this problem resides in differentiating between two broad principles that appear to underlie statistical practice generally, namely: the *likelihood* principle and the *confidence* principle, as they have been labelled by Birnbaum [1962] and Birnbaum [1969]. Birnbaum's investigation of these principles scored a major triumph in 1962 in the form of a demonstration that the principles of *conditionality* and of *sufficiency* (which conditionality itself entails) obtain if and only if the principle of *like-

lihood also obtains. This result was hailed by Savage as "a landmark in statistics", largely because, as Giere ([1977a], p. 6) remarks, "Birnbaum claimed to have shown that two principles widely held by non-Bayesian statisticians (sufficiency and conditionality) jointly imply an important consequence of Bayesian statistics (likelihood)". Although Birnbaum has been viewed as a leading advocate of the likelihood principle ever since, his position changed radically within less than a decade, when he discovered the incompatibility of the likelihood principle with ordinary statistical practice in the form of (what he referred to as) "the confidence concept" (which underlies Neyman-Pearson hypothesis tests), a result he reported with some dismay (Birnbaum [1969], p. 131). Yet this incompatibility may harbor the basic difference between principles of inference and procedures of decision.

All of the concepts with which Birnbaum has been concerned may be said to represent aspects of alternative interpretations of "evidential meaning", i.e., as delineating those features of data acquired by means of experiment and observation that "make a difference" to the evidential support provided to appropriate hypotheses. Informally formulated, for example, the principle of *conditionality* asserts that, "the evidential meaning of any outcome of any mixture experiment is the same as that of the corresponding outcome of the corresponding component experiment", i.e., however an experiment is selected for performance, its "evidential meaning" depends exclusively upon the results actually obtained (Birnbaum [1962], p. 271). The principle of *sufficiency*, in turn, states that the "evidential meaning" of an experiment is completely conveyed by sufficient statistics, i.e., by those summaries of the data that preserve likelihood ratios between competing hypotheses, given that evidence (Birnbaum [1962], p. 270). The *likelihood* principle, finally, asserts that, "the evidential meaning of any outcome x of any experiment E is fully characterized by giving the likelihood function $cf(x, \theta)$", which is specifiable up to an arbitrary positive constant c (Birnbaum [1962], p. 271). Thus, as Birnbaum expressed it, these principles assert, respectively, the "irrelevance of (component) experiments not actually performed", the "irrelevance of observations independent of a sufficient statistic", and "the irrelevance of outcomes not actually observed" (Birnbaum [1962], pp. 270–271).

What Birnbaum subsequently ascertained, however, is that these principles of "evidential meaning" are incompatible with the assignment of *error probabilities* within the context of Neyman-Pearson hypothesis tests, since likelihood ratios *per se* "cannot be construed so as to allow useful appraisal, and thereby possible control, of probabilities of erroneous interpretations" (Birn-

ACCEPTANCE AND REJECTION RULES 239

baum [1969], p. 128). He later formulated this desideratum as the following *confidence principle*, namely:

(Conf): A concept of statistical evidence is not plausible unless it finds 'strong evidence for H_2 as against H_1' with small probability (α) when H_1 is true, and with much larger probability $(I - \beta)$ when H_2 is true (Birnbaum [1977], p. 24).

He considered this principle to be more adequate than the likelihood notion, which he formulated by the following *likelihood principle:*

(L'): If an observed sample point has very small probability (density) under H_1, relative to its probability (density) under H_2, then it provides strong statistical evidence for H_2 as against H_1 (Birnbaum [1977], p. 46, n. 8);

presumably on the grounds that (Conf) "refines and assimilates" what Birnbaum considered to be "the traditional concept" of statistical tests, i.e.,

(P): A concept of statistical evidence is not plausible unless it finds 'strong evidence against H_1' with very small probability when H_1 is true (Birnbaum [1977], p. 46, n. 8).

Indeed, according to Giere, Birnbaum had come to this realization as early as 1964 in the form of the following "principle of unbiasedness", namely:

(U): Systematically misleading or inappropriate interpretations shall be impossible; that is, under no θ shall there be high probability of outcomes interpreted as 'strong evidence against θ' (Giere [1977a], p. 9).

Although this principle is obviously too strong so long as 'misleading results' are *possible* by chance alone, likelihood ratios themselves *are* statistics which fail to provide for the appraisal of probabilities of error.

If these are Birnbaum's arguments, however, then their weaknesses are strikingly apparent. For Birnbaum first discovers that conditionality and sufficiency entail likelihood (Birnbaum [1962]), that is:

(1) if conditionality and sufficiency, then likelihood, necessarily.

Birnbaum later discovers that likelihood and confidence are not compatible (Birnbaum [1969]), that is:

(2) if confidence, then not likelihood, and conversely, necessarily.

He subsequently formulates "the traditional concept" of statistical tests,

which presumably entails the confidence concept (Birnbaum [1977]), that is:
    (3) if unbiasedness, then confidence, necessarily.
Birnbaum finally assumes the adequacy of "the traditional concept" with an endorsement of unbiasedness (Birnbaum [1977]), that is:
    (4) unbiasedness.
From these premises, of course, Birnbaum is entitled to deduce,
    (5) confidence;
from premises (3) and (4), which with premise (2) entails,
    (6) not likelihood;
which, in turn, with premise (1) entails the conclusion,
    (7) not both conditionality and sufficiency.

Birnbaum's argument, of course, is perfectly *valid*, insofar as (5), (6), and (7) all follow from the premises provided. The difficulties, therefore, such as they may be, concern this argument's *soundness* — in particular, the issue of whether these premises all happen to be true. As Hacking ([1973], p. 487) has observed,

> ... deductive logic settles nothing outside pure mathematics. The connection between p, $p \supset q$, and q is a pivot on which reasonable discussion turns. One man uses it to conclude q, another to reject p. A third who cannot tolerate q and yet is convinced of p and $p \supset q$ goes away in a quandary; so long as he grants that his position is unclear, he is still being reasonable.

The principal problem, of course, is premise (4), the assertion of which may well be supposed to 'beg the question'; for, with at least equal enthusiasm, one might endorse the likelihood principle (on the basis of Birnbaum [1962], for example; cf. Pratt [1977], p. 62) and assert instead,
    (4′) likelihood;
which, of course, together with premise (2) entails,
    (5′) not confidence;
which, in turn, with premise (3) likewise entails,
    (6′) not unbiasedness.
The difficulties thus encountered, however, are not restricted to the need for independent arguments in support of premise (4) but extend to the claim that unbiasedness *per se* entails confidence as premise (3) alleges. For if likelihood principles applied to large numbers of trials over a wide variety of conditions would satisfy the unbiasedness desideratum, for example, then premise (3) is false and Birnbaum's argument fails.

If Birnbaum's position appears to leave these matters unsettled, however, perhaps a somewhat different approach may prove beneficial. During the

ACCEPTANCE AND REJECTION RULES 241

course of his defense of Neyman-Pearson procedures, for example, Giere has introduced a potentially valuable distinction between "exploratory inquiry" and "confirmatory inquiry" which appears to be relevant to the issues involved here. He suggests that 'acceptance' and 'rejection' are context-dependent notions that assume importance only in relation to *the purposes of inquiry*; in particular,

> In exploratory inquiry there are two purposes for which the truth of an accepted hypothesis may be assumed: (i) as an auxiliary hypothesis in the design of later experiments, and (ii) as a 'fact' to be explained by any proposed theory of the phenomena under investigation. Thus there are two ways a mistaken acceptance or rejection can impede 'the progress of inquiry' or 'the growth of knowledge'. It can lead to mistaken or confusing results in later experiments, and it can mislead theoretical speculation. (Giere [1975a], p. 127)

The results of exploratory inquiries within the designated domain, therefore, not only provide guidelines for subsequent research but also generate 'facts' for future explanation. The differences between "exploratory" and "confirmatory" inquiry, moreover, arise in great measure due to the consideration that large numbers of alternative hypotheses will still be compatible with the evidence acquired by specified tests *even with significance levels fixed at relatively low* α, e.g., .05 (Giere [1975a], pp. 125–126). The range of compatible alternatives can usually be improved, of course, by adopting successively lower and lower bounds for α and β, but their existence can never be completely eliminated by any inquiry.

The entire class of hypotheses that would not be rejected for a specified number of trials n and significance level α, therefore, may be referred to as "Δh", where (following Giere) outcomes within a rejection region r will *not* occur (with a probability of .95, for example) when h is true, but *will* occur (with a probability of .95, once again) when *not-*Δh is true: Neyman-Pearson procedures, in other words, are directed toward *classes of hypotheses*, consisting of h and its "near rivals", let us say, rather than toward *single specific hypotheses*, such as h apart from Δh, *per se* (Giere [1979a], pp. 220–223). The practical difference between "exploratory" and "confirmatory" inquiries, therefore, is mainly (though not entirely) determined by the discriminatory capabilities of *large* numbers of trials rather than *small* numbers of trials, where exploratory testing employing limited resources provides indications of promising directions for additional investigations with less-limited resources, that is:

Research Results:

Type of Test	Data is α-significant	Data is not α-significant	Context
Small Samples (naive state)	Abandon h (grossly wrong)	Retain Δh (roughly right)	Exploratory Inquiry
Large Samples (critical state)	Reject h' (as "false")	Accept h' (as "true")	Confirmatory Inquiry

Investigative Payoff-Matrix

Scientific inquiry thus proceeds by tentatively entertaining hypotheses in a certain class Δh, which are subject to preliminary exploration requiring only limited allocations of resources. Once this naive state has been overcome by acquiring the results of these investigations, however, hypotheses, such as h', which fall within the range of Δh, may receive eventual acceptance as a consequence of further critical scrutiny (Giere [1979a], pp. 75–86; cf. Seidenfeld [1979], pp. 81–83).

An account of this kind captures at least one basic feature of scientific inquiry, namely: that it proceeds by a process of *successive approximation*. The benefits of changes from α, β, and Δ that are rather large to others that are relatively small as the number of trials n increases does not diminish merely because it is a process which may be endlessly repeated, with ever smaller α, β, and Δ, unless the potential for further inquiry must eventually expire. Nevertheless, while Giere's reconstruction ought to be admired as an attempt to defend Neyman-Pearson procedures as *principles of inference*, at least three crucial aspects of these methods appear to undermine the propriety of this interpretation; for (i) the choice of an hypothesis of primary interest, (ii) the choice of a significance level α, and, (iii) the choice of the appropriate power β, respectively, reflect implicit reliance upon three decision-making elements, successively, as follows:

(i) selecting the hypothesis of "primary" interest requires a decision to minimize Type I errors in conformity with the *minimax loss* principle;

(ii) selecting the significance level (or "size") α that is appropriate requires a quantitative formulation of the *dis-utility* of Type I errors; and,

(iii) selecting the "power" of that test at $1 - \beta$ requires a decision to minimize Type II errors in conformity with the *minimax regret* principle.

Apart from the formulation of the hypotheses $h, h^*, \ldots,$ themselves and the acquisition of the empirical evidence e by conducting an appropriate sequence of trials, therefore, Neyman-Pearson procedures actually presuppose *prin-*

ciples of decision (including, of course, subjective evaluations of relative utility).

The importance of these ingredients for the application of Neyman-Pearson procedures, moreover, may be illustrated by considerations such as these:

(a) with fixed size α and number of trials n, a broad range of different hypotheses would all have been accepted (as Giere's use of 'Δh' reflects) if they had been subjected to test in lieu of the hypothesis actually selected;

(b) with variable n and fixed hypotheses h and h^* (or even with fixed n and fixed hypotheses), whether h or h^* is accepted (or is rejected) will vary with the specific significance level α that is employed; and, furthermore,

(c) with fixed n, fixed α, and fixed β, whether one or the other member of a fixed pair of alternatives, h and h^*, is accepted or rejected will often depend upon which of those hypotheses is selected as of "primary" interest (an objection overcome by alternative formulations; cf. Seidenfeld [1979], especially Ch. 2).

Although Giere and Seidenfeld have recommended "professional judgements" and "discrepancy measures" as partial remedies for these difficulties, it is not easy to dismiss their weight as evidence supporting a "behavioral" interpretation (Giere [1976], p. 79; Seidenfeld [1979], pp. 79–80; cf. Kiefer [1977]). And while Birnbaum himself supposed that an "evidential" interpretation could be sustained on the basis of the confidence principle, even he acknowledged the necessity to qualify that conception as follows (Birnbaum [1977], pp. 26–27):

a conclusion reached in a scientific investigation... requires not only
(1) statistical evidence of sufficient strength concerning the statistical hypothesis of interest.
In addition the investigator (or community of investigators) must judge
(2) the adequacy of the mathematical-statistical model, which serves as the conceptual frame of reference for the interpretation of the statistical evidence, to represent the research situation in relevant respects; and,
(3) the compatibility with other knowledge and evidence of a conclusion that may be supported by statistical evidence provided by the current investigation (for example, strong statistical evidence against the [alternatives under consideration]).

Indeed, Giere has gone so far as to describe the assumption that one or the other of the alternatives under consideration, h and h^*, must be true as a *necessary condition* for the adequacy of the "evidential" interpretation, a difficulty partially overcome through 'Δh' (Giere [1975], pp. 231–235; cf. Rogers [1971], pp. 403–412).

So long as the available alternatives consist of "forced choices" between specific actions (purchasing or not purchasing tomatoes, marketing or not marketing wrenches, etc.) in which subjective evaluations of their relative utility appears to be theoretically unavoidable, these features are not "defects"; but to the extent to which *principles of inference* are intended to secure the desideratum of "epistemic objectivity" by supplying standards whose systematic application to any logically consistent, finite class of accepted beliefs *b* or evidence sentences *e* would warrant assigning all and only the same measures of evidential support to all and only the same hypotheses, they appear inadequate. Thus, a striking advantage of likelihood principles over Neyman-Pearson tests as *principles of inference* is that the former, but not the latter, reflect the "evidential meaning" (in Birnbaum's sense) of the evidence which is *actually acquired* as a result of conducting particular empirical tests. Thus, as Pratt ([1977], p. 62; cf. Smith [1977], pp. 73–74) has remarked,

Even for simple hypotheses, the question arises whether the tail probabilities cited [for specific tests] are to correspond to the particular data observed, or are to be fixed in advance with only 'accept' or 'reject' determined by the data. The latter seems to me clearly a very inadequate expression of the evidence.... Dichotomous reports seem less satisfactory the less behaviorally and more evidentially the 'decisions' are interpreted.

The "preference procedures" Neyman and Pearson have proposed, in other words, may be perfectly suitable for decision-making between restricted alternatives without also fulfilling the appropriate conditions for drawing inferences on the basis of empirical evidence.

AN INDUCTIVE ACCEPTANCE RULE

The conception that underlies the likelihood principle is the existence of a quantitative measure of the strength of evidential support with respect to designated hypotheses, where this measure of support may assume any value from zero to one inclusively (or any ratio of those values when alternatives are compared). Pratt ([1977], p. 62) has suggested a condition of adequacy which is compatible with the likelihood principle but incompatible with the confidence principle, namely:

(CA): A concept of statistical evidence is (very) inadequate if it does not distinguish evidence of (very) different strengths.

The results that arise from orthodox statistical hypothesis tests conducted with a significance level of .05, for example, do not discriminate within a rejection region between outcomes with probabilities of .04, .03, .02, etc.

nor within an acceptance region between outcomes with probabilities of 1.0, 0.9, 0.8, etc. Consequently, *how near* or *how far* from rejection or acceptance an hypothesis may have come does not constitute any part of the "evidential meaning" of Neyman-Pearson procedures. In this respect, likelihood ratios provide more "evidential meaning" than do error probabilities, which in turn provide more "evidential meaning" than do dichotomous reports (such as "significant" or "not significant", "accept" or "reject", and so forth). Pratt's condition of adequacy, therefore, supports the likelihood principle.

In an important paper published in 1973 which has not received the attention it deserves, Hacking has related the differences between the likelihood and the confidence principles to general issues of explanation and prediction on the ground that confidence pertains to *predictive* desiderata, while likelihood pertains to *explanatory* desiderata:

> I propose, ..., that the incompatibility of sufficiency and confidence concepts is no passing phase in the evolution of statistical ideas. On the contrary, it brings to light a deep divergence between two of our most important bases of reasoning, explanation and prediction. Sufficiency is a concept concerned with inferring to the best explanation. Confidence is a concept concerned with inferring to the best prediction. That in practice the concepts often coincide reminds us of the familiar fact that explanation and prediction are very often symmetric. But they do not always coincide. (Hacking [1973], p. 498)

Thus, in the chapter that follows, I shall argue that Hacking is correct as far as he goes, but that the difference involved here extends beyond issues of explanation and prediction to those of inference and decision; for confidence appears to be fundamental both for *prediction* and for *decision*, while likelihoods appear to be fundamental both for *explanation* and for *inference*, which we shall now consider.

There are important relations between likelihoods and explanations, on the one hand, and explanation and acceptance, on the other. As Giere ([1976], pp. 69–70) has emphasized, likelihood principles, such as sufficiency,

> ... are clearly incompatible with the use of error probabilities which are defined relative to some sample space of possible outcomes, only one of which is actually obtained. The question is whether ... it is possible to justify conditionalizing on the result of the randomizer without going so far as to conditionalize on the actual data, thus abandoning error probabilities altogether. (Cf. Seidenfeld [1979], pp. 36–47 and pp. 119–120.)

But even though "conditionalizing on the actual data" entails abandoning error probabilities altogether, the direct connection between likelihoods and explanations is by no means difficult to discern; for the likelihood of h, given

e, reflects not only the probability of e, given h, but also represents "the degree of nomic expectability" for e as an *explanandum*, given h as its *explanans* (as that concept was introduced in Chapter 5 and elaborated upon in Chapter 8). "The actual data" is important, therefore, since no explanation is called for unless its explanandum occurs. If explanations are arguments, moreover, then their assertion as adequate entails their acceptance as true: *insofar as science aims at explanation, therefore, it cannot dispense with acceptance*. But the theory of explanation does not need or require error probabilities at all.

Giere's distinction between "exploratory" and "confirmatory" inquiry may serve a useful function, nevertheless. In particular, it is difficult to deny that scientists themselves have to *decide* which hypotheses to subject to test, where coarse differences between hypotheses as "roughly right" or as "grossly wrong" require preliminary detection. Consequently, Neyman-Pearson procedures appear to play an important role with respect to the preliminary stages of scientific inquiry, more or less as Giere claims, namely:

Research Projects:

Type of Test	Data is α-significant	Data is not α-significant	Context
Small Samples (naive state)	Abandon h (grossly wrong)	Retain Δh (roughly right)	Exploratory Inquiry

Stage I: Neyman-Pearson Procedures

So long as science itself aims at establishing general hypotheses of broad scope and systematic power, however, more sensitive methods of inquiry are required, i.e., those which likelihood principles supply:

Research Projects:

Type of Test	Hypothesis Eliminated	Hypothesis Corroborated	Context
Large Samples (critical state)	Reject h' (as "false")	Accept h' (as "true")	Explanatory Inquiry

Stage II: Maximum-Likelihood Principles

Instead of extending Neyman-Pearson procedures within the "confirmatory" domain, therefore, I would suggest that a deeper division separates these stages in the conduct of inquiry. For Neyman-Pearson procedures appear to be appropriate for "pilot studies" involving small samples within the context of *exploratory* inquiry, while maximum-likelihood principles appear to be appropriate for "extensive research" involving large samples within the context of *explanatory* inquiry.

It should be emphasized, therefore, that Neyman-Pearson procedures in exploratory inquiry function as *principles of decision*, which result in decisions concerning the course of inquiry, but maximum-likelihood principles in explanatory inquiry function as *principles of inference*, which result in inferences concerning what is true. In order to warrant the conception of maximum-likelihood principles as principles of inference, of course, it is essential to provide some *more formal* characterization, especially because these principles are supposed to fulfill this function only when applied to large numbers of trials under a wide variety of conditions. The *large number* condition, moreover, seems to be susceptible to more or less successful formalization; but the *wide variety* condition poses somewhat more difficult problems with respect to this desideratum. All lawlike sentences that are *true*, of course, are maximally specific, necessarily; that is, their reference class descriptions, 'K' or '$K \cdot T^i$', specify complete sets of nomically relevant predicates relative to their attribute phenomena, 'χ' or 'O^i' (as explained in Chapter 3). Consequently, for all other predicates 'F', such that (a) neither 'F' nor '$-F$' entails the presence or the absence of these outcome phenomena, and, (b) 'K' or '$K \cdot T^i$' entails neither 'F' nor '$-F$', it *must* be the case that 'F' is nomically irrelevant to the truth of that lawlike sentence *if* that lawlike sentence is true. The *wide variety* condition is therefore intended to promote the search for (other) nomically relevant predicates if they exist.

The measure of the severity of e as a test of h, relative to b, advanced in Chapter 8, of course, represents a different notion, namely: the extent to which the hypothesis h exceeds or contravenes accepted background knowledge b. The measure of the severity of e as a test of h^2, relative to h^1, analogously, compares differences in the degree of nomic expectability for outcomes of that kind given those hypotheses, respectively. The conception under consideration within the present context, however, involves *pragmatic elements* not reflected by those notions; for, although the severity of a test (in this sense) will be *relative to accepted background knowledge b*, the members of the scientific community conducting these tests are morally obligated to attempt to refute those hypotheses under investigation by discovering or by inventing some property or process which would establish their falsity, if they are false. Otherwise, in the absence of these procedural safeguards, the practice of science would fail to promote the aim of science in providing for the acceptance, rejection, and modification of hypotheses and theories which may be employed for the purposes of explanation and prediction — for explanations are not adequate unless their explanans are true. Attempting to eliminate alternative hypotheses by means of severe tests of this kind thus requires *not only* that the members of the scientific community should inventory their background knowledge in order to isolate and identify those properties $F, G, \ldots$ most promising as potential falsifiers with respect to these hypotheses *but also* that they must exercise their imaginative capacities in order to insure no such properties are overlooked or ignored — to the best of their ability!

Thus, as Popper ([1965], p. 418) has emphasized, "if e is *not* a report about the results of our sincere attempts to overthrow h, then we shall simply deceive ourselves if we think we can interpret $C(h/e)$ as degree of corroboration, or anything like it". But, although *sincerity* cannot be formalized, there appears to be a fundamental connection between the large numbers and the wide variety conditions, on the one hand, and the difference between extensional distributions and intensional generalizations, on the other. An *extensional distribution*, after all, describes (some segment of) the world's history, without claiming to constrain the course of other world's histories as well; while *intensional generalizations* not only restrict the history of this world, but also restrain the histories of other possible worlds as well. While large numbers of trials might be sufficient for *confirming* extensional distributions, therefore, a wide variety of conditions is also necessary for *corroborating* intensional generalizations: the purpose of this condition, in other words, is to promote the corroboration of lawlike sentences, which have the form of subjunctive conditionals, by tests designed to instantiate their antecedents (which might otherwise remain counterfactual) over a wide variety of conditions, i.e., through repeated attempts to refute them!

In the exploration of principles of induction, of course, it is extremely important to distinguish clearly between the world itself and our beliefs about the world; for although our beliefs will be true just in case the world possesses the properties our beliefs attribute to it, the only available measure of the truth of our beliefs is the extent to which the evidence at our disposal lends them support. Thus, although the concept of a random sample may be provided an ontic definition *with respect to the world itself* as follows,

> (D-1) A frequency distribution $\mathscr{F}_n$ is a random sample (of size n) if and only if it records the results of n random trials; where random trials are equal and independent (as defined in Chapter 5);

ACCEPTANCE AND REJECTION RULES

the formulation of principles of induction requires reliance upon the derivative concept of a sample that is random *with respect to our beliefs about the world*. Then the concept of a random sample may be provided an epistemic definition relative to the knowledge context $\mathcal{K}zt$ as follows:

(D-2) A frequency distribution statement, '$\mathcal{F}_n(O^i xt/Kxt \cdot T^i xt) = f$', records a random sample (of size n) within the knowledge context $\mathcal{K}zt$ if and only if any predicate, say, 'M', that is a maximally specific reference predicate relative to the outcome predicate '$O^i xt$' for each member m of that n sample is logically equivalent to '$Kxt \cdot T^i xt$' in $\mathcal{K}zt$.

A frequency distribution statement belonging to $\mathcal{K}zt$ records a random sample of size n, therefore, when and only when each of those n trials is obtained under precisely the same relevant conditions "to the best or our knowledge". Thus, let us assume that our belief that this is the case may be measured by an *a priori* degree of confidence in data d reporting such a sample relative to $\mathcal{K}zt$ that is equal to one; and that, in the absence of such a belief, our *a priori* degree of confidence (in this new distinctive sense) is not known.

Although we have focused primarily upon *simple* frequency distributions, which reflect outcomes of one particular kind, we are obviously concerned with *complex* frequency distributions, which reflect all possible outcomes, as well. Let us assume that a *random variable X* is any outcome of any singular trial of any experimental arrangement such that, if the possible outcomes of that trial are $O^1, O^2, \ldots, X$ may assume any of the corresponding values $X = O^1$, $X = O^2, \ldots$ (Feller [1950], pp. 212–213). A complex frequency distribution for a sequence of n trials of such a random variable X will thus record the relative frequency f_i with which each possible outcome O^i occurred within that sequence, where the frequency statements '$\mathcal{F}_n(X = O^1) = f_1$', '$\mathcal{F}_n(X = O^2) = f_2$', ..., satisfy the following three conditions, namely:

(a) each such frequency f_i is a non-negative real number;
(b) the numerical values of $f_1, f_2, \ldots$ sum to one; and,
(c) for any pair of outcomes O^i and O^j, $f_{i \vee j} = f_i + f_j$.

Analogously, a complex probability distribution for a random variable X will report the probability r_i for each possible outcome O^i of that random variable as, say, '$P(X = O^1) = r_1$', '$P(X = O^2) = r_2$', ..., where, as before,

(a) each probability r_i is a non-negative real number;
(b) the numerical values of $r_1, r_2, \ldots$ sum to one; and
(c) for any pair of outcomes O^i and O^j, $r_{i \vee j} = r_i + r_j$.

Furthermore, the corresponding *frequency distribution function* $\mathscr{F}_n{}^*$ for the random variable X (whose values may be represented by non-negative real numbers x) and the corresponding *probability distribution function* $\mathscr{P}^*$ may be defined as follows:

$$\mathscr{F}_n{}^*(x) = \sum_{i \leq [x]} f_i\text{; and, } \mathscr{P}^*(x) = \sum_{i \leq [x]} r_i\text{.}$$

[Cf. Hodges and Lehmann [1964], pp. 16–17; and Feller [1950], pp. 212–213.]

Since our *a priori* degree of confidence in the reliability of a frequency distribution function for the purpose of determining the likelihoods of various probability distribution functions is based upon our beliefs both with respect to (i) the set of all properties that are nomically relevant to these O^i-outcome kinds and (ii) the sub-set of those properties that were jointly present on each of these n trials, it should be obvious that, even when our *a priori* degree of confidence happens to have the value of one, that degree of confidence represents a *conjecture*, namely: that each and every member of the trial sequence was actually conducted under the same relevant conditions. It would be useful, therefore, to have a means for measuring posterior as well as prior degrees of confidence on the basis of particular features of that frequency distribution function itself; and, indeed, there appears to be suitable theoretical support for such a conception in the form of *the central limit theorem*. For let $X_1, X_2, \ldots, X_n$ be a sequence of n trials of a random variable X with probabilities that are equal and independent, i.e., this is a sequence of n independent identically distributed random variables. Let θ be the theoretical mean and σ^2 the theoretical variance of the experimental population. Then if $S_n = X_1 + X_2 + \ldots + X_n$ and $\bar{X} = S_n/n$, as n increases without bound, the distribution function of $\sqrt{n}(\bar{X} - \theta)/\sigma$ converges toward a normal distribution with mean zero and variance one; that is, $\sqrt{n}(\bar{X} - \theta)/\sigma \rightarrow \mathscr{N}(0,1)$ (cf. Feller [1950], p. 244). What is most remarkable about this theorem, moreover, is that it imposes no requirements upon the *form* of the underlying distribution functions: "The importance of this theorem, as far as practical applications are concerned, is the fact that the mean $\bar{X}$ of a random sample from *any* distribution with finite variance σ^2 and mean θ is approximately distributed as a normal variate with mean θ and variance σ^2/n" (Mood and Graybill [1963], p. 150).

Consequently, a suitable measure of our posterior degree of confidence in the data d would be the extent to which the frequency distribution function $\mathscr{F}_n{}^*$ diverges from our *a priori* expectation on the assumption that this was a sequence of n independent identically distributed random variables; for sure-

ACCEPTANCE AND REJECTION RULES 251

y our confidence in the reliability of our data should be shaken if the distribution function $\mathscr{F}_n^*$ does not converge toward a normal distribution $\mathscr{N}$. As it happens, Paul Lévy has provided a promising measure for this purpose; for a necessary and sufficient condition for the (weak) convergence of a distribution function $\mathscr{F}_n^*$ toward a normal distribution function $\mathscr{N}$ is the following: $D(\mathscr{F}_n^*, \mathscr{N}) \to 0$, where the distance $D(\mathscr{G}, \mathscr{H})$ between two distribution functions $\mathscr{G}$ and $\mathscr{H}$ is defined as the infimum of all ϵ such that for all values of x,

$$\mathscr{H}(x - \epsilon) - \epsilon \leqslant \mathscr{G}(x) \leqslant \mathscr{H}(x + \epsilon) + \epsilon.$$

[Cf. Gnedenko and Kolmorgorov [1954], p. 33.] The distance $D(\mathscr{G}, \mathscr{H})$, moreover, satisfies the conditions for a metric; that is,
(a) $D(\mathscr{G}, \mathscr{H}) = 0$ if and only if $\mathscr{G} = \mathscr{H}$;
(b) $D(\mathscr{G}, \mathscr{H}) = D(\mathscr{H}, \mathscr{G})$; and,
(c) $D(\mathscr{G}, \mathscr{I}) \leqslant D(\mathscr{G}, \mathscr{H}) + D(\mathscr{H}, \mathscr{I})$.

Let us take as our measure of divergence from the normal distribution that ϵ which is the smallest ϵ satisfying the specified conditions, such that the Lévy distance $D(\mathscr{F}_n^*, \mathscr{N}) = \epsilon$. Then as our measure of confidence in the reliability of a frequency distribution function $\mathscr{F}_n^*$ as relevant evidence for determining the likelihoods of alternative probability distribution functions, we might embrace the following principle, namely:

> *The Principle of Confidence:* When our prior degree of confidence in a frequency distribution function $\mathscr{F}_n^*$ is equal to one, then our posterior degree of confidence in that same distribution function $\mathscr{F}_n^*$ is equal to $1 - \epsilon$, i.e., $C^*(\mathscr{F}_n^* / \mathscr{K}zt) = 1 - \epsilon$; otherwise, $C^*(\mathscr{F}_n^* / \mathscr{K}zt)$ is not known.

The point of the final provision, therefore, is to explicitly acknowledge that when these conditions are not satisfied the degree of reliability of data $d = \mathscr{F}_n^*$ for the purpose of determining the likelihoods of hypotheses within a knowledge context $\mathscr{K}zt$ cannot be measured and epistemic caution is clearly required. [The Lévy distance is employed here as an illustration of the kind of measure that might appropriately be relied upon for this purpose. Other measures of potential interest are those discussed by Gnedenko [1962], pp. 444–458, and by Hájek [1969], Chs. 1 and 2.]

Let us further assume a set of j mutually exclusive hypotheses, $h^1, h^2, \ldots, h^j$, that is jointly exhaustive in the following sense: with respect to random variable X, this set includes at most twelve hypotheses for a possible outcome O^i such that these hypotheses attribute probabilities to the first decimal place for outcomes of that kind, i.e., 0.0, 0.1, ..., 0.9, 1.0 (or, perhaps, universal strength u); or, if the first decimal place has been ascertained in $\mathscr{K}zt$, attribute probabilities to the second decimal place for outcomes of that kind, i.e., 0.10, 0.11, ..., 0.19, 0.20; or, ...; and so on. The appropriate level of specificity of the members of such sets, therefore, is determined on the basis of empirical and theoretical considerations within $\mathscr{K}zt$; for example, the number of possible values of that variable and the frequencies with which those outcomes have occurred in the past to the best of our knowledge. Provided the data at our disposal has enabled us to ascertain the first decimal place with a suitable degree of confidence, we may proceed to determine the second decimal place with a suitable degree of confidence, and so forth, by what is essentially a method of successive approximation, eventually advancing to any degree of specificity desired. A principal benefit of this procedure, of course, is that it allows for the systematic investigation of an arbitrarily large number of hypotheses of arbitrary specificity a finite number at a time.

Since every member of each such set of hypotheses will be homogeneous in the sense of ascribing a probability of different strength for the same outcome O^i to the same conditions $K \cdot T^i$, for example, the incapacity to satisfy the desideratum reflected by the principle of confidence, i.e., that a sequence consisting of independent identically distributed random variables will almost always exhibit convergence tendencies (as *The Short Run Principle* declares), should suggest the necessity to consider alternative sets of hypotheses which are also homogeneous in ascribing probabilities of different strengths for the same outcome O^i to the same conditions but for which *those conditions themselves* are subjected to modification. Thus, if a sequence of $K \cdot T^i$ trials has failed to satisfy this desideratum, perhaps a sequence of $K \cdot T^j$ trials, of $K \cdot T^k$ trials, and so on, might prove to be more adequate as a specification of the appropriate reference class for O^i outcome phenomena. A suitable combination of the principle of confidence with the likelihood principle, therefore, promises to provide an inference rule for inductive acceptance, such as the following:

> *Inference Rule for Inductive Acceptance:* If h^i is the best supported member of a set of mutually exclusive hypotheses (which is jointly exhaustive in the sense defined), given the evidence e within the knowledge context $\mathscr{K}zt$, then h^i may be accepted in $\mathscr{K}zt$ if the likelihood of h^i, given e, in $\mathscr{K}zt$, exceeds the degree of divergence ϵ of the data e from a normal distribution, i.e., if $L(h/e \ \& \ \mathscr{K}zt) > \epsilon$, where $e = \mathscr{F}_n^*$ in $\mathscr{K}zt$; otherwise, no member of that set has to be accepted into $\mathscr{K}zt$.

The degree of divergence ϵ of the data $e = \mathscr{F}_n^*$ from a normal distribution $\mathscr{N}$ thus functions as an index of epistemic caution, since no hypothesis may be accepted from such a set unless its likelihood is greater than the degree of divergence of that data from a normal distribution; hence, hypotheses with a high likelihood on the data may be acceptable even when our measure of confidence happens to be low, while hypotheses with a low likelihood on that data will be acceptable only when our measure of confidence happens to be high.

Several features of this inference rule for inductive acceptance, however, require further elaboration. The *first* is that it is intended to be applicable to general hypotheses concerning extensional distributions, i.e., limiting frequencies and constant conjunctions, as well as to those concerning intensional generalizations, i.e., universal and statistical lawlike sentences. Consequently, the term "probability" as it occurs in the discussion of the principle of confidence must be construed in a very broad sense, i.e., in the case of value u, the "probabilistic" conditions may be only trivially fulfilled. When this inference rule is applied to intensional alternatives, moreover, it is understood that the *wide variety* desideratum has to be satisfied, where "acceptance" is warranted only by virtue of repeated failures to refute over large numbers of trials and a wide variety of conditions. Otherwise, although this acceptance rule is responsive to the *large numbers* desideratum, the only hypotheses whose acceptance is warranted will be extensional distributions.

The *second* is that this inductive rule is intended to apply within the context of the logical consistency, deductive closure, and available evidence conditions formulated by (CR-1), (CR-2), (CR-3), and (CR-4). In particular, when any member of a set of mutually exclusive alternative hypotheses is accepted into a knowledge context $\mathscr{K}zt$, the negations of those incompatible alternatives must also be accepted into $\mathscr{K}zt$, in conformity with (CR-1) and (CR-2). When the evidence on the basis of which an hypothesis h has been accepted or rejected changes by virtue of acquiring additional observations, performing further experiments, or constructing alternative theories, "new" hypotheses such as h^* may deserve acceptance and "old" hypotheses such as h may have to be rejected, in conformity with (CR-3) and (CR-4). It thus provides for the acceptance, rejection, and modification of the members of the set of beliefs, Szt, accepted by an individual z at a time t; that is, the knowledge context, $\mathscr{K}zt$, must always reflect those beliefs best supported by the evidence available at any time, relative to this principle.

The *third* is that it is intended to be applicable within the context of Stage II (or "explanatory") inquiries, which involve large numbers of trials, rather than within the context of Stage I (or "exploratory") inquiries, which involve small numbers of trials. Consequently, even though those hypotheses ascribing higher probabilities to the evidence will therefore possess higher likelihoods and confront less demanding conditions of acceptance, in general, with respect to the confidence desideratum, their appraisal by means of this inference principle presupposes that they have previously withstood scrutiny by means of those decision procedures. Moreover, the confidence desideratum itself affords intuitively satisfying theoretical resolutions of "the paradox of minimal evidence", which we have already considered, and of "the paradox of perfect evidence", which Popper ([1965], pp. 406–408) has described. For even though the likelihood of an hypothesis on scanty or partial evidence might equal its likelihood on abundant or extensive evidence, that added evidence is not therefore irrelevant or insignificant precisely because, as Popper claims and the confidence desideratum discloses, *evidence may affect the rationality of a belief in more than one way* (Popper [1965], p. 408).

The *fourth* (and final) is that this principle of inference is intended to be applicable on the basis of frequency distribution data concerning the relative frequencies with which different outcomes actually occur in finite sequences during the course of this world's history, where discovering that $\mathscr{F}_n(O^i xt/Kxt \cdot T^i xt \cdot Fxt) = \mathscr{F}_n(O^i xt/Kxt \cdot T^i xt \cdot -Fxt)$, for example, over large number of trials would ordinarily establish that a property F is *statistically irrelevant* to the occurrence of O^i, relative to $K \cdot T^i$, in the world W; while that same discovery over large numbers of trials conducted under a wide variety of conditions would ordinarily establish that the property F is *nomically irrelevant* to the occurrence of that same outcome under the same conditions, in any world W — provided, of course, that these tests were *severe*. A point worth making, therefore, is that the inductive principle advanced here is intended to serve as a *theoretical representative* of a general conception of scientific procedures rather than as a *practical requirement* for individual instances of scientific reasoning: indeed, even the inference that a certain sequence is *random* will ordinarily be established by evidence suggesting (i) a tendency toward insensitivity to ordinal selection and (ii) a tendency toward freedom from aftereffect, i.e., evidence that it is *normal*. It would be a mistake, therefore, to suppose that the specific details of this proposal are more important than the general conception of scientific procedures it represents. And should the objection be advanced that this inference rule seems to be somewhat complex, let us bear in mind that complex problems may require complex solutions: simple theories should be preferred, after all, only when they are also adequate.

ACCEPTANCE AND REJECTION RULES   255

IN DEFENSE OF THIS CONCEPTION

Considerations such as these, of course, may establish the *prima facie* plausibility of this inference rule for inductive acceptance, but they leave important issues yet to be resolved, especially the evaluation of this rule in comparison with its own competitors. Traditional principles of induction, probabilistic conceptions of confirmation advocated by logical Bayesians, by empirical Bayesians, and by subjective Bayesians, and "orthodox" statistical hypothesis testing procedures, for example, have all encountered substantial objections to their adoption within the context of scientific inference. It might serve a useful purpose, therefore, to appraise the adequacy of a rule of this kind in relation to difficulties such as these; for, even though the results of this comparison cannot be *conclusive* so long as other principles are possible, it should nevertheless afford a foundation for appraising the comparative preferability of these alternative conceptions, as a measure of their relative adequacy. Let us therefore attempt to subject this inductive principle to a severe test of its own by ascertaining the extent to which it promises to overcome the obstacles that have been discovered to confront alternative conceptions.

The traditional principles of induction embraced by Reichenbach [1949] and by Salmon [1967], for example, encountered at least three different difficulties, namely:

(a) traditional principles of induction afford no solution to problems of inductive inconsistency, whether with regard to the infinite class of limiting frequencies that are logically compatible with any finite frequency or with regard to the corresponding classes of possible Goodmanian alternatives; but these problems are overcome within the context of the present conception insofar as, first, the application of this principle is restricted to finite classes of hypotheses of appropriate specificity, as we have just discovered, and, second, the class of permissible predicates for the formulation of intensional generalizations is restricted to universals, as explained in Chapter 7;

(b) traditional principles of induction would have the consequence, whether intentional or otherwise, of eliminating hypotheses and theories couched in dispositional, theoretical, or continuous mathematical predicates from the language of empirical science, at least to the extent to which those concepts and relations are not translatable into an observational vocabulary; but this difficulty does not arise within the context of the present conception, since intensional generalizations are subject to test by means of their extensional implications, which entail no such translations — indeed, as Chapter 6 relates, this approach does not assume an observational/theoretical distinction at all;

(c) traditional principles of induction afford no basis for distinguishing between genuinely lawlike and merely accidental generalizations, not only because their application is directed toward extensional distributions rather than toward intensional generalizations, but also because discriminating between those claims that may be true of the actual word and those that may be true of all possible worlds dictates the subtle changes in attitude attending an exchange of the conception of *confirmation* as the successful search for instances of an hypothesis or theory for the conception of *corroboration* as the *un*successful search for instances of their *negations*, which our rule requires.

Perhaps it should be emphasized, therefore, that the incapacities of traditional principles of induction are rooted less in their restricted application to relative frequencies within finite sequences during the history of the actual world — an incapacity no other empirical procedure could possibly overcome — than in their inadequate reflection of the character of nomological relations. Nevertheless, their employment in combination with more satisfactory conceptions of the differences between genuinely lawlike and merely accidental generalizations could not completely compensate for their inherent limitations; for, unless the large numbers and wide variety conditions are supplemented by a "critical attitude", i.e., a commitment to sincere attempts at falsification, which is the core of the Popperian perspective, the shortcomings of the traditional conception cannot be overcome. It would be possible (obviously) to restrict the application of traditional principles to finite sets of incompatible hypotheses, to refrain from constructing lawlike sentences from non-universal predicates, to abandon the distinction between observational and theoretical language, and to embrace the conception of permanent properties underlying the intensional construction. The account that would result from these alterations, no doubt, would be more defensible than the traditional account, but it would still be insufficient.

The probabilistic conception of confirmation supported by the logical Bayesian approach, represented by Carnap [1962] and by Hintikka [1966], for example, also encountered at least three principal difficulties, namely:

(a) the application of Bayes' theorem within the context of a logical interpretation presupposes some method of assigning values to the *a priori* probabilities without which no *a posteriori* probabilities may be obtained, where the latitude of variation in the choice of these parameters suggests that their justification is logically arbitrary and empirically vacuous, a difficulty overcome by the inference rule for inductive acceptance, which, of course, requires no *a priori* probabilities at all (although it does depend upon prior degrees of confidence, whose values, however, are readily ascertainable, empirically testable, and comparatively uncontroversial);

(b) to the extent to which this approach envisions continuous progress toward hypotheses and theories that are true with the accumulation of observational and experimental evidence, it appears to rest upon an indefensible presumption, while the very conception of probabilistic appraisals of varied kinds of tests by means of a linear ordering raises problems of its own; but, the acceptance principle under consideration here takes for granted that the evidence available at any specific time will depend upon then-existing technology, changes in which may afford access to previously unexplored data domains leading to refutations of "old" hypotheses and corroborations of "new", where the *wide variety* desideratum fulfills a crucial but informal function;

(c) the logical Bayesian approach, like traditional principles of induction, confronts the extraordinary difficulty of distinguishing between genuinely lawlike and merely accidental generalizations — indeed, if anything, it would appear even more problematical on the logical approach to differentiate between them, where, for example, the work of Burks [1977] and of Uchii [1973] cannot even cope with the difference between logical and nomological truths; while the inference rule recommended above recognizes important differences between the process of *confirming* extensional distributions and the process of *corroborating* intensional generalizations.

The probabilistic conceptions supported by the empirical Bayesians such as Jaynes [1976] and Rosenkrantz [1977] and by the subjective Bayesians such as Savage [1954] and de Finetti [1964], moreover, appear even less appealing:

(a) in attempting to exploit certain relations between "information" and "energy", especially in the form of the second law of thermodynamics, for example, the empirical Bayesian conception tends to take for granted something inductive procedures are presumably required to ascertain, namely: the maximum entropy tendency itself; while the principle of inference elaborated here not only makes no unwarranted empirical assumptions but also does not depend upon the existence of symmetries and averages which do not invariably obtain (although it does, of course, exploit certain mathematical relations, such as the central limit theorem, and depend upon empirical assumptions, in the form of testable conjectures);

(b) within the context of the subjective Bayesian approach, moreover, as de Finetti sought to emphasize, there is no necessary connection, in principle, between personal probabilities and relative frequencies in the actual world W, reflecting the lack of any determinate systematic connection between the principles of Bayesian inference, relevant empirical evidence, and the occurrence of physical phenomena, which promotes the convergence of personal opinion between different investigators as a matter of theoretical necessity, rather than as the occasional outcome of fortuitous conditions; whereas the inference rule recommended above explicitly specifies the evidence relations required for the acceptance and the rejection of alternative hypotheses and theories;

(c) the subjective conception, in particular, violates the principle of empiricism (by permitting the acceptance and the rejection of hypotheses and theories without reliance upon observation and experimentation) and even disregards the objective of inquiry (namely, the discovery of hypotheses and theories of broad scope and systematic power that are applicable for the purposes of explanation and prediction, i.e., laws as properties of the physical world); while the acceptance principle advocated here especially promotes the acceptance and the rejection of hypotheses and theories of broad scope and systematic power on the basis of evidence acquired by means of empirical procedures.

Even apart from the specific problems encountered by alternative versions of Bayesian principles, however, there appear to be general grounds for doubting the suitability of the *probabilistic framework* itself (to which all Bayesian approaches are committed) with respect to *acceptance rules* themselves (to which *not* all Bayesian approaches are committed) in the form of "the lottery paradox" and "conjunctivitis", both of which were identified by Kyburg [1964] and [1970b], respectively. As explained in Chapter 8, the systematic development of a probabilistic account of confirmation might yield a "probabilistic acceptance and rejection rule" such as the following, namely:

Tentative Rule for Inductive Acceptance: when $m(h/e)$ is greater than .5, accept h; when $m(h/e)$ is less than .5, reject h; and when $m(h/e)$ equals .5, h may be accepted, rejected, or left in suspense.

Although Hempel [1962] proposed this rule within the context of an account of epistemic utilities, it will serve the purpose of illustrating certain difficulties encountered by principles of this general kind. This specific rule, for example, does not appear to be completely adequate, insofar as, when combined with (CR-1) and (CR-2), for example, it leads to logical contradictions (as Kyburg [1970b], p. 61, points out), provided there are jointly exhaustive sets of alternative hypotheses $\{h^1, h^2, ..., h^n\}$ consisting of three or more members, where each member of those sets has a probability less

ACCEPTANCE AND REJECTION RULES

than .5, relative to e; for this rule would require rejecting $h^1, h^2, \ldots,$ and h^n, which, in turn, entails accepting $-h^1, -h^2, \ldots,$ and $-h^n$, which contradicts the assumption that $h^1, h^2, \ldots,$ or h^n must be true.

While Kyburg has suggested several possible solutions for problems such as these [some of which involve weakening conditions (CR-1) and (CR-2)], let us consider whether any less drastic measures might suffice instead. As one potential option, for example, this tentative rule might be revised to serve simply as a "probabilistic acceptance rule" such as the following:

Alternate Rule for Inductive Acceptance: when $m(h/e)$ is greater than .5, accept h; when $m(h/e)$ equals .5, h may be accepted or left in suspense.

However promising this maneuver may initially appear, however, it cannot fulfill its intended function; for, since probabilities are additive, i.e., when h and h^* are incompatible, $m(h \vee h^*/e) = m(h/e) + m(h^*/e)$, and sum to one, i.e., if $h, \ldots, h^n$ are exclusive and exhaustive, $m(h \vee \ldots \vee h^n/e) = 1$, if the probability of h, given e, is low, then the probability of $-h$, given e, will be high, necessarily. Consequently, if there are jointly exhaustive sets of three or more alternative hypotheses as before, where each member of those sets has a probability less than .5, relative to e, then the negations of each of those hypotheses must be equal to or greater than .5. This alternative rule therefore requires or permits accepting $-h^1, -h^2, \ldots,$ and $-h^n$, which once again contradicts the assumption that $h^1, h^2, \ldots,$ or h^n must be true. This problem, moreover, cannot be overcome by increasing the acceptance requirement to higher probabilities, say, .9, .95, or .999, for example, as Kyburg has shown; indeed, the generalization of this difficulty is "the lottery paradox" itself.

An illustration of this "paradox" would be a 1000 ticket lottery with a single winning ticket, such that for each of the 1000 individual tickets $T_1, T_2, \ldots, T_{999}, T_{1000}$, the probability that that ticket will win that lottery is exactly .001. Consequently, the probability that the same ticket will *not* win that lottery is exactly .999, leading to the acceptance of the conclusion that ticket n will not win for every one of the 1000 tickets, thereby contradicting the assumption that one of them will win — even with an acceptance requirement of .999! In his analysis of this difficulty, therefore, Kyburg has pointed out what he takes to be an ambiguity in Hempel's formulation of conditions analogous to (CR-1) an (CR-2) above, namely: the option of taking the closure condition as applying to each member of $\mathcal{K}zt$ *individually*, rather than to their collective *conjunction*. He then offers the following recommendation:

Let us take [Hempel's (CR-1)] to mean *not* that every logical consequence of the conjunction of the statements of $\mathscr{X}$ belong to $\mathscr{X}$, but only that every logical consequence of each single element of $\mathscr{X}$ belongs to $\mathscr{X}$. Thus if P and Q belong to $\mathscr{X}$, their conjunction may not belong to $\mathscr{X}$ unless it is included in $\mathscr{X}$ on independent grounds. In conjunction with a probabilistic rule of detachment this interpretation of [Hempel's (CR-1)] is very natural – it is clear that P can be overwhelmingly probable and Q can be overwhelmingly probable without their conjunction being overwhelmingly probable. (Kyburg [1964], p. 307)

Presumably, therefore, in the situation before us, Kyburg recommends that we accept that ticket T_1 will not win, that ticket T_2 will not win, and so forth for all 1000 tickets, and yet not accept that *no* ticket will win – unless that conclusion follows *on independent grounds*!

While this could appear to be a reasonable suggestion under these circumstances, it leads, in effect, to the imposition of restrictions upon the *logical apparatus* of the language $\mathfrak{L}zt$ itself, rather than to specific limitations upon the *conditions of acceptance* for individual hypotheses into the knowledge context $\mathscr{X}zt$, i.e., it entails weakened conditions of consistency or of closure (cf. Kyburg [1970b]). Nevertheless, however arbitrary it might seem to be, no "lottery paradox" is generated by the acceptance of 999 sentences asserting of each of 999 tickets that ticket T_1 will not win, that ticket T_2 will not win, and so forth; nor does any "lottery paradox" arise from accepting the conjunction of any number of these 999 sentences. Once 999 such belief-sentences have been accepted into the knowledge context $\mathscr{X}zt$, however, inconsistencies would arise *either* from accepting the sentence, 'Ticket T_{1000} will not win', *or* from assigning a probability other than *one* to any sentence asserting that ticket T_{1000} will win or a probability other than *zero* – for example, .999 – to any sentence asserting that ticket T_{1000} will lose! When this situation has been clarified, the semblance of paradox tends to disappear; for surely the acceptance of an hypothesis h into a knowledge context $\mathscr{X}zt$ requires consideration of the logical consequences attending its acceptance.

Rather than restricting the principles of deduction, therefore, it seems to be far preferable to delimit the conditions of acceptance, such that an hypothesis h may be accepted into a knowledge context $\mathscr{X}zt$ *only if* its acceptance is logically compatible with the truth of the other members of $\mathscr{X}$. The solution to "the lottery paradox", in other words, does not consist in weakening either the condition of consistency or the condition of closure but in their rigorous enforcement. The implicit conception, therefore, is once again that according to which an individual z at a time t may be in a knowledge context $\mathscr{X}zt$ *only if* the set of belief-sentences Szt representing the set of beliefs $\mathscr{B}zt$ accepted by z at t forms a (minimally) rational set of beliefs, i.e., by satisfying (CR-1) and (CR-2). If this approach provides a resolution for "the lottery paradox", however, it does not also resolve "conjunctivitis", namely: that the probability for a conjunction of two or more hypotheses, $h^1 \& h^2 \& \ldots \& h^n$, on evidence e, will tend to be lower than the probability of any of its singular conjuncts, $h^1, h^2, \ldots, h^n$, on evidence e, as a consequence of the multiplication axiom, i.e., $m(h \& h^*/e) = m(h/e) \cdot m(h^*/h \& e)$. It is this relationship, of course, which entails the difficulty that the degree of confirmation for any universal generalization will always equal *zero* for any consistent, finite set of evidence — a problem overcome by Hintikka within the context of a "slightly more optimistic" conception than Carnap's (as explained in Chapter 8).

Kyburg campaigns against "conjunctivitis" in the form of the principle of conjunction, i.e., the deductive inference from 'p' and 'q' to '$p \& q$', primarily on the strength of the following argument, namely: that the retention of this principle within the context of inductive procedures entails the consequence

... that given a language and a body of evidence, there is essentially just one strongest statement that can be accepted. This approach to induction is global with a vengeance. It suggests that as scientists or even as people we do not induce hypothesis by hypothesis, but that induction consists in principle of inducing at each stage of inquiry — i.e., for each body of evidence e — a single monumentally complex conjunctive statement. (Kyburg [1970b], p. 76; see also Smokler [1971], pp. 332–333)

But it appears as though Kyburg may have become somewhat carried away in the defense of an alternative conception: what, after all could be more rational than to accept a deductive principle violations of which are demonstrably inconsistent? Moreover, when Kyburg claims that this approach is "global with a vengeance", because induction does not then occur "hypothesis by hypothesis", what he must mean is "*simple* hypothesis by *simple* hypothesis"; otherwise, since a single monumentally complex conjunctive statement is an "hypothesis" if there ever is one, he would commit the fallacy of self-refutation. Consequently, it seems reasonable to assume that the issue becomes instead whether or not there are any suitable grounds for alleging that induction does or does not consist of inducing at each stage, i.e., for each body of evidence e, some single hypothesis of perhaps momental complexity.

It is fascinating to observe, therefore, that the conception of scientific procedure as essentially a process of "conjectures and refutations" formalized by the inference rule for inductive acceptance could be viewed as instantiating Kyburg's conditions; indeed, it would not be mistaken to affirm that any method of inquiry in which likelihood principles are applied to lawlike hypotheses with respect to an evidential domain might be entertained as satisfying this desideratum: since the likelihood of h, given e, is equal to the probability of e, given h, so long as (CR-1), (CR-2), (CR-3), and (CR-4) apply, there will ordinarily be only *one* strongest statement that can be accepted. Indeed, if more than one *strongest* statement could be accepted on the basis of inductive principles, then either those principles would violate the condition of consistency (CR-2) or they would fail to fulfill the conception of objectivity, which should not be ignored within the context of normative epistemology. To the extent to which "conjectures and refutations" reflects an intuitively satisfying and theoretically illuminating conceptual foundation for understanding scientific inquiries, therefore, this approach represents a striking counterexample to Kyburg's claims. And if further evidence is called for, it should be remarked that the most crucial difficulties encountered by "probabilistic" principles of acceptance are not problems for "likelihood" principles: unlike probabilities, likelihoods do not obey axioms of addition, multiplication, and summation.

The "orthodox" statistical hypothesis testing procedures Neyman and Pearson [1966] recommend, which Giere [1975], [1976], and [1977] seeks to defend, require additional attention. The principal questions, of course, concern the distinction between *evidential* and *behavioral* interpretations of these principles, where the underlying difference between drawing-an-inference and making-a-decision is not easily defined; nevertheless, it seems evident that Neyman-Pearson procedures, which are based upon error probabilities, are incompatible with likelihood principles, which depend upon the evidence actually acquired. So long as the available alternatives consist of "forced choices" in relation to which subjective evaluations of relative utilities are clearly unavoidable, these procedures possess attractive features; but as principles of inference intended to satisfy the desideratum of "epistemic objectivity", they appear inadequate. Likelihood ratios provide more "evidential meaning" than do error probabilities, which in turn provide more "evidential meaning" than do dichotomous reports (of "significant" or not). In this respect, it seems, likelihood ratios are not only more sensitive to the evidence available, but likelihood principles, applied to large numbers of trials over a wide variety of conditions within the context of severe tests, also satisfy the unbiasedness desideratum, partially relieving the tension between Birnbaum's conception of confidence (Conf) and the likelihood procedures elaborated above.

Giere's distinction between "exploratory" and "confirmatory" inquiry is potentially valuable, nevertheless. It appears perfectly correct to say that scientists have to *"decide"* which hypotheses to subject to test, where differ-

ACCEPTANCE AND REJECTION RULES 263

ences between hypotheses as "roughly right" and as "grossly wrong" do require preliminary detection. Consequently, Neyman-Pearson procedures should fulfill an important role with respect to the preliminary stages of scientific inquiry. So long as science itself aims at establishing general hypotheses and theories of broad scope and systematic power, however, likelihood principles seem to be required. Instead of extending Neyman-Pearson procedures within the "confirmatory" domain, therefore, the suggestion has been made that a deeper division separates these stages in the conduct of inquiry. Thus, Neyman-Pearson procedures appear to be appropriate for "pilot studies" involving small samples within the context of *exploratory* inquiry, while maximum-likelihood principles are appropriate for "extensive research" involving large samples within the context of *explanatory* inquiry, as this chapter has explained. In the sections that remain, therefore, we shall have occasion to finally consider the general character of the theoretical account of scientific procedures that emerges from these investigations, with particular concern for understanding the extent to which scientific knowledge properly qualifies as objective.

10. RATIONALITY AND FALLIBILITY

The conception of scientific procedure elaborated in Chapters 7, 8, and 9, of course, stands in striking contrast to the familiar presumption that science proceeds by way of Observation, Classification, Generalization, Derivation, and Experimentation, which yields probable, if not demonstrable, knowledge. Hempel [1966] has emphasized the indefensibility of this popular account, which he refers to as "the narrow inductivist conception of scientific inquiry", on a number of counts, including, for example, the following:

(a) the imperative, "Observe", unguided by the consideration of hypotheses deserving exploration, does not afford an appropriate point of departure for a scientific investigation, since observation may be carried out indefinitely as an indiscriminate and directionless, but scientifically insignificant, activity;

(b) the injunction, "Classify", similarly, overlooks unlimited possibilities for arranging and rearranging the results of observation by a succession of unilluminating classification schemes, since the explanatory connections which they may implicitly display acquire their importance in relation to hypotheses;

(c) the directive, "Generalize", moreover, cannot be adequately construed as a mechanical procedure of deducing or deriving hypotheses and theories from empirical evidence; for, as Hempel remarks, "Scientific hypotheses and theories are not *derived* from observed facts, but *invented* in order to account for them" (Hempel [1966], p. 15).

A more adequate alternative to this popular conception, therefore, would be to entertain its initial stages as those of Hypothesis-Formation, Derivation, and Experimentation, i.e., as a process of "conjectures and refutations", requiring ingenuity and imagination in the development and construction of hypotheses and theories, more or less along the lines developed in Chapter 7.

An account of scientific inquiry bearing comparison to this conception has been advanced by none other than Neyman [1977], who suggests an analysis involving three distinct stages, in particular:

> (i) Empirical establishment of apparently stable long-run relative frequencies (or frequencies' for short) of events judged interesting, as they develop in nature.
> (ii) Guessing and then verifying the 'chance mechanism', the repeated operation of which produces the observed frequencies. This is a problem of 'frequentist probability

theory'. Occasionally, this step is labeled 'model building'. Naturally, the guessed chance mechanism is hypothetical.

(iii) Using the hypothetical chance mechanism of the phenomenon studied to *deduce* rules of adjusting our actions (or 'decisions') to the observations so as to ensure the highest 'measure' of 'success'. (Neyman [1977], p. 99)

In certain respects, of course, this conception harmonizes well with, say, "the broad deductivist conception of scientific inquiry", especially in its acknowledgement of the essential function fulfilled by hypothetical conjectures, which represents an important advance over the popular conception. In other respects, however, Neyman's analysis appears to require further refinement: the idea of "chance mechanisms" whose repeated operation produces observed frequencies, for example, invites analysis by means of 'propensity probability theory', in which "hypothetical" is unambiguously interpreted as meaning theoretical and conjectural rather than as, perhaps, purely imaginary and instrumentalistic. The relationship between *drawing-inferences* and *making-decisions* which Neyman endorses, i.e., his conception of "inductive behavior", moreover, does not necessarily reflect the most suitable construction. In order to promote more valuable comparisons, therefore, let us consider the general conception of scientific procedure introduced in Chapters 7, 8, and 9, in somewhat greater detail.

SCIENTIFIC RATIONALITY

The conception of scientific procedure recommended here is based upon the assumption that the program of science is one of providing for the acceptance, rejection, and modification of hypotheses and theories of broad scope and systematic power which may be employed for the purposes of explanation and prediction. Since scientific explanations presuppose general premises in the form of lawlike sentences, scientific procedure cannot dispense with inductive acceptance rules; indeed, since scientific explanations are adequate only when they are true, these rules must support inductive acceptance as *true*, i.e., as descriptive of the actual world, W, rather than as *useful*, i.e., as instrumental for dealing with the world (in van Fraassen's sense of "empirically adequate", for example, discussed in Chapter 6). Since lawlike sentences are intensional generalizations imposing constraints upon the histories of physically possible worlds, W, moreover, they must be distinguished *ontologically* from (merely) extensional distributions which do not constrain the histories of other possible worlds, where the former, but not the latter, attribute permanent dispositional properties, χ, to every member of appropriately specified reference classes, K. And lawlike sentences must be further distinguished *epistemically* from (merely) extensional distributions insofar as generalizations which are not both lawlike and true are theoretically vulnerable to the eventual discovery of incompatible evidence — where the class of quasi-accidental (or, quasi-lawlike) sentences as extensional implications of lawlike generalizations poses the crucial case that "proves the rule" (as explained in Chapter 7).

The distinction between "exploratory" and "explanatory" inquiry introduced in Chapter 9 corresponds in a partial and approximate fashion to the difference between the "context of discovery" and the "context of justification" as it has come to be known, that is: the difference between formulating an hypothesis or a theory *as a potential object of inquiry*, on the one hand, and accepting that hypothesis or theory *as a warranted object of belief*, on the other. The process of discovery, of course, is essentially psychological, while that of justification is essentially logical. As Hempel [1966], p. 16, has observed,

> In his endeavor to find a solution to his problem, the scientist may give free rein to his imagination, and the course of his creative thinking may be influenced even by scientifically questionable notions. Kepler's study of planetary motion, for example, was inspired by his interest in a mystical doctrine about numbers and a passion to demonstrate the music of the spheres. Yet, scientific objectivity is safeguarded by the principle that while hypotheses and theories may be freely invented and *proposed* in science, they can be *accepted* into the body of scientific knowledge only if they pass critical scrutiny, which included in particular the checking of suitable test implications by careful observation or experiment.

Thus, the function of Neyman-Pearson procedures within the context of Stage I inquiries falls within the context of discovery, in general, because the results of these investigations are *not* supposed to be employed for drawing any conclusions concerning what is true and what is false; instead, their role is to provide guidance in making some decisions concerning potentially promising directions for further investigation. The function of likelihood principles within the context of Stage II inquiries falls within the context of justification, by contrast, because the results of these investigations *are* to be employed for drawing conclusions concering what is true and what is false: the purpose of "explanatory" inquiry, unlike that of "exploratory" inquiry, *is* to promote the acceptance of hypotheses and theories as objects of belief.

In terms of the analysis of scientific procedures presented in previous chapters, therefore, the division of inquiry into Stage I (or, "exploration") and Stage II (or, "explanation") appears to provide an important but partial account of the elements of induction; in particular, it seems clear that the initial phases of scientific investigations involve hypothesis-formation, *per se*, as well as "pilot studies", since the application of Neyman-Pearson methods presupposes the specification of well-formed hypothetical alternatives, such as h and h^*, for designated reference classes R and attributes A. This first level of inquiry might best be envisioned as one of Discovery and Exploration rather than merely as one of Exploration; for the creative and imaginative efforts required for the formation of hypotheses and theories deserves explicit recognition. The second level of inquiry, of course, likewise represents two distinct phases of empirical analysis, insofar as the establishment of 'stable relative frequencies' (as Neyman would describe them), as extensional distributions, involves the application of the inference rule for inductive acceptance to large numbers of trials over a wide variety of conditions; while the establishment of 'single-case propensities', by contrast, as intensional generalizations, involves the application of the inference rule for inductive acceptance to large numbers of trials over a wide variety of conditions *within the context of severe (or "critical") tests* — as Confirmatory and as Corroboratory modes of investigation, respectively. Moreover, although these stages of inquiry appear to present a more complete and less inadequate conception of scientific procedures in general, they nevertheless still omit a third level of inquiry, namely: Stage III (or, let us say, the "application" stage), which draws upon the contents of the knowledge context $\mathcal{K}zt$ for purposes of Explanation-and-Prediction, on the one hand, and of Practical Decision-Making, on the other.

There thus appear to be at least thee distinct stages of scientific inquiry; specifically, Stage I (or, "exploration"), Stage II (or, "explanation") and Stage III (or, "application"), each of which consists of at least two different phases, including *Discovery* and *Exploration*, *Confirmation* and *Corroboration*, and *Explanation-and-Prediction* and *Practical Decision-Making*. Indeed, on the reasonable assumption that practical decision-making under these conditions is dominated by the principle of maximizing expected utility, each such stage reflects the prevalence of distinctive principles of procedure, namely: Neyman-Pearson procedures, maximum likelihood principles, and maximizing expected utility strategies, respectively. The position offered here, therefore, supports the conclusion that all of these approaches have been successful in identifying important, if partial, aspects of scientific practice, where the comprehensive characterization of this multi-faceted phenomenon requires their integration and refinement. Scientific inquiries generally may be entertained accordingly as falling (more or less) into three broad stages:

SCIENTIFIC INQUIRIES

	Stage		Objective	Methods
I.	*The Exploration Stage:*			
	Phase 1:	*Discovery*	Hypothesis-Formation	Imagination and Ingenuity
	Phase 2:	*Exploration*	Pilot Studies	Neyman-Pearson Procedures
II.	*The Explanation Stage:*			
	Phase 3:	*Confirmation*	Extensive Research	Maximum Likelihood Principles
	Phase 4:	*Corroboration*	Inference to Laws	Severe Tests and Critical Attitudes
III.	*The Application Stage:*			
	Phase 5:	*Explanation-and-Prediction*	Understanding the World	Deductive Inference
	Phase 6:	*Practical Decision-Making*	Influencing History	Maximizing Expected Utility

The Three Broad Stages of Scientific Investigations

The relations between these three stages of inquiry no doubt warrant further elaboration. Insofar as Phase 1 and Phase 2 are both concerned with the formulation of hypotheses or theories as potential objects of inquiry through creative hypothesis-formation and preliminary pilot studies, it would not be seriously misleading to envision Stage I inquiries as falling within the context of discovery, so long as Neyman-Pearson procedures are being employed as *principles of decision* rather than as *principles of inference*. The subjective ingredients that invariably occur as ineliminable features of Phases 1, 2, and 6, moreover, tend to differentiate those aspects of inquiry from other phases, although it is important to recognize that the attitudinal difference distinguishing Phase 4 from Phase 3 cannot be completely formalized. Since Stage II is referred to as "The Explanation Stage" while Phase 5 is referred to as the "Explanation-and-Prediction" phase, perhaps it should be emphasized that Stage II represents inductive inference — and corroborative inference, especially — as "inference to the best explanation" (or, as Harman [1970], p. 89, suggests, as "inference to the best of competing explanatory statements"); Phase 5, however, represents the application of explanatory statements *which have been accepted into the knowledge context* $\mathscr{K}zt$ to the occurrence of singular events (in relation to lawlike sentences) and to the occurrence of quasi-lawlike or lawlike phenomena (in relation to general theories), as described in Chapters 5 and 6. Stage II reflects the context of justification employing likelihood principles, while Phase 5 is a deductive phase in which probabilistic principles apply (in concert with maximal specificity considerations) — although, as reflection upon "the lottery paradox" has displayed in Chapter 9, predictive inferences involving events that occur only with some probability may dictate arbitrary choices between competing hypotheses, which undermines the desideratum of objectivity.

The conception that gains support from these arguments at last, therefore, is that probabilities and likelihoods fulfill radically different functions in "the inferential scheme of things", where *probabilities* dominate "direct inference", i.e., reasoning from populations to samples, from propensities to frequencies, or from intensional generalizations to extensional distributions (as aspects of Phase 5 inquiries), while *likelihoods* dominate "inverse inference", i.e., reasoning from samples to populations, from frequencies to propensities, or from extensional distributions to intensional generalizations (as elements of Stage II inquiries). Moreover, the principles of "direct inference", thus construed, appear to satisfy the *special consequence* condition, while those of "inverse inference" satisfy the *converse consequence* condition; in other words, any evidence which 'confirms' a probabilistic inference h also 'confirms' any consequence of h ("the special consequence condition"), whereas any evidence which 'confirms' a likelihood inference h also 'confirms' any hypothesis that entails h ("the converse consequence condition"). These conditions cannot be satisfied simultaneously, as Hempel [1965], pp. 32–33, has reported, since otherwise evidence which 'confirms' an hypothesis h, such as, "Ravens exist", would also 'confirm' the conjunction of h with any other statement h^*, such as Hooke's law (by the converse consequence condition), with the result that the observation of a raven would 'confirm' Hooke's law (by the special consequence condition). Hempel therefore abandons the converse consequence condition; but here it has to be retained, for when a theory h^* entails a lawlike hypothesis h, which in turn explains a phenomenon e, then the likelihood of h^*, given e, cannot be less than the likelihood of h, given e. Since likelihood principles thus satisfy the converse consequence condition, it is important to understand how this debilitating consequence is supposed to be avoided.

Indeed, this difficulty would appear to be acute within the context of the present construction, since conditions (CR-1) and (CR-2) have the effect of guaranteeing that any consequence of any belief accepted into $\mathcal{H}zt$ must be accepted into $\mathcal{H}zt$ as well, strongly suggesting that the special consequence condition cannot easily be avoided. The observation of a sample of sodium s which burned yellow in a Bunsen flame, for example, might appropriately 'confirm' the hypothesis h, "All samples of sodium burn yellow in a Bunsen flame", but it would be difficult to imagine that it would also 'confirm' the hypothesis h^*, "The absolute is perfect", even though hypothesis h^* is a consequence of the conjunction of h and h^*, "All samples of sodium burn yellow in a Bunsen flame and the absolute is perfect". But an hypothesis such as h^* would not receive empirical support from the observation of a sample of sodium which burned yellow in a Bunsen flame, since that hypothesis is not thereby subjected to *any* number of tests under *any* variety of conditions; consequently, the only hypothesis whose acceptance could be warranted would be h itself — provided of course that the evidence available satisfied the conditions specified by the inference rule for inductive acceptance. Notice, in particular, moreover, that since the likelihood $L(h/e)$ of an hypothesis h relative to evidence e is equal to the nomic expectability $E(e/h)$ of e as an explanandum with h as its explanans (as Chapter 8 has explained), both the requirement of strict maximal specificity (RSMS) and the conception of scientific theories introduced in Chapters 5 and 6 should be invoked to establish the appropriate relationship between a particular hypothesis h and specified evidence e. Explanatory relevance and irrelevance relations, therefore, provide the foundation for determining evidential relevance and irrelevance relations, respectively, in relation to lawlike hypotheses and theories h within the context of this intensional explication, thereby reinforcing their significance.

The combination of a "critical attitude" with likelihood principles thus forestalls the criticism which Hempel directs against the converse consequence condition without abandoning that conception; indeed, it seems plain that the utilization of likelihood principles, in general, entails a commitment to some version of that particular condition. The clearest formulation of this connection, moreover, may be that which Smokler [1968], p. 310, has advanced, namely:

> If a theory explains an evidential statement (and correspondingly is confirmed by that evidential statement), then in turn any second theory from which the first is logically derivable is also confirmed by that evidential statement. But the special-consequence condition is not satisfied by this notion. For a theory may be confirmed by evidence and yet the logical consequence of that theory not be confirmed by the evidence. To claim that, in the latter case, the evidence is indirect evidence for the hypothesis does not, I think, provide us with a lucid picture of the situation.

Smokler refers to this approach as "abductive inference", in which evidence is taken to 'confirm' an hypothesis if that hypothesis (together with a specification of initial conditions and auxiliary hypotheses) "explains" that evidence; and he also suggests that confirmation functions satisfying the special consequence condition tend to be probabilistic, while those that are non-probabilistic tend to satisfy the converse consequence condition (Smokler [1968], p. 312). For reasons such as these, therefore, it appears to be essential to distinguish probabilistic principles from likelihood principles with respect to their range of intended application; for the differences involved here extend beyond issues of explanation and prediction to those of inference and decision, encompassing the utilization of likelihood principles for inductive inference within Stage II and of probabilistic principles for deductive inference within Stage III.

Strictly speaking, however, the definition of an "inductive argument" is applicable to *statistical predictions* as well as to *likelihood inferences* even though they happen to rely upon probabilistic principles; for predicandum-phenomena, unlike explanandum-phenomena, are not presumed *known to have occurred within the knowledge context* $\mathcal{K}zt$, i.e., the description of an explanandum-outcome (but not of a predicandum-outcome) must be entailed by $\mathcal{K}zt$ for its explanans (but not for its predicans) to be "epistemically adequate" (in relation to what is *taken* to be the case), as opposed to 'ontically adequate" (in relation to what *is* the case). The criteria of adequacy for nomically significant scientific explanations advanced in Chapters 5 and 6, in particular, require that the sentences constituting the explanation – both the explanans and its explanandum – must be true, relative to the language framework, $\mathcal{L}zt$; in application to our *beliefs* about the world, therefore, the sentences constituting the explanation – both the explanans and its explanandum – must be *believed* to be true, relative to the knowledge context $\mathcal{K}zt$. To the extent to which a prediction is intended to provide grounds for anticipating the occurrence of the event described by its predicandum, it would be self-defeating, if not absurd, to impose the requirement that *predicandum-sentences* must be entailed by $\mathcal{K}zt$ for their predicans to be "epistemically adequate", since that would represent the claim that arguments intended to provide grounds for anticipating the occurrence of an event are adequate for their purpose *only if* they are completely dispensable in principle – because their predicandum-phenomena must then not only already have occurred but already be *known* to have occurred within the knowledge context $\mathcal{K}zt$. Since a predicandum may be false even when its predicans happens to be true in probabilistic inferences, application of the results of Stage II inquiries within Phase 5 encounters the "inductive uncertainty" that lottery problems display.

These considerations thus reinforce the observation that features which qualify as inadequacies with respect to explanations are not necessarily also inadequacies with respect to predictions, as Chapters 4, 5, and 8 have previously explained. But it should not be overlooked that likelihood principles employed in Stage II inquiries avoid altogether those "inductive infirmities" which arise from the simultaneous satisfaction of the special consequence and the converse consequence conditions; for, although (CR-1) and (CR-2) require that any deductive consequence of any belief *already accepted into* $\mathcal{K}zt$ must also be accepted into $\mathcal{K}zt$, they do not therefore serve an inductive function analogous to that fulfilled by the special consequence condition. Inductive inference within the context of Stage II inquiries takes place in accordance with the procedures of confirmation and of corroboration supported by the inference rule for inductive acceptance, which do not satisfy the special consequence condition at all. Thus, (CR-1) and (CR-2) are applicable to the accepted contents of $\mathcal{K}zt$ and dictate the acceptance of logical truths and of other analytical sentences relative to the language framework $\mathcal{L}zt$, but they do not, in general, determine the *acceptability* of empirical propositions (other than prohibiting the acceptance of incompatible beliefs, of course). So long as a sentence such as, "The absolute is perfect", is not a necessary truth, therefore, its acceptance is not supported by (CR-1) and (CR-2) merely because it happens to be conjoined with some testable, well-tested, or otherwise acceptable hypothesis, even though the inference rule for inductive acceptance satisfies the converse consequence condition; and, if such a sentence were a necessary truth, its acceptance would be supported by (CR-1) and (CR-2), whether or not the inference rule for inductive acceptance satisfied such a condition. This objection thus poses no insurmountable obstacles for this interpretation.

Comparisons with alternative conceptions are even more instructive. For to the extent to which Bayesians, such as Jeffrey [1970], tend to eschew "acceptance and rejection", they appear to be unable to sustain the construction of scientific inquiries as providing for the acceptance and the rejection of hypotheses and theories which may be employed for the purpose of explanation; and to the extent to which "orthodox" statisticians, including Neyman [1977], embrace "principles of decision" in lieu of "principles of inference", their approach likewise promotes the practical objective of influencing the course of events in preference to the theoretical task of understanding the world's structure. This difference in attitude, moreover, is exhibited by Jeffrey's remarks on the relationship between (Bayesian) theory and practice, that is:

It interests me as a way of making sense of scientific methodology within a framework which serves for the analysis of decision-making under conditions of uncertainty. Indeed, like most Bayesians today, I came to it from the latter, 'practical', direction. The great challenge which I think Bayesianism more likely than Popperianism to meet is that of relating theory to practice — explaining why and how scientific theories can be expected to improve the quality of decision-making. (Jeffrey [1975], p. 111)

It is striking to realize, therefore, that Neyman would endorse precisely the same conception, with appropriate substitutions, *mutatis mutandis*. Their emphasis upon Phases 1, 2, and 6 (with "Bayesian" rather than "Neyman-Pearson" procedures as appropriate) and de-emphasis upon Phases 3, 4, and 5 (with respect to explanation) strongly suggest that at least *two* distinct conceptions of "scientific rationality" are involved here; for, Neyman and Jeffrey appear to be concerned with (let us say) *applied* scientific rationality, while Popper is preoccupied instead with *pure* scientific rationality.

The distinction between "pure" and "applied" scientific rationality, of course, is intended to be analytical rather than evaluative: which would be of greater utility obviously depends upon pragmatical conditions. But it is enormously important to discover that the principal competitor with Jeffrey's Bayesian analysis of the relation between theory and practice is not Popper's conception of "conjectures and refutations" but Neyman's conception of "inductive behavior", i.e., alternative "principles of decision" rather than alternative "principles of inference"! Indeed, from this point of view, the subjective features of Neyman-Pearson procedures, on the one hand, and of Bayesian personalism, on the other, no longer appear to pose serious objections to their utilization as "principles of decision". For, so long as these procedures are employed for *deciding upon courses of action* instead of *drawing conclusions concerning what is true*, their inadequacies with respect to Stage II inquiries assume only secondary importance; conversely, moreover, if intended to function as "principles of inference" rather than as "principles of decision", these same inadequacies assume decisive significance. Neither Bayesian nor "orthodox" approaches clarify and illuminate the essential properties of scientific rationality, i.e., in relation to Stage II inquiries. Consequently, although Hacking [1973] is correct as far as he goes in associating Birnbaum's confidence principle with *predictions* and principles of likelihood with *explanations* as related in Chapter 9, these considerations reinforce the conclusion that these differences also underlie the distinction between *inference* and *decision*, which, we have found, is by no means simple to unravel. In relation to Stage II inquiries, however, the Popperian approach to pure scientific rationality appears to have no serious rivals.

PERSONAL PROBABILITIES

In his important investigation of philosophical problems of statistical inference, Seidenfeld [1979], pp. 7–10, adopts five "conditions of adequacy" for characterizing (what he refers to as) an *inductive logic*, most of which are familiar from Chapter 8, namely:

(CA-1) the *coherence* condition, i.e., 'confirmation functions' must obey the mathematical properties of the calculus of probability;

(CA-2) the *conditionalization* condition, i.e., that changes in measures of evidential support occur in conformity with Bayes' theorem;

(CA-3) the *total evidence* condition, i.e., that the confirmation of any hypothesis must be based upon all the available relevant evidence;

(CA-4) the *direct inference* condition, i.e., measures of evidential support should equal physical probabilities when those happen to be known;

(CA-5) the *objectivity* condition, i.e., there exists exactly one correct confirmation function, which remains invariant over persons and over times.

It is interesting to observe, therefore, that the conception of scientific inquiry recommended here violates both (CA-1), i.e., the thesis of a probabilistic framework, and (CA-2), i.e., the thesis of conditionalization, while satisfying (CA-3), i.e., in the form of (CR-3) and (CR-4), (CA-4), i.e., within the context of Phase 5, and (CA-5), i.e., as discussed in Chapter 8. Because this account does not satisfy (CA-1) and (CA-2), it does not qualify as an *inductive logic*, even though it appears to provide a plausible interpretation of the logical structure of scientific inference.

The significance of this circumstance, of course, depends upon the function these "conditions of adequacy" are intended to fulfill. If they are supposed to serve as essential requirements that *any* acceptable construction of statistical inference and scientific procedure *must* invariably satisfy, then "inductive logic" is intended in a broad sense and (CA-1) through (CA-5) represent important claims. If they are supposed to serve as definitional conditions for particular constructions of statistical inference and scientific procedure to qualify *as accounts of a certain special variety*, however, then "inductive logic" is intended in a narrow sense and (CA-1) through (CA-5) do not represent important claims — unless there are independent grounds supporting each of these conditions, in which case, presumably, they are supposed to be perfectly general requirements, after all. To the extent to which (CA-1) through (CA-5) exclude *non-Bayesian* accounts of statistical inference and of scientific procedure, therefore, they appear to beg-the-question; and to the extent to which *non-Bayesian* accounts of statistical inference and scientific procedure are determined to be defensible, these requirements also appear to be unwarranted. Consequently, these "conditions of adequacy" are either (a) perfectly general requirements imposing important but unwarranted constraints upon alternative conceptions or (b) definitional conditions defining accounts of a certain special variety known as "inductive logics". And if these conditions are interpreted as analytically true of any "inductive logic", then the problem becomes whether or not statistical inference and scientific procedure fall within the domain of *inductive logic* thus defined. If the arguments presented in Chapters 7, 8, 9, and 10, are sound, therefore, the answer appears to be, "No"; but the appropriate standard is the extent to which alternative constructs clarify and illuminate the object of inquiry, which is science itself. [It is intriguing to observe, moreover, that Bayesian interpretations have not met with great success in satisfying (CA-5), as explained in Chapter 8; indeed, Seidenfeld himself has recently abandoned this particular condition.]

Perhaps it should be emphasized, therefore, that although Popperians reject (CA-1), the *coherence* condition, as an adequacy condition for inductive principles, they do not therefore permit incoherent *belief-contexts* to qualify as "rational"; (CR-1) and (CR-2), on the contrary, require probabilistic beliefs, like other beliefs, to be logically consistent and deductively closed within any *knowledge context* $\mathcal{K}zt$. The condition of coherence within the Bayesian scheme of things, however, emanates instead from the Bayesian *static* assumption, namely: the requirement that, for any person z, there exists a set of propositions A (forming a Boolean algebra) such that, for any hypothesis h belonging to A and any fact f known to z included in A, a prior probability function, $P_f(h)$, is defined and satisfies the probability axioms. The axioms, as we all know, dictate that the probabilities of mutually exclusive and jointly exhaustive alternative hypotheses must sum to one; and that, so long as hypotheses h^i and h^j are incompatible, their probabilities must be additive. These conditions are conveniently illustrated by games of chance and by physical set-ups, such as drawing a card from a standard deck or firing a rifle at a target under well-controlled conditions, in which their applicability (apart from difficult issues of determinism) appears to be practically impossible to question. But while the consistency of a set of beliefs, such as f, accepted by z at t, must conform to the principles of probability as a necessary condition *with respect to beliefs concerning the probabilities of events*, it would be unwarranted and question-begging to take for granted that similar constraints obtain *with respect to measures of support for hypotheses as such* — apart from those direct inference contexts within Phase 5 in which they necessarily coincide. What Popperians deny is that these same constraints apply to inverse inference contexts within Stage II!

To the extent to which Seidenfeld's "conditions of adequacy" (CA-1) to (CA-4) appear perfectly reasonable when applied to the prediction of singular events (whether or not accompanied by a "probabilistic acceptance rule" such as those discussed in Chapter 9), moreover, it seems plausible to suppose that the Bayesian conception of rationality represents an unwarranted generalization of conditions appropriate for *direct inference* to *inverse inference* as well in pursuit of a unified theory. This position is reinforced not only by the inadequacy of Bayesian accounts to contend with the context of justification reflected by Stage II inquiries, but also by the implausibility of probabilistic principles in application to *general hypotheses* as objects of inquiry, which the following questions tend to reflect, that is:

(i) since there will always be an *infinite number* of logically possible mutually exclusive hypotheses $h^1, h^2, \ldots$, on any given finite evidence e, is it *reasonable* that their measures of evidential support *should sum to one*?;

(ii) since there will always be an infinite number of logically possible mutually exclusive hypotheses $h^2, h^3, \ldots$, *arbitrarily similar* to any other h^1, is it *reasonable* that measures of evidential support *should be additive*?;

(iii) since the available relevant evidence e supporting an hypothesis h does not change merely because that hypothesis has "qualified for acceptance", it is *reasonable* that h's measure of evidential support *should change to one*?

The answers to questions such as these, I believe, confirm my conclusion that *the Bayesian conception of rationality should no longer be taken for granted*. Nevertheless, it may serve a useful purpose to compare the Bayesian notion of "degrees of belief" or of "strengths of conviction" (in which object-language and meta-language principles appear to be conflated) with its object-language counterpart of "probabilistic beliefs" as elements of a knowledge context $\mathscr{K}zt$.

In relation to the discussion of Chapter 1, the difference involved here approximately corresponds to the distinction between "personal probabilities" and "epistemic probabilities" (assuming (CR-1) and (CR-2) are both satisfied) as follows:

(b) for any sentence S in $\mathfrak{L}zt$, S describes a *personal probability* for z at t, given facts f, if and only if z accepts f at t and it is the case that z at t believes that S is probable;

which has been represented by '$P_f(S)$', i.e., z's probability for S, given f; while "epistemic probabilities", by contrast, are completely objective in also assuming the specification of a set of "rules of inference" which properly apply within the language framework $\mathcal{L}zt$ to produce the knowledge context $\mathcal{K}zt$ as follows:

(b*) for any sentence S in $\mathcal{L}zt$, S describes an *epistemic probability* for z at t, relative to evidence $\{E\}$ and $\mathcal{K}zt$, if and only if z accepts $\{E\}$ at t and it is the case that $\{E\}$-and-$\mathcal{K}zt$ entails or implies S is probable;

where, in the case of predictive inference involving the occurrence of a singular event which S describes, the evidence $\{E\}$ should contain information concerning the *kind of trial* (or the "initial conditions") of which that event is supposed to be an instance (to the best of our knowledge). Thus, if it is the case that $\{E\}$-and-$\mathcal{K}zt$ entails or implies S is probable in relation to accepted lawlike sentences or extensional distributions (by satisfying appropriate maximal specificity conditions), then S will describe an epistemic probability for z at t; moreover, if this prediction is warranted by an argument from premises including lawlike sentences, then this inference will possess potential explanatory significance (provided the corresponding conditions of adequacy are subsequently fulfilled within the knowledge context $\mathcal{K}zt$), but otherwise will not. And there appears to be little reason to doubt that "probabilistic acceptance rules" are provisionally applicable here in accordance with (CR-3) and (CR-4).

Once again, however, the distinction between pure scientific rationality and applied scientific rationality is relevant; for, although the application of "probabilistic acceptance rules" appears to be appropriate within the context of Phase 5 as an aspect of *pure* scientific rationality for understanding the world, their application within the context of Phase 6 as an aspect of *applied* scientific rationality in influencing events is not obviously required. Decision-making in accordance with the strategy of maximizing expected utility, for example, presupposes "probabilities of hypotheses" but not their "acceptance" and "rejection", a difference further substantiating the importance of the distinction between these alternative conceptions of "scientific rationality". The definition of "epistemic probability", moreover, clarifies the relationship between *probabilistic beliefs* within the knowledge context $\mathcal{K}zt$, on the one hand, and *physical probabilities* (as frequencies or propensities), on the other; for the requirement formulated by Seidenfeld's condition (CA-4) is intended to ensure that subjective expectations should equal objective expectations, to the best of our knowledge. Just as Carnap supposed that epistemic probabilities (or, "probability$_1$") should provide 'the best estimate' of ontic probabilities (or, "probabilitity$_2$"), which in turn, represent "rational betting quotients" for corresponding singular events, it is tempting to conclude that personal probabilities should be entertained as estimates of single-case propensities (Mellor [1971]), which is certainly not unreasonable so long as the corresponding single-case propensities happen to exist (compare Salmon [1979a]). The position to be developed here, however, suggests that personal probabilities are themselves a distinct species of single-case propensity and that single-case propensities as personal probabilities may exist even if the world is otherwise deterministic, which we shall now consider.

With reference to formulae (I) and (III) presented in Chapter 8, therefore, those schemata should be entertained as elliptical representations of *sentential functions* attributing appropriate "degrees of belief" (or, alternatively, "strengths of conviction") to particular person-at-times. Formula (I), for example, tacitly represents

(I*) $[P_f(h) = r]_{zt}$,

which is an instance of the general form, 'φzt', where 'φ' is '$P_f(h) = r$'; while formula (III) analogously represents,

(III*) $(Et^*)[P_{f \cup \{e\}}(h) = r]_{zt^*} \supset (Et)[P_f(h/e) = r]_{zt}$,

which is an instance of the form, '$(Et^*)\varphi zt^* \supset (Et) \psi zt$'. Although the property of coherence (or its absence) will be a characteristic of individual z's "state of mind", let us say, *at a specific time t*, therefore, conditionalization (or its absence) will characterize a certain relationship between such an individual's "state of mind" *on two or more occasions*, $t^1, t^2, \ldots$; for example, "before" and "after" the acquisition of some experience or observation e, where (presumably) whatever causes these changes qualifies as 'evidence'. Although Bayesian accounts sometimes seem to suggest that all of us — or, perhaps, all who are not either babes-in-arms, brain-damaged, or mentally retarded — are *expected* to fulfill these conditions, they are properly understood as normative requirements, like consistency and closure, which are seldom if ever completely satisfied, as explained in Chapter 8.

The important point, of course, is not that these functions are false if construed as descriptive generalizations, but that sentences such as these attribute particular *psychological properties* to persons-at-times, e.g., having the degree of belief r that the sentence 'H' is true (or, having a strength of conviction r that proposition h is the case). In view of their sentential (or propositional) objects, moreover, there are substantial grounds for conceiving these psychological properties as *propositional attitudes* of a special variety; in particular, they appear to satisfy the criteria Fodor has proposed, namely:

(i) degree-of-belief sentences may be analyzed as relations between people and specific sentences (or particular propositions);

(ii) the sort of thing people have degrees-of-belief in are the same sort of thing they can be said to affirm, to assert, to deny, etc.;

(iii) the objects of degrees-of-belief are referentially opaque; that is, they are not interchangeable *salva veritate* even when they are equivalent;

(iv) the objects of degrees-of-belief themselves have logical form, i.e., they are non-arbitrarily related as *tokens* of designated *types;* and,

(v) degrees-of-belief as propositional attitudes reflect familiar mental states of uncertainty, of skepticism, of doubt, etc. (Fodor [1978], pp. 502—509)

Even though "degrees of belief" are not also *true if and only if their objects of belief are true*, therefore, they clearly qualify as a particular variety of propositional attitude, nevertheless. (Fodor [1978], pp. 506—507) [See also Jeffrey [1965], pp. 48—59 and pp. 145—152; and Mellor [1971], esp. pp. 12—16. Although he interprets "partial beliefs" as propositional attitudes and dispositional in character, Mellor does not imply that they are themselves a particular species of propensity (or, perhaps, of "chance", in his sense).]

If *degrees-of-belief* are propositional attitudes like *beliefs*, there are grounds for inferring that they must fulfill an analogous role with respect to *behavior*; for persons who *believe* a sentence '*H*' is true (or who believe that a proposition *h* is the case) are influenced by that belief to *act as though it were true under appropriately specified conditions* (encompassing motivational, circumstantial, and auxiliary-belief variables). Similarly, persons having a certain *degree-of-belief* that sentence '*H*' is true (or who tend to believe that proposition *h* is the case) are thereby influenced to *tend to act as though it were true under appropriately specified conditions* (encompassing motivational, circumstantial, and auxiliary-belief parameters, once again). The probabilistic structure of these propositional attitudes thus supports the possibility that *degrees-of-belief are psychological propensities*, i.e., dispositional tendencies of statistical strength to exhibit specific kinds of behavior under relevant test conditions (specifying motivational, circumstantial, and auxiliary, belief variables), which encompass "betting behavior" among their single-case displays. [Mellor, by contrast, appears preoccupied with "betting behavior":

In the case of partial belief, moreover, it is a great merit of personalist theory that it is able to prescribe a single kind of action to display dispositional belief in events of every kind, e.g. choosing odds for a bet on its occurrence. The homogeneity that can thus be imposed on the most diverse partial beliefs indeed justifies treating them as a single family of attitudes. It is easier with partial belief than with belief itself to specify what can count as a display of the alleged disposition and the dispositional account of partial belief is accordingly the less contested.

(Mellor [1971], p. 15) He seems not even to consider the possibility of alternative conceptions, such as the one being proposed (cf. Skyrms [   ] also).]

The general conception under consideration here, therefore, is that betting behavior *per se* is by no means the only manifestation of these probabilistic properties, which will also be displayed by the frequencies with which individuals exhibit appropriate responses over sequences of relevant trials. Tversky [1974], for example, has described several situations which appear to satisfy this conception. The first involves "probability learning", namely:

...a subject had to predict on each trial which of two lights will turn on. In a typical experiment, the light on the right is lit on 2/3 of the trials and the light on the left is lit on the remaining 1/3 of the trials. The subject is instructed to make as many correct predictions as he can and is rewarded for his correct predictions. Contrary to the optimal policy of predicting the more frequent light on every trial, people usually probability-match, that is, predict the right on 2/3 of the trials and the left on 1/3 of the trials.

Since it could be maintained that this situation is "excessively artificial" as an indication of behavior under *serious* conditions, Tversky also reports a second situation involving "more realistic" circumstances, specifically:

Fighter pilots in the Pacific during World War Two encountered situations requiring incendiary shells about 1/3 of the time and armour-piercing shells about 2/3 of the time. Since there was no general procedure for predicting on every mission which type of shells would be required, the optimal policy was clearly to use armour-piercing shells on every mission. It was observed, however, that when left to their own devices, pilots armed themselves with incendiary and armour-piercing shells in the proportion of 1 to 2.

Since in each specific trial one or the other outcome would occur, it seems reasonable to assume that the individuals under consideration had acquired "degrees of belief" in the propositions *that the left light would come on, that incendiary shells would be required*, and so on, which were exhibited by the relative frequencies with which corresponding behavior was displayed.

If "degrees of belief" and "strength of conviction" designate single-case propensities, of course, then they should certainly satisfy the *definitional conditions* for dispositional predicates advanced in Chapter 2. Provided the notion of "an actual physical state" is construed sufficiently broadly to encompass *psychological* properties (as Maxwell [1970], p. 189, for example, suggests), there would appear to be suitable grounds for regarding these predicates as designating tendencies to bring about appropriate outcome responses when subject to appropriate singular trials, which are actual mental states of particular individuals (should they happen to be satisfied at all). The "single-case" aspect, perhaps, requires additional elaboration, since these properties are theoretically measured by the limiting frequencies they would display if they were subject to infinite sequences of trials under identical test conditions, as explained in Chapter 3. An an informal account, however, the following may suffice; for, if there is a *fixed set* of conditions which is a *complete set* of conditions (such that no other factors are *causally relevant* to the class of outcomes, given those fixed factors), then the property displayed is *probabilistic*, i.e., a propensity, if its displays are *statistical* and, in principle, for each single case *unpredictable*. So long as these requirements are fulfilled by Tversky's illustrations, therefore, "degrees-of-belief" certainly appear to be a species of single-case propensity.

If personal probabilities are psychological propensities, of course, then people must be "chance set-ups" or "experimental arrangements" in the sense of being objects (or arrangements of objects) upon which trials may be conducted, where each trial results in one or more possible outcomes. Furthermore, when z believes more strongly in h^1 than in h^2, z's tendency to act on h^1 is greater than z's tendency to act on h^2; consequently, z should display h^1-behavior more frequently than z would display h^2-behavior, under the appropriate test conditions — provided the preceding account is correct. In spite of its intuitive plausibility, however, there are at least three grounds upon which objections might be raised to this general conception, namely:

(a) within a decision-theoretical context, the subjects under consideration should conform to the policy of maximizing their expected utility, which would influence the behavior displayed under relevant test conditions;

(b) the paradigm exhibit of degrees of belief is ordinarily assumed to be "betting behavior", which this account appears to minimize in favor of a frequency measure of the strength of an individual's conviction; and,

(c) if personal probabilities satisfy Bayes' theorem, while single-case propensities do not (as certain authors have recently alleged), then how can personal probabilities possibly be a species of single-case propensity?

The first objection might be referred to as *Jeffrey's complaint*, since Jeffrey has elsewhere criticized the notion of "acting on hypotheses" thus:

This notion is foreign or peripheral to Bayesian decision theory, which sees us rather as acting on our degrees of belief and preference: on judgemental probabilities and utilities. I take it that to act on a hypothesis in a decision problem is to opt for the very act one would choose in that problem if one were sure of the truth of the hypothesis. Then where high utilities are associated with low probabilities, to act on the most probable hypothesis may be to flout the Bayesian principle, *act so as to maximize expected utility*. (Jeffrey [1975a], p. 149)

Although not all Bayesians endorse personal probabilities, (virtually) all who endorse personal probabilities are Bayesians. The examples that Tversky cites appear to be of moment here, for the situations they represent are remarkably simple in their decision-theoretic character; for instance,

States of Nature

	Left light lights	Right light lights	Expectation
Pick Left:	reward	no reward	1/3
Pick Right:	no reward	reward	2/3

Payoff Matrix

Since the probabilities for left and right are 1/3 and 2/3, respectively, but the rewards for picking correctly are the same in either case, the optimal solution on Bayesian principles is to predict the right light *every single time*!

In a case of this kind, therefore, the Bayesian strategy of acting so as to maximize expected utility *entails* acting as if the more probable hypothesis were true. Since the subjects in question did not satisfy this conception, it illustrates another aspect of the normative character of Bayesian rationality: those who fail to maximize expected utility are not merely "bad Bayesians" but *irrational individuals*. When presented with Tversky's examples, for instance, Lindley exhorted his colleagues to devote their energies to teaching others the principles of maximizing expected utility, remarking: "Why do you spend your time studying how people make decisions, when we know how they *should* make decisions?" (Lindley [1974], p. 181). Notice especially, however, that if individuals did uniformly conform to Bayesian principles, the choice behavior they would then display under conditions such as these could not be construed as a measure of their "degrees of belief" or "strengths of conviction", since, apart from the special case of equal degrees of belief requiring a tie-breaker, their choice would always be the same, thereby affording no foundation for discriminating between *specific degrees of belief*. Even Jeffrey's own procedures, moreover, afford no solution to this problem, since his "preference orderings" are compatible with whole families of probability and utility functions themselves. [See Jeffrey [1965], Chapters 6–9; and compare Kyburg [1978a], pp. 169–172.]

The second objection, of course, could qualify as *de Finetti's criticism*, since he especially has championed the role of "betting quotients" as measures of "degrees of belief". Personal probabilities are supposed to satisfy the axioms of probability, therefore, since otherwise it would be possible to have a so-called "Dutch Book" made against oneself, i.e., a set of bets in relation to which one is guaranteed to suffer a net loss, regardless of the outcome (Heilig [1978], pp. 330–332). So long as an individual's degrees of belief qualify as *coherent*, however, it will not be possible for a "Dutch Book" to be fulfilled; protection against such an outcome, therefore, seemingly provides the principal motivation for coherence. Nevertheless, the "Dutch Book" rationale envisions a rather complicated and highly artificial betting situation; for, as Heilig has lucidly explained, an individual z's "betting quotients" will be immune from a "Dutch Book" provided the following conditions are satisfied, namely:

If an individual is willing to:
(Condition 1) make bets on all events at betting quotients that he himself has fixed;
(Condition 2) have the amounts of the gross gains and the stakes
(Condition 3) as well as their signs, and thereby the direction of the betting, arbitrarily fixed by the opponent,
then a Dutch Book against the individual is excluded only if the sum of the betting quotients is exactly one,
because otherwise the opponent can make a Dutch Book against the individual.
If the sum of the betting quotients is *less* than one, he can do this through *positive* stakes; if it is *more* than one, through *negative* stakes. (Heilig [1978], p. 339)

It is interesting to observe, moreover, that it is no condition of coherence on this conception that the betting quotients z happens to fix necessarily have to *correspond* to the degrees of belief z happens to have with respect to the relevant class of outcomes; indeed, for z to secure a rational expectation of gain within the context of a betting situation, z's interests will be served by *violating* the condition of coherence. [See Heilig [1978], pp. 332–340; Kyburg [1978a], pp. 159–164; and compare Mellor [1971], pp. 23–25, regarding "sincerity".] Since "betting behavior" clearly falls within Stage III as an aspect of *applied* scientific rationality, moreover, while (CR-1) and (CR-2) provide a warrant for coherence as an aspect of *pure* scientific rationality, it seems inappropriate to impose a greater burden upon "betting rationales" than they may reasonably bear.

The third objection, finally, should be described as "*Humphreys' paradox*", since it counts against the construction of propensities as probabilities even *without* the analysis of personal probabilities as single-case propensities (as discovered by Paul W. Humphreys and related to Salmon in private correspondence; Salmon [1979a], pp. 213–214). Salmon has formulated this criticism as follows:

...there is an important limitation upon identifying propensities with probabilities, for we do not seem to have propensities to match up with "inverse" probabilities. ... Propensity can, I think, be a useful causal concept in the context of a probabilistic theory of causation, but if it is used in that way, it seems to inherit the temporal asymmetry of the causal relation. (Salmon [1979a], pp. 213–214)

Among propensity advocates, neither Mellor nor Giere appear to have appreciated this point, since both take for granted that propensities (or "chances") satisfy the usual mathematical relations (cf. Giere [1976a] and Mellor [1971], p. 30). In the probabilistic causal calculus advanced in Chapter 3, by contrast, p may bring about q with the strength n (where p occurs prior to or simultaneous with q), whether or not q brings about p with *any* strength m (which *can* happen in the case of causal relations of co-existence or of co-variation, an issue discussed in Chapters 3 and 6). Salmon elaborates this point with a simple illustration:

Given suitable "direct" probabilities we can, for example, use Bayes' theorem to compute the probability of a particular cause of death. Suppose we are given a set of probabilities from which we can deduce that the probability that a certain person died as a result of being shot through the head is 3/4. It would be strange, under these circumstances, to say that this corpse has a propensity ... of 3/4 to have had its skull perforated by a bullet. (Salmon [1979a], pp. 213–214)

Thus, if single-case propensities are causal tendencies which are temporally asymmetrical and therefore fail to satisfy Bayes' theorem (except, perhaps, with respect to laws of co-existence or of co-variation, as Chapters 3 and 6 tend to suggest), are they then not *probabilities* after all?

Now whether or not propensities satisfy Bayes' theorem, they are nevertheless "probabilistic" in their character, as axioms A22–A26 of Chapter 3 clearly represent. Since single-case propensities obey the probabilistic principles of addition, summation, and multiplication, therefore, declining to refer to them as "probabilities" would appear to be somewhat arbitrary, especially since, as Ackermann has remarked, there is no single unique axiomatization of probability: "*The* probability calculus as it is often referred to is actually a collection of closely related systems" (Ackermann [1976], p. 66). Moreover, even if propensities *per se* fail to satisfy Bayes' theorem, the *frequencies* which they probabilistically imply do fulfill those relations. The situation thus appears to be analogous with the relations between universal lawlike sentences and corresponding material conditionals which they deductively entail, since these extensional distributions also differ in their syntactical properties from those intensional generalizations; indeed, in the case of lawlike sentences of causal form, their general failure to satisfy transposition clearly qualifies as a theoretical necessity, since the absence of this restriction would undermine the adequacy of those interpretations. For the same reasons,

I take it, frequency constructions do not qualify as causal conceptions, unless they are supplemented by further constraints in order to achieve "causal directedness", in which case they too will fail to satisfy both direct and inverse probability relations — in precisely the same fashion! But if these considerations are correct, then *either* there is no such thing as "probabilistic causation" *or else* "probabilistic causation" can be *probabilistic* without invariably satisfying inverse as well as direct probability relations.

According to Bayes' theorem, of course, the probability of h, given e, is equal to the probability of e, given h, multiplied by the prior probability of h divided by the prior probability of e; that is, $P(h/e) = P(e/h) \cdot P(h)/P(e)$. In application to causal constructions of probability relations, therefore, it must be the case that, in general, if an h-phenomenon 'tends to bring about' an e-phenomenon with some strength n, then that e-phenomenon must similarly 'tend to bring about' that h-phenomenon with some strength m, or Bayes' theorem cannot possibly be satisfied thereby. Although single-case propensity interpretations may satisfactorily interpret *direct* probabilities (from 'h' to 'e', say), they cannot generally simultaneously satisfactorily interpret *inverse* probabilities (from 'e' to 'h') with the same substitution instances. This means that even when there does exist some value n representing the propensity for p, say, to bring about q, there does not usually, if not always, also exist some value m representing the propensity for q to bring about p, since these "inverse propensities" in general do not exist. Since relations of correlation invariably do exist for the frequency with which q occurs relative to p when p occurs with some frequency relative to q, however, it appears reasonable to conjecture that any interpretation of "probabilities" as causal relations invariably satisfying *both* direct *and* inverse probabilistic relations could not possibly be adequate. Perhaps this means that the propensity construction has to be classified as a *non-standard* conception of probability, which does not preclude its importance *even as an interpretation of probability*! Non-Euclidean geometry first emerged as a *non-standard* conception of geometry, but its significance is none the less for that. Perhaps, therefore, the propensity construction of probability stands to standard accounts of probability just as non-Euclidean constructions of geometry stood to standard accounts of geometry before the advent of special and of general relativity.

Even though single-case propensities fail to satisfy Bayes' theorem, moreover, personal probabilities may still be construed as single-case propensities, since the tendency to revise beliefs in accordance with conditionalization is a *deterministic*, rather than a *statistical*, disposition: if z happens to acquire "new" evidence 'e' at time t^*, then the redistribution of his beliefs will have a single unique Bayesian solution relative to his "old" beliefs and distribution of probabilities at some time t. (III*) thus attributes an enduring disposition to an individual z over time, where t is prior to or simultaneous with t^*, i.e., the relation between the values involved is *deductive*, as Salmon [1979a], p. 213, remarks in passing. The statistical dispositional character of personal probabilities as single-case propensities, therefore, arises solely from the requirement of coherence *per se* and not from conditionalization by means of this method, since the Bayesian theorem simply fixes the values of some beliefs (or degrees of belief, strengths of conviction, etc.) as functions of the values of other beliefs (degrees of belief, strengths of conviction, etc.). Since Popperians endorse coherence as a necessary condition for probabilistic beliefs belonging to the knowledge context $\mathcal{K}zt$, moreover, while rejecting the method of Bayesian conditionalization, it is important to recognize that, although the process of revising "old" beliefs on the basis of "new" evidence in accordance with Bayes' theorem presupposes that these beliefs are distributed in accordance with the axioms of the calculus of probability, the converse is not true. [See Carnap [1962a] for an analogous dispositional interpretation of Bayesian principles.]

Perhaps I should emphasize that, with respect to the interpretation of degrees of belief as psychological propensities, whether or not a person z were a Bayesian in the sense of following Bayes' rule of maximizing expected utility as well as a Bayesian in the sense of revising his degrees of belief in accordance with Bayes' theorem would certainly be explanatorily and predictively relevant to his behavior; consequently, at least *some* of the manifestations of having specific degrees of belief in a proposition p would assume classical Bayesian character under those specific circumstances. Indeed, when the conditions Heilig specifies happen to be satisfied by a person's (other) motives and beliefs, then the betting behavior he would display would *also* fulfill classical Bayesian desiderata. The point under consideration, therefore, is not that these conceptions are altogether irrelevant or theoretically superfluous in principle, but rather that the Bayesian account thus defined is clearly incomplete, since there are innumerable *other* manifestations, in principle, of degrees of belief interpreted as psychological propensities. The Tversky illustrations, moreover, may be misleading in certain respects, since it is by no means necessary for there to be *any* objective probabilities in the world mirrored by degrees of belief; but, nevertheless, they do exemplify situations in which there is no reason to doubt that the persons in question have acquired corresponding degrees of belief, which makes them illuminating — if partially problematic — examples. The behavioral consequences of having degrees of belief, therefore, may only be comprehensively characterized by indefinitely large sets of test-condition/outcome-response causal conditionals.

The problem that lingers on, therefore, concerns the relationship between personal probabilities, physical probabilities, and rational beliefs. Let us say that a *degree of belief* in 'h' is *rational* if and only if (a) it is coherent and (b) it equals the measure of support for 'h' (relative to the relevant available evidence); and that a *degree of belief* in 'h' is *fair* if and only if (a) it is coherent and (b) it equals the probability for h to occur (under either a limiting frequency or a single-case propensity construction). Then a person's beliefs may be fair *without* being rational or rational *without* being fair. Moreover, insofar as single-case propensities entail but are not entailed by corresponding frequencies, those beliefs could be *fair* on frequency criteria, even if the world were otherwise deterministic. To the extent to which *measures of support* exist apart from personal probabilities, degrees of belief may be characterized as "optimistic", "pessimistic", or "rational", while coherent sets of beliefs maintained by conditionalization will otherwise always be *equally* "rational". But, "rational" or not, so long as there are personal probabilities as such, there will be at least *some* single-case propensities. These reflections, I might add, are of more than hypothetical interest from the perspective developed here, since *probabilistic beliefs* within the knowledge context $\mathcal{K}zt$ also qualify as psychological properties and as propositional attitudes of a similar variety; indeed, each claim advanced above concerning "degrees of belief" and "strength of conviction" also appears to be true of "probabilistic beliefs", *mutatis mutandis* — with the crucial difference, of course, that these beliefs are 'accepted as true' in $\mathcal{K}zt$, i.e., they are not beliefs about which z is unsure, except for the "inductive uncertainty" which attends probabilistic predictions of singular events within Phase 5 inquiries.

SCIENTIFIC FALLIBILITY

According to the conception introduced in Chapter 1, of course, a person z is in a *knowledge context* $\mathcal{K}zt$ at time t if and only if the set of belief sentences Szt representing the set of beliefs $\mathcal{B}$ accepted by z at t forms a (maximally) rational set of beliefs, i.e., Szt must satisfy (CR-1), (CR-2), (CR-3), and (CR-4), relative to the inductive, deductive, and perceptual rules that z should properly apply in language $\mathcal{L}zt$. The epistemic resources upon which we all may draw in attempting to ascertain the truth about ourselves and the world around us, therefore, include the following:

(1) the language framework $\mathfrak{L}$ which each person z accepts at t;
(2) the deductive rules of inference upon which z relies at t;
(3) the inductive rules of inference upon which z relies at t;
(4) the experiential findings available to z at t, relative to $\mathfrak{L}$; and,
(5) the powers of imagination and conjecture which z can exercise at t.

To the extent to which a collection of individuals $z^1, z^2, \ldots$ relies upon the same language framework $\mathfrak{L}$ and similar rules of inference, they constitute a *community* C; and to the extent to which the members of such a community are committed to the program of providing for the acceptance, rejection, and modification of hypotheses and theories of broad scope and systematic power which may be employed for the purposes of explanation and prediction (by means of a suitable set of inductive, deductive, and perceptual rules of inference), it appears, they will also constitute a *scientific community* Z.

The conception of a "scientific community Z" represented here, moreover, is one in which every member of Z employs the same "rules of inference" relative to a common language framework, $\mathfrak{L}$, and consequently would derive *all and only* the same inferential consequences under the same evidential conditions — except insofar as those rules themselves afforded latitude for variation (in particular, in the case of probabilistic predictions discussed above), i.e., it is an "impersonal" conception to the extent to which it reflects the pursuit of *objective knowledge*. "Objective knowledge" consists of conjectures which have survived our best efforts to refute them by large numbers of trials over a wide variety of conditions, thereby qualifying for tentative and provisional acceptance within the knowledge context $\mathscr{K}$ in conformity with (CR-3) and (CR-4). Nevertheless, although this process of knowledge acquisition is envisioned as an objective process, the members of community Z fulfill indispensable roles, since the success of that community depends upon its members' ability to contribute toward its scientific goals. For their efforts are essential to establishing beliefs $b_1, b_2, \ldots$ warranting acceptance by the standards of the community through creative hypothesis-formation, inductive and deductive inferences, observation and experimentation — and especially imagination and ingenuity with respect to changing and improving the language framework $\mathfrak{L}$ itself, upon which Z ultimately depends for the description and interpretation of its experiential findings and its scientific theories.

In order for a scientific community to successfully pursue its intended program, therefore, it must not only have *access to appropriate standards of inquiry* (in the form of suitable "rules of inference"), but it must also have *the capacity to apply those standards under relevant experiential conditions* (as a function of its members perceptual and inferential capabilities). For the community's epistemic resources are dependent upon the contributions made by each of its members in pursuit of their common epistemic goals. The role of the theory of science in relation to this objective, therefore, is to attempt to ascertain which standards of inquiry are most appropriate for this purpose (through a process of *vindication*), which in turn presupposes an attempt to ascertain which program of inquiry most adequately illuminates scientific investigations (through a process of *exoneration*). If such an undertaking were to succeed, therefore, it would have the effect of making appropriate standards of inquiry more accessible to the members of the scientific community; conversely, if such an undertaking were to lack that effect, then it could hardly succeed. The problem, of course, is to tell the difference, where absence of agreement does not guarantee a lack of success; but failure has doubtless been quite common. As relevant evidence, moreover, a variety of illustrations come to mind, most notably, perhaps, "the narrow thesis of empiricism" (discussed in Chapter 2), which would require the vocabulary of science to be exclusively observational, the consequences of which have been displayed in Chapters 7 and 8. Its successor, "the liberalized thesis of empiricism", by contrast (as Hempel [1952], p. 31, explains it), would require that every term of empirical science be supplied with observational criteria of application by means of reduction sentence procedures.

Neither of these theses concerning the language of science has survived critical scrutiny (advanced, for example, by Hempel [1952]); but still other theses have been advanced in their place. The most encompassing of these, I believe, is van Fraassen's formulation, namely: that surely the language of science is strictly extensional (in some appropriate sense), as discussed in Chapter 6. But if the considerations developed in this investigation are at all well-founded, van Fraassen's thesis cannot be sustained, not only because first-order extensional languages which disavow quantification over possible objects and events are unable to cope with some of the most fundamental problems under consideration here, including, for example,

(a) the analysis of counterfactual and of subjunctive conditionals;
(b) the distinction between lawlike and accidental generalizations;
(c) the discovery of adequate criteria for scientific explanations;

but also because even second-order extensional languages that accommodate it nevertheless encounter difficulties with problems such as these:

(d) explicating the meaning of single-case dispositional predicates;
(e) providing an ontological foundation for nomological conditionals;

(f) supplying non-teleological solutions to problems (a), (b), and (c).

It is intriguing to observe, therefore, that Carnap himself did not object, in principle, to 'strongly intensional' formulations such as those advocated here; on the contrary, Carnap (Schilpp, ed. [1963], p. 952) asserted instead,

> Once the problem of the explication of nomic form has been solved and a logic of causal modalities has been constructed, it will be possible to use these modalities for the explication of subjunctive and, in particular, of counter-factual conditionals. Presumably it will then also be possible to introduce disposition terms by explict definitions.

This brief paragraph, moreover, virtually describes the program of Chapter 3.

Although the conception of "intensional realism" recommended throughout represents a relatively liberal position in dispensing with an observational/theoretical language distinction (which appears to be unwarranted from a dispositional point of view), it remains comparatively conservative in retaining the analytic/synthetic language distinction (which appears to fulfill a theoretically significant function). In particular, the distinction between the class of *narrowly* analytical sentences (whose truth is a function of the syntactical rules of the language framework ℒ alone) and the class of *broadly* analytical sentences (whose truth is a function of the syntactical and semantical rules of the language framework ℒ alone) is one that requires cautious adherence in order to theoretically differentiate between the logical and the physical modalities. A sentence such as, "Titanium melts at a temperature of 1675°C", for example, might assert a physical necessity within one language framework ℒ* and a logical necessity within another ℒ, relative to the class of broadly analytical sentences, where the only difference between them is that "titanium" is defined as "the element with atomic number 22" in ℒ* and as "the element with atomic number 22 and melting point of 1675°C" in ℒ.

While the definitions that have been provided for the various modalities in Chapter 3 implicitly maintain the fundamental distinction involved here by tacitly restricting the class of logical possibilities to just those sentences whose truth does not contradict any of the syntactical truths of ℒ, it would be useful to establish a general procedure that might be invoked in selecting an appropriate language framework for the purpose of maximizing the empirical content of hypotheses and theories which is consistent with these reflections. Let us therefore adopt the following principle, namely:

> *The Principle of Maximizing Empirical Content:* If a sentence S asserting that every member of a reference class K possesses the

property χ is true within the framework $\mathfrak{L}$ (where x satisfies 'K' if and only if x satisfies 'F^1 & ... & F^n') and the corresponding sentence S^* asserting that every member of K^* possesses the property χ is true within the framework $\mathfrak{L}^*$ (where x satisfies 'K^*' if and only if x satisfies 'F^1 & ... & F^m') and the only difference between $\mathfrak{L}$ and $\mathfrak{L}^*$ is that $\{F^1, ..., F^m\} \subset \{F^1, ..., F^n\}$, then framework $\mathfrak{L}^*$ should be adopted in preference to framework $\mathfrak{L}$.

Adherence to the *Principle of Maximizing Empirical Content* (which displaces analytical truths by empirical claims whenever appropriate), therefore, promotes the objective of securing hypotheses and theories of broad scope and systematic power (by maximizing empirical content without obfuscating underlying modal distinctions), while affording a corresponding principle for replacing "logically trivial" scientific explanations by their "nomically significant" counterparts (by substituting lawlike sentences for meaning postulates whenever appropriate), which were discussed in Chapters 5 and 6.

Reliance upon meaning postulates in place of lawlike sentences, moreover, warrants elaboration, for "logically trivial" and "nomologically significant" scientific explanations otherwise have to satisfy the same conditions. Since a dispositional predicate (of universal or statistical strength), 'χxt', may be defined by a (possibly infinite) set of relevant test and outcome response 'causal' conditionals, '$T^1 xt \ni_m O^1 xt^*$', '$T^2 xt \ni_n O^2 xt^*$', and so forth, any such intensional definition accepted into the language framework $\mathfrak{L}zt$ will entail a (possibly infinite) set of "meaning postulates" of the forms, '$(x)(t) [\chi xt \ni (T^1 xt \ni_m O^1 xt^*)]$', '$(x)(t)[\chi xt \ni (T^2 xt \ni_n O^2 xt^*)]$', and so on, as explained in Chapter 6. Each of these sentences will represent a partial specification of the meaning of that predicate, 'χxt', whose meaning is equivalent to their (possibly infinite) conjunction within the language framework $\mathfrak{L}zt$. A sentence of this sort, moreover, may serve as an "explanatory premise" in formulating a "logically trivial" (causal or non-causal) explanation for the occurrence of a singular event described by an explanandum-sentence, *ES*, regardless of whether that event occurs as a manifestation of a dispositional property of universal or of statistical strength. Consider, for example, the *universal-deductive* explanation of the following explanandum-phenomenon:

(MP) For all x and all t, if x has a melting point of $1063°C$ at t, then heating x to $1063°C$ at t would invariably bring about its melting at t^*;

(C1)   Jan's bracelet b has a melting point of $1063°C$ at t;
(C2)   Jan's bracelet b is heated to $1063°C$ at t;
(ES)   Jan's bracelet b melts at t^*.

Thus, the "logical triviality" of explanations of this kind follows from their employment of "meaning postulates" (MP) in lieu of "covering laws" (CL).

"Logically trivial" scientific explanations, of course, will typically qualify as the strongest possible with respect to manifestations of transient properties of things of a specified kind, while "nomologically significant" scientific explanations will be indispensable with respect to manifestations of permanent properties of things of those same kinds. Since motives, beliefs, and character traits, for example, are among the transient properties of individual human beings as members of this species, these reflections suggest the tentative but plausible conjecture that the extended controversy over the relative status of the *natural* (or "physical") sciences and the *social* (or "behavioral") sciences with respect to scientific methods may ultimately represent a difference in degree emanating from their relative emphasis upon permanent and transient properties, respectively. Although all of the sciences are concerned with dispositional properties and with manifest behavior, perhaps the widespread variation in behavior patterns among human beings partially displays the influence of habits, customs, and traditions as contributors to the acquisition of *transient* dispositions: psychology, sociology, and anthropology might therefore be viewed as (at least in part) concerned with personal, interpersonal, and social patterns of behavior that frequently but not exclusively represent the manifestation of transient dispositions, while physics, chemistry, and biology, from this same perspective, are preoccupied instead with persistent and enduring properties of things, i.e., properties no members of such kinds can be without. Yet even if this speculation were to prove to be well-founded, whatever insight it might afford concerning the character of an assuredly artificial division of labor in the pursuit of scientific knowledge should not be over-emphasized: as Malinowki's account of cultural determinants aptly illustrates, the social and the natural sciences alike aim at establishing hypotheses and theories of broad scope and systematic power which may be employed for the purposes of explanation and prediction within their respective domains.

Although "the broad deductivist conception of scientific inquiry" that has been developed here undoubtedly improves upon "the narrow inductivist conception of scientific inquiry" discussed in Chapter 10, this account does not entail or imply the *success* of scientific inquiries. The limited abilities of actual human beings $z^1, z^2, \ldots$, with respect to inductive, deductive, and perceptual modes of reasoning creates an imposing obstacle to maintaining the capacity to apply those standards appropriate to scientific inquiries under relevant experiential conditions. The difficulty of exercising imagination and ingenuity in the discovery and invention of hypotheses and theories of broad scope and systematic power compound an already difficult situation. And the necessity for *nature's cooperation* in attempting to arrange this world's history to include sets of events potentially relevant for testing alternative hypotheses and theories, on the one hand, and for *technological innovations* which may be required for it to be merely technically possible to perform appropriate experiments or to obtain essential observations, on the other, establish formidable barriers to successful scientific inquiries. It should therefore come as no surprise that the belief-sentences S belonging to *the knowledge context $\mathcal{K}$ of the scientific community Z at t* are not necessarily true; for, as Carnap explained, we must admit — at least with respect to all empirical knowledge — that imperfect knowledge is the best we can have. Yet imperfect knowledge is not sufficent for us to determine with absolute assurance the truth or the falsity of any belief about ourselves or the world around us: the risk of error is indeed the price of "scientific knowledge".

Our hopes for success, therefore, are rooted in our ontology, for it is the existence of permanent properties that provides a theoretical foundation for understanding nomological conditionals and lawlike sentences. This ontological conception affords the basis for the realization that falsification practices provide crucial opportunities for discriminating between accidental and lawlike generalizations, for only statements which are not both *lawlike* and *true* are theoretically vulnerable to the eventual discovery of refuting evidence, where the role of background knowledge and auxiliary hypotheses reflects the consideration that even the results of apparent refutations must be envisioned as tentative and fallible, relative to conditions (CR-3) and (CR-4) — which serves to emphasize the fallibilistic character of the conception of 'scientific knowledge' as the outcome of a process consisting essentially of "conjectures and refutations". Nevertheless, to the extent to which our conception of scientific laws entails that *when they are true they do not change*, a conditional argument relating the past and the future is possible after all, with respect to the members of the class of presumptive scientific laws; for,

> Even if the world *is* as we believe it to be in these specific respects, it remains *logically* possible that they might still change tomorrow; but *if* the world is as we believe it to be in these specific

respects, then it is not *physically* possible that they actually will change tomorrow.

And this, no doubt, *is* as much as we should ask of our "inductive" procedures.

From an ontological point of view at last, therefore, the physical world is a world of dispositions that must be construed as a continuous sequence of atomic events, each of which itself consists of the instantiation of arrangements of objects which are themselves instantiations of dispositions. Since there are no causal connections, i.e., no 'causal and effect' or *brings about* relations, between simultaneous happenings (apart from the special classes of laws of co-existence and of co-variation), events that are described by occasion sentence conjunctions that are all true together do not cause one another; however, if an event of kind K and an event of kind T^1, or kind T^2, ..., are all instantiated as features of a single atomic event t by a single individual thing, a, it will (invariably or probably) be the case that some event of kind O^1, or kind O^2, ..., occurs as a feature of a subsequent event t^* (where, in principle, there are no *a priori* boundaries to the variety of antecedent test trials or outcome response consequents that may occur together as features of a single atomic event − except, of course, those imposed by logical and physical impossibilities). As a result, the continuous sequence of instantiations of arrangements of sets of dispositions that constitutes an individual object's history not only records that historical sequence *per se* but implicitly reflects those features of its past theoretically relevant to its explanation (where any feature of an atomic event t that contributes to bringing about the occurrence of some feature of atomic event t^* is necessarily relevant to its explanation).

Rendering these *implicit* features *explicit*, of course, requires access to the set N of lawlike sentences true of the physical world as well as access to the set H of historical descriptions true of that individual's history, relative to which the occurrence of specific features of that history may be subject to systematic explanation; indeed, on the basis of those historical descriptions true of an individual a at a time t, the occurrence of subsequent events as features of that individual's history may be subject to systematic prediction as well (with logical certainty, if all of the relevant laws are universal; or with merely probabilistic confidence, if they are not). Hence, given the set N of all laws and the set H of true descriptions of any atomic event t during the course of the world's history, every feature of the subsequent atomic event t^* may be systematically predicted or explained as a manifestation of some underlying (universal or statistical) dispositional property of the world. The history of an object thus imposes 'causal constraints' upon its subsequent development in the form of historically determined possibilities, necessities, and impossibilities for that object's future course.

An illustration of the significance of this theoretical conception is posed by the relationship between bodies and minds, which appears to be as follows: as a product of heredity and gestation, each human being enters the world possessing a neurophysiological apparatus, i.e., a 'brain', with some determinate structure, K. Among the permanent properties of every brain of structure K is a disposition to acquire other dispositions as outcome responses to particular kinds of trial tests, which may be referred to as its 'capacity to learn' or, for short, *intelligence*. The characteristics of this specific disposition, of course, vary as a function of the underlying structure; but, in any case, as that individual thing undergoes multifarious experiences during the course of its life history, it will acquire, as invariable or probable outcomes, innumerable complex tendencies to respond to various kinds of environmental variables with specific kinds of outcome behavior. Since every event of this individual's history happens to be unique, such a thing may acquire behavioral dispositions of distinctive kinds, although, to the extent to which things of this kind are exposed to similar — though not *exactly* similar — happenings, their dispositional acquisitions will tend to be the same in kind, if not in strength. Indeed, as an ontological perception, things of many kinds are analogous in structural characteristics; for things that are *gold* are like things that are *people*, insofar as gold has its characteristic malleability, melting point, and boiling point, and people too have characteristic degrees of flexibility, and boiling points, and melting points, when appropriately conceived. Yet each and every instance of either of these kinds is a unique individual thing.

From this point of view, therefore, every atomic event that occurs during the course of the world's history is a manifestation of some dispositional property of the world and every physical object that exists is an instantiation of some set of dispositions; as a result, every structural property of the world is dispositional in kind. On the basis of the preceding considerations, taken altogether, it appears to be reasonable inference that a dispositional ontology provides a logically elegant and theoretically illuminating analysis of the basic kinds of relations that may obtain in the physical world; indeed, to the extent to which the concepts of *object*, of *event*, of *property*, of *natural kind*, of *lawlike sentence, subjunctive* and *causal conditional*, of *logical, physical*, and *historical possibility*, of *names* and *descriptions* for objects and events, of *space* and of *time*, and of *explanation* and *prediction*, are both philosophically sound and theoretically derivable on the basis of definitions for *dispositions* and for *permanent properties* of things of a certain kind, the philosophical benefits of a dispositional explication of all of these conceptions appear to be enormously appealing. There will always be grounds for dissent, to be sure, and the price of intensionality may be thought too high a price to pay, even for these benefits; but surely the burden of proof is upon those who would deny the theoretical potential of a dispositional construction. Whether "philosophy of science is philosophy enough" (as Quine [1953a] suggests), of course, depends upon the emphasis one chooses to place upon the world's contingent history as opposed to its physical structure; but, there appear to be no obvious problems involving names and definite descriptions that lie beyond its scope or would warrant its rejection. The issue underlying any distinction between scientific and philosophical ontologies, after all, is whether there may be more in heaven and earth than is dreamt of in our philosophy: for the ontology of science, a world of dispositions is world enough.

REFERENCES

Achinstein, P. [1971], *Law and Explanation*, Oxford: Oxford University Press.
Ackermann, R. [1965], 'Discussion: Deductive Scientific Explanation', *Philosophy of Science* 32 (1965), 155–167.
Ackermann, R. [1976], *The Philosophy of Karl Popper*, Amherst: University of Massachusetts Press.
Ackermann, R. and A. Stenner [1966], 'Discussion: A Corrected Model of Explanation', *Philosophy of Science* 33 (1966), 168–171.
Anderson, A. and N. Belnap [1975], *Entailment: The Logic of Relevance and Necessity*, Princeton: Princeton University Press.
Austin, J. [1950], 'Truth', *Proceedings of the Aristotelian Society, Supplementary Volume* 24 (1950), 111–128.
Ayer, A. J. [1946], *Language, Truth, and Logic*, New York: Dover Publications.
Ayer, A. J. [1956], 'What is a Law of Nature?' *Revue Internationale de Philosophie* 10 (1956), 144–165.
Ayer, A. J., ed. [1959], *Logical Positivism*, New York: The Free Press.
Bennett, J. [1974], 'Counterfactuals and Possible Worlds', *Canadian Journal of Philosophy* 4 (1974), 381–402.
Beth, E. W. [1969], 'Semantic Entailment and Formal Derivability', in *The Philosophy of Mathematics*, J. Hintikka, ed., Oxford: Oxford University Press, 1969, pp. 9–41.
Birnbaum, A. [1962], 'On the Foundations of Statistical Inference', *Journal of the American Statistical Association* 57 (1962), 269–306.
Birnbaum, A. [1969], 'Concepts of Statistical Evidence', in *Philosophy, Science and Method*, S. Morgenbesser, *et al.*, eds., New York: St. Martin's Press, 1969, pp. 112–143.
Birnbaum, A. [1974], 'Discussion of the Papers by Professor Tversky and by Professor Suppes', *The Journal of the Royal Statistical Society*, Series B (Methodological) 36 (1974), 183–184.
Birnbaum, A. [1977], 'The Neyman-Pearson Theory as Decision Theory, and as Inference Theory: with a Criticism of the Lindley-Savage Argument for Bayesian Theory'. *Synthese* 36 (1977), 19–49.
Braithwaite, R. B. [1953], *Scientific Explanation*, Cambridge: Cambridge University Press.
Braithwaite, R. B. [1957], 'On Unknown Probabilities', in *Observation and Interpretation in the Philosophy of Physics*, S. Körner, ed., New York: Dover Publications, 1957, pp. 3–11.
Buchler, J., ed. [1955], *Philosophical Writings of Peirce*, New York: Dover Publications.
Burks, A. [1977], *Chance, Cause, Reason*, Chicago: University of Chicago Press.
Carnap, R. [1936–37], 'Testability and Meaning', *Philosophy of Science* 3 (1936), 419–471; and *Philosophy of Science* 4 (1937), 1–40.

REFERENCES

Carnap, R. [1939], *Foundations of Logic and Mathematics*, Chicago: University of Chicago Press.
Carnap, R. [1947], *Meaning and Necessity*, Chicago: University of Chicago Press.
Carnap, R. [1949], 'Truth and Confirmation', in *Readings in Philosophical Analysis*, H. Feigl and W. Sellars, eds., New York: Appleton-Century-Crofts, Inc., 1949, pp. 119–127.
Carnap, R. [1949a], 'The Two Concepts of Probability', in *Readings in Philosophical Analysis*, H. Feigl and W. Sellars, eds., New York: Appleton-Century-Crofts, Inc., 1949, pp. 330–348.
Carnap, R. [1952], *The Continuum of Inductive Methods*, Chicago: University of Chicago Press.
Carnap, R. [1955], 'Meaning and Synonymy in Natural Language', *Philosophical Studies* 7 (1955), 33–47.
Carnap, R. [1956], 'On the Methodological Character of Theoretical Concepts', in *Minnesota Studies in the Philosophy of Science*, Vol. I, H. Feigl and M. Scriven, eds., Minneapolis: University of Minnesota Press, 1956, pp. 38–76.
Carnap, R. [1962], *Logical Foundations of Probability*, 2nd ed., Chicago: University of Chicago Press.
Carnap, R. [1962a], 'The Aim of Inductive Logic', in *Logic, Methodology and Philosophy of Science*, E. Nagel, et al., eds., Stanford: Stanford University Press, 1962, pp. 303–318.
Carnap, R. [1963], 'Replies and Systematic Expositions', in *The Philosophy of Rudolf Carnap*, P. Schilpp, ed., La Salle, Illinois: Open Court, 1963, pp. 859–1013.
Carnap, R. [1966], *Philosophical Foundations of Physics*, New York: Basic Books.
Cartwright, N. [1979], 'Causal Laws and Effective Strategies', *Noûs* 13 (1979), 419–437.
Chalmers, A. F. [1976], *What is this thing called Science?* New York: The Humanities Press.
Coffa, A. [1974], 'Hempel's Ambiguity', *Synthese* 28 (1974), 141–163.
Coffa, A. [1974a], 'Randomness and Knowledge', in *Boston Studies in the Philosophy of Science*, Vol. XX, K. Schaffner and R. Cohen, eds., Dordrecht, Holland: D. Reidel, 1974, pp. 103–115.
Copi, I. [1973], *Symbolic Logic*, 4th ed., New York: Macmillan.
Cramér, H. [1946], *Mathematical Methods of Statistics*, Princeton: Princeton University Press.
de Finetti, B. [1964], 'Foresight: Its Logical Laws, Its Subjective Sources', in *Studies in Subjective Probability*, H. Kyburg and H. Smokler, eds., New York: John Wiley & Sons, 1964, pp. 93–158.
Dias, P. and A. Shimony [   ], 'A Critique of Jaynes' Maximum Entropy Principle' (unpublished manuscript).
Dretske, F. I. [1977], 'Laws of Nature', *Philosophy of Science* 44 (1977), 248–268.
Eberle, R., D. Kaplan and R. Montague [1961], 'Hempel and Oppenheim on Explanation', *Philosophy of Science* 28 (1961), 418–428.
Feigl, H. [1963], 'De Principiis Non Disputandum...?' In *Philosophical Analysis*, M. Black, ed., Englewood Cliffs, N.J.: Prentice-Hall, 1963, pp. 113–147.
Feller, W. [1950], *An Introduction to Probability Theory and Its Applications*, Vol. I. New York: John Wiley & Sons.

REFERENCES

Fetzer, J. H. [1971], 'Dispositional Probabilities', in *Boston Studies in the Philosophy of Science*, Vol. VIII, R. Buck and R. Cohen, eds., Dordrecht, Holland: D. Reidel, 1971, pp. 473–482.
Fetzer, J. H. [1972], 'Philosophy of Science versus Psychology of Science', *American Psychologist* 27 (1972), 662–665.
Fetzer, J. H. [1974], 'Grünbaum's "Defense" of the Symmetry Thesis', *Philosophical Studies* 25 (1974), 173–187.
Fetzer, J. H. [1974a], 'On "Epistemic Possibility"', *Philosophia* 4 (1974), 327–335.
Fetzer, J. H. [1974b], 'Statistical Explanations', in *Boston Studies in the Philosophy of Science*, Vol. XX, K. Schaffner and R. Cohen, eds., Dordrecht, Holland: D. Reidel, 1974, pp. 337–347.
Fetzer, J. H. [1974c], 'Statistical Probabilities: Single Case Propensities vs. Long Run Frequencies', in *Developments in the Methodology of Social Science*, W. Leinfellner and E. Köhler, eds., Dordrecht, Holland: D. Reidel, 1974, pp. 387–397.
Fetzer, J. H. [1974d], 'A Single Case Propensity Theory of Explanation', *Synthese* 28 (1974), 171–198.
Fetzer, J. H. [1975], 'On the Historical Explanation of Unique Events', *Theory and Decision* 6 (1975), 87–97.
Fetzer, J. H. [1975a], 'Discussion Review: Achinstein's *Law and Explanation*', *Philosophy of Science* 42 (1975), 320–333.
Fetzer, J. H. [1976], 'Elements of Induction', in *Local Induction*, R. Bogdan, ed., Dordrecht, Holland: D. Reidel, 1976, pp. 145–170.
Fetzer, J. H. [1976a], 'The Likeness of Lawlikeness', in *Boston Studies in the Philosophy of Science*, Vol. XXXII, R. Cohen, C. Hooker, A. Michalos, and J. van Evra, eds., Dordrecht, Holland: D. Reidel, 1976, pp. 377–391.
Fetzer, J. H. [1977], 'Reichenbach, Reference Classes, and Single Case "Probabilities"', *Synthese* 34 (1977), 185–217. Errata, *Synthese* 37 (1978), 113–114.
Fetzer, J. H. [1977a], 'A World of Dispositions', *Synthese* 34 (1977), 397–421.
Fetzer, J. H. [1978], Review of Ackermann's *The Philosophy of Karl Popper*, *Philosophy of Science* 45 (1978), 491–493.
Fetzer, J. H. [1978a], 'On Mellor on Dispositions', *Philosophia* 7 (1978), 651–660.
Fetzer, J. H. [1979], 'Discussion Review: Chalmers' *What is this thing called Science?*' *Erkenntnis* 14 (1979), 393–404.
Fetzer, J. H. [   ], 'Critical Study: Wolgast's *Paradoxes of Knowledge*', *Philosophia* (forthcoming).
Fetzer, J. H. and D. E. Nute [1979], 'Syntax, Semantics, and Ontology: A Probabilistic Causal Calculus', *Synthese* 40 (1979), 453–495.
Fetzer, J. H. and D. E. Nute [1980], 'A Probabilistic Causal Calculus: Conflicting Conceptions', *Synthese* 44 (1980), 241–246. Errata, *Synthese* 48 (1981), 493.
Fisher, R. A. [1956], *Statistical Methods and Scientific Inference*, New York: Hafner Press, 1956, 3rd ed., 1973.
Fodor, J. [1978], 'Propositional Attitudes', *Monist* 61 (1978), 500–523.
Gettier, E. [1963], 'Is Justified True Belief Knowledge?' *Analysis* 23 (1963), 121–123.
Giere, R. N. [1973], Review of Mellor's *The Matter of Chance*, *Ratio* 15 (1973), 149–155.
Giere, R. N. [1973a], 'Objective Single Case Probabilities and the Foundations of

Statistics', in *Logic, Methodology and Philosophy of Science*, P. Suppes *et al.*, eds., Amsterdam: North-Holland, 1973, pp. 467–483.
Giere, R. N. [1975], 'The Epistemological Roots of Scientific Knowledge', in *Minnesota Studies in the Philosophy of Science*, Vol. VI, G. Maxwell and R. Anderson, eds., Minneapolis: University of Minnesota Press, 1975, pp. 212–261.
Giere, R. N. [1975a], 'Popper and the Non-Bayesian Tradition: Comments on Richard Jeffrey', *Synthese* 30 (1975), 119–132.
Giere, R. N. [1976], 'Empirical Probability, Objective Statistical Methods and Scientific Inquiry', in *Foundations of Probability Theory, Statistical Inference, and Statistical Theories of Science*, Vol. II, W. Harper and C. Hooker, eds., Dordrecht, Holland: D. Reidel, 1976, pp. 63–101.
Giere, R. N. [1976a], 'A Laplacean Formal Semantics for Single-Case Propensities', *Journal of Philosophical Logic* 5 (1976), 321–353.
Giere, R. N. [1977], 'Testing vs. Information Models of Statistical Inference', in *Logic, Laws and Life*, R. Colodny, ed., Pittsburgh: University of Pittsburgh Press, 1977, pp. 19–70.
Giere, R. N. [1977a], 'Allan Birnbaum's Conception of Statistical Evidence', *Synthese* 36 (1977), 5–13.
Giere, R. N. [1979], 'Propensity and Necessity', *Synthese* 40 (1979), 439–451.
Giere, R. N. [1979a], *Understanding Scientific Reasoning*, New York: Holt, Rinehart, and Winston.
Giere, R. N. [1979b], 'Foundations of Probability and Statistical Inference', in *Current Research in Philosophy of Science*, P. Asquith and H. Kyburg, eds., East Lansing: Philosophy of Science Association, 1979, pp. 503–533.
Gillies, D. A. [1973], *An Objective Theory of Probability*, London: Methuen.
Gnedenko, B. V. [1962], *The Theory of Probability*, New York: Chelsea.
Gnedenko, B. V. and A. N. Kolmogorov [1954], *Limit Distributions for Sums of Independent Random Variables*, Reading, Massachusetts: Addison-Wesley.
Good, I. J. [1976], 'The Bayesian Influence, of How to Sweep Subjectivism Under the Rug', in *Foundations of Probability Theory, Statistical Inference, and Statistical Theories of Science*, Vol. II, W. Harper and C. Hooker, eds., Dordrecht, Holland: D. Reidel, 1976, pp. 125–174.
Goodman, N. [1965], *Fact, Fiction, and Forecast*, Indianapolis: Bobbs-Merrill.
Hacking, I. [1965], *Logic of Statistical Inference*, Cambridge: Cambridge University Press.
Hacking, I. [1967], 'Slightly More Realistic Personal Probability', *Philosophy of Science* 34 (1967), 311–325.
Hacking, I. [1967a], 'Possibility', *Philosophical Review* 76 (1967), 143–168.
Hacking, I. [1968], 'One Problem About Induction', in *The Problem of Inductive Logic*, I. Lakatos, ed., Amsterdam: North-Holland, 1968, pp. 44–59.
Hacking, I. [1968a], 'On Falling Short of Strict Coherence', *Philosophy of Science* 35 (1968), 284–286.
Hacking, I. [1973], 'Propensities, Statistics, and Inductive Logic', in *Logic, Methodology, and Philosophy of Science*, P. Suppes *et al.*, eds., Amsterdam: North-Holland, 1973, pp. 485–500.
Hacking, I. [1975], *The Emergence of Probability*, Cambridge: Cambridge University Press.

REFERENCES

Hacking, I. [1980], 'The Theory of Probable Inference: Neyman, Peirce, and Braithwaite', in *Science, Belief, and Behavior*, D. H. Mellor, ed., Cambridge: Cambridge University Press, 1980, pp. 141–160.
Hájek, J. [1969], *Non-Parametric Statistics*, San Francisco: Holden-Day.
Harman, G. [1970], 'Induction', in *Induction, Acceptance, and Rational Belief*, M. Swain, ed., Dordrecht, Holland: D. Reidel, 1970, pp. 83–99.
Heilig, K. [1978], 'Carnap and de Finetti on Bets and the Probability of Singular Events: The Dutch Book Argument Reconsidered', *British Journal for the Philosophy of Science* 29 (1978), 325–346.
Hempel, C. G. [1950], 'Problems and Changes in the Empiricist Criterion of Meaning', *Revue Internationale de Philosophie* 11 (1950), 41–63.
Hempel, C. G. [1951], 'The Concept of Cognitive Significance: A Reconsideration', *Proceedings of the American Academy of Arts and Science* 80 (1951), 61–77.
Hempel, C. G. [1952], *Fundamentals of Concept Formation in Empirical Science*, Chicago: University of Chicago Press.
Hempel, C. G. [1962], 'Deductive-Nomological vs. Statistical Explanation', in *Minnesota Studies in the Philosophy of Science*, Vol. III, H. Feigl and G. Maxwell, eds., Minneapolis: University of Minnesota Press, 1962, pp. 98–169.
Hempel, C. G. [1965], *Aspects of Scientific Explanation*, New York: The Free Press.
Hempel, C. G. [1966], *The Philosophy of Natural Science*, Englewood Cliffs, N.J.: Prentice-Hall.
Hempel, C. G. [1968], 'Maximal Specificity and Lawlikeness in Probabilistic Explanation', *Philosophy of Science* 35 (1968), 116–133.
Hempel, C. G. [1970], 'On the "Standard Conception" of Scientific Theories', in *Minnesota Studies in the Philosophy of Science*, Vol. IV, M. Radner and S. Winokur, eds., Minneapolis: University of Minnesota Press, 1970, pp. 142–163.
Hempel, C. G. [1976], 'Nachwort 1976', in *Aspekte wissenschaftlicher Erklärung*, Berlin: Walter de Gruyter, 1977. This is a German edition of Hempel [1965].
Hempel, C. G. and P. Oppenheim [1948], 'Studies in the Logic of Explanation', *Philosophy of Science* 15 (1948), 135–175.
Hilpinen, R. [1968], *Rules of Acceptance and Inductive Logic*, Amsterdam: North-Holland.
Hintikka, J. [1966], 'A Two-Dimensional Continuum of Inductive Methods', in *Aspects of Inductive Logic*, J. Hintikka and P. Suppes, eds., Amsterdam: North-Holland, 1966, pp. 113–132.
Hintikka, J. [1968], 'Induction by Enumeration and Induction by Elimination', in *The Problem of Inductive Logic*, I. Lakatos, ed., Amsterdam: North-Holland, 1968, pp. 191–216.
Hintikka, J. [1971], 'Unknown Probabilities, Bayesianism, and de Finetti's Representation Theorem', in *Boston Studies in the Philosophy of Science*, Vol. VIII, R. Buck and R. Cohen, eds., Dordrecht, Holland: D. Reidel, 1971, pp. 325–341.
Hodges, J. L. and E. L. Lehmann [1964], *Basic Concepts of Probability and Statistics*, San Francisco: Holden-Day.
Humphreys, P. [1977], 'Randomness, Independence, and Hypotheses', *Synthese* 36 (1977), 415–426.
Jaynes, E. T. [1967], 'Foundations of Probability Theory and Statistical Mechanics', in

Delaware Seminar in the Foundations of Physics, M. Bunge, ed., New York: Springer-Verlag, 1967, pp. 77–101.
Jeffrey, R. [1965], *The Logic of Decision*, New York: McGraw-Hill.
Jeffrey, R. [1970], 'Dracula Meets Wolfman: Acceptance vs. Partial Belief', in *Induction, Acceptance, and Rational Belief*, M. Swain, ed., Dordrecht, Holland: D. Reidel, 1970, pp. 157–185.
Jeffrey, R. [1975], 'Probability and Falsification: A Critique of the Popper Program', *Synthese* 20 (1975), 95–117.
Jeffrey, R. [1975a], 'Replies', *Synthese* 20 (1975), 149–157.
Kaplan, D. [1961], 'Explanation Revisted', *Philosophy of Science* 28 (1961), 429–436.
Kemeny, J. G. [1955], 'Fair Bets and Inductive Probabilities', *Journal of Symbolic Logic* 20 (1955), 263–273.
Kiefer, J. [1977], 'The Foundations of Statistics – Are There Any?' *Synthese* 36 (1977), 161–176.
Kim, J. [1963], 'Discussion: On the Logical Conditions of Deductive Explanation', *Philosophy of Science* 30 (1963), 286–291.
Kolmogorov, A. N. [1956], *Foundations of the Theory of Probability*, New York: Chelsea.
Koopman, B. O. [1940], 'The Axioms and Algebra of Intuitive Probability', *Annals of Mathematics* 41 (1940), 269-292.
Kripke, S. [1971], 'Identity and Necessity', in *Identity and Individuation*, M. Munitz, ed., New York: New York University Press, 1971, pp. 135–164.
Kripke, S. [1972], 'Naming and Necessity', in *Semantics of Natural Language*, D. Davidson and G. Harman, eds., Dordrecht, Holland: D. Reidel, 1972, pp. 253–355 and pp. 763–769.
Kripke, S. [1979], 'Speaker's Reference and Semantic Reference', in *Contemporary Perspectives in the Philosophy of Language*, P. French et al., eds., Minnesota: University of Minnesota Press, 1979, pp. 6–27.
Krüger, L. [1976], 'Are Statistical Explanations Possible?' *Philosophy of Science* 43 (1976), 129–146.
Kuhn, T. S. [1964], *The Structure of Scientific Revolutions*, Chicago: University of Chicago Press.
Kyburg, H. E. [1964], 'Probability, Rationality, and a Rule of Detachment', in *Logic, Methodology, and Philosophy of Science*, Y. Bar-Hillel, ed., Amsterdam, North-Holland, 1965, pp. 301–310.
Kyburg, H. E. [1965], 'Comment', *Philosophy of Science* 32 (1965), 145–151.
Kyburg, H. E. [1970], *Probability and Inductive Logic*, New York: Macmillan.
Kyburg, H. E. [1970a], 'Discussion: More on Maximal Specificity', *Philosophy of Science* 37 (1970), 295–300.
Kyburg, H. E. [1970b], 'Conjunctivitis', in *Induction, Acceptance, and Rational Belief*, M. Swain, ed., Dordrecht, Holland: D. Reidel, 1970, pp. 55–82.
Kyburg, H. E. [1974], 'Propensities and Probabilities', *British Journal for the Philosophy of Science* 25 (1974), 25–42.
Kyburg, H. E. [1974a], 'Randomness', in *Boston Studies in the Philosophy of Science*, Vol. XX, K. Schaffner and R. Cohen, eds., Dordrecht, Holland: D. Reidel, 1974, pp. 137–149.

REFERENCES

Kyburg, H. E. [1974b], *The Logical Foundations of Statistical Inference*, Dordrecht, Holland: D. Reidel.
Kyburg, H. E. [1976], 'Chance', *Journal of Philosophical Logic* 5 (1976), 355–393.
Kyburg, H. E. [1977], 'Decisions, Conclusions, and Utilities', *Synthese* 36 (1977), 87–96.
Kyburg, H. E. [1978], 'Propensities and Probabilities', in *Dispositions*, R. Tuomela, ed., Dordrecht, Holland: D. Reidel, 1978, pp. 277–301. This is a slightly revised version of Kyburg [1974].
Kyburg, H. E. [1978a], 'Subjective Probability: Criticisms, Reflections, and Problems', *Journal of Philosophical Logic* 7 (1978), 157–180.
Kyburg, H. E. [1979], 'Tyche and Athena', *Synthese* 40 (1979), 415–438.
Kyburg, H. E. [1979a], 'The Application of Formal Methods in the Philosophy of Science', in *Current Research in the Philosophy of Science*, P. Asquith and H. Kyburg, eds., East Lansing: Philosophy of Science Association, 1979, pp. 28–39.
Kyburg, H. E. and H. Smokler, eds. [1980], 'Introduction', in *Studies in Subjective Probability*, 2nd ed., Huntington, New York: R. Krieger, 1980, pp. 3–18.
Lakatos, I. [1968], 'Changes in the Problem of Inductive Logic', in *The Problem of Inductive Logic*, I. Lakatos, ed., Amsterdam: North-Holland, 1968, pp. 315–417.
Lakatos, I. [1970], 'Falsification and the Methodology of Scientific Research Programmes', in *Criticism and the Growth of Knowledge*, I. Lakatos and A. Musgrave, eds., Cambridge: Cambridge University Press, 1970, pp. 91–195.
Lakatos, I. [1971], 'History of Science and Its Rational Reconstruction', in *Boston Studies in the Philosophy of Science*, Vol. VIII, R. Buck and R. Cohen, eds., Dordrecht, Holland: D. Reidel, 1971, pp. 91–136.
Lehmann, R. S [1955], 'On Confirmation and Rational Betting', *Journal of Symbolic Logic* 20 (1955), 251–261.
Lehrer, K. [1974], *Knowledge*, Oxford: Oxford University Press.
Lehrer, K. [1976], 'Induction, Consensus, and Catastrophe', in *Local Induction*, R. Bogdan, ed., Dordrecht, Holland: D. Reidel, 1976, pp. 115–143.
Levi, I. [1967], *Gambling with Truth*, New York: Alfred A. Knopf.
Levi, I. [1969], 'Are Statistical Hypotheses Covering Laws?' *Synthese* 20 (1969), 297–307.
Levi, I. [1970], 'Probability and Evidence', in *Induction, Acceptance, and Rational Belief*, M. Swain, ed., Dordrecht, Holland: D. Reidel, 1970, 134–156.
Levi. I. [1971], 'Certainty, Probability, and the Correction of Evidence', *Noûs* 5 (1971), 299–312.
Levi, I. [1973], '... But Fair to Chance', *Journal of Philosophy* 70 (1973), 52–55.
Levi, I. [1974], 'On Indeterminate Probabilities', *Journal of Philosophy* 71 (1974), 391–418.
Levi, I. [1976], 'Acceptance Revisited', in *Local Induction*, R. Bogdan, ed., Dordrecht, Holland: D. Reidel, 1976, pp. 1–71.
Levi, I. [1977], 'Subjunctives, Dispositions, and Chances', *Synthese* 34 (1977), 423–455.
Levi, I. [1979], 'Inductive Appraisal', in *Current Research in the Philosophy of Science*, P. Asquith and H. Kyburg, eds., East Lansing: Philosophy of Science Association, 1979, pp. 339–351.

REFERENCES

Lewis, D. [1973], *Counterfactuals*, Cambridge: Harvard University Press.
Lindley, D. V. [1971], *Bayesian Statistics: A Review*, Philadelphia: Society for Industrial and Applied Mathematics.
Lindley, D. V. [1974], 'Discussion on the Papers by Professor Tversky and by Professor Suppes', *The Journal of the Royal Statistical Society*, Series B (Methodological) 36 (1974), 181–182.
Lindley, D. V. [1977], 'The Distinction between Inference and Decision', *Synthese* 36 (1977), 51–58.
Maxwell, G. [1970], 'Structural Realism and the Meaning of Theoretical Terms', in *Minnesota Studies in the Philosophy of Science*, Vol. IV, M. Radner and S. Winokur, eds., Minneapolis: University of Minnesota Press, 1970, pp. 181–192.
Mellor, D. H. [1969], 'Chance', *Proceedings of the Aristotelian Society*, Supplementary Volume (1969), pp. 11–36.
Mellor, D. H. [1971], *The Matter of Chance*, Cambridge: Cambridge University Press.
Mellor, D. H. [1974], 'In Defense of Dispositions', *Philosophical Review* 83 (1974), 157–181.
Merrill, G. H. [1978], 'Formalization, Possible Worlds and the Foundations of Modal Logic', *Erkenntnis* 12 (1978), 305–327.
Miller, D. [1975], 'Making Sense of Method: Comments on Richard Jeffrey', *Synthese* 30 (1975), 139–147.
Mood, A. M. and F. A. Graybill [1963], *Introduction to the Theory of Statistics*, 2nd ed., New York: Mc Graw-Hill.
Moore, D. S. [1979], *Statistics: Concepts and Controversies*, San Francisco: W. H. Freeman.
Morgan, C. G. [1972], 'Discussion: On Two Proposed Models of Explanation', *Philosophy of Science* 39 (1972), 74–81.
Morgenbesser, S. [1969], 'The Realist-Instrumentalist Controversy', in *Philosophy, Science, and Method*, S. Morgenbesser et al., eds., New York: St. Martin's Press, 1969, pp. 200–218.
Nagel, E. [1939], *Principles of the Theory of Probability*, Chicago: University of Chicago Press.
Nagel, E. [1963], 'Carnap's Theory of Induction', in *The Philosophy of Rudolf Carnap*, P. Schilpp, ed., La Salle, Illinois: Open Court, 1963, pp. 785–825.
Neyman, J. [1977], 'Frequentist Probability and Frequentist Statistics', *Synthese* 36 (1977), 97–131.
Neyman, J. and E. Pearson [1966], *Joint Statistical Papers*, Berkeley: University of California Press.
Nickles, Thomas [1971], 'Covering Law Explanation', *Philosophy of Science* 38 (1971), 542–561.
Niiniluoto, I. [1976], 'Inductive Explanation, Propensity, and Action', in *Essays on Explanation and Understanding*, J. Manninen and R. Tuomela, eds., Dordrecht, Holland: D. Reidel, 1977, pp. 335–368.
Niiniluoto, I. [1977], 'On a K-dimensional System of Inductive Logic', in *PSA 1976*, F. Suppe and P. Asquith, eds., East Lansing: Philosophy of Science Association, 1977, pp. 425–447.
Niiniluoto, I. [1978], 'Truthlikeness: Comments on Recent Discussion', *Synthese* 38 (1978), 281–329.

REFERENCES

Niiniluoto, I. [1981], 'Statistical Explanation', in *Contemporary Philosophy 1966–1978*, G. Fløistad, ed., The Hague: Martinus Nijhoff, 1981.

Niiniluoto, I. [1981a], 'Statistical Explanation Reconsidered', *Synthese* 48 (1981), 437–472.

Niiniluto, I. [   ], 'Inductive Logic as a Methodological Research Programme' (unpublished manuscript).

Niiniluto, I. and R. Tuomela [1973], *Theoretical Concepts and Hypothetico-Inductive Inference*, Dordrecht, Holland: D. Reidel.

Nute, D. E. [1975], 'Counterfactuals', *Notre Dame Journal of Formal Logic* 16 (1975), 475–482.

Nute, D. E. [1975a], 'Counterfactuals and the Similarity of Worlds'. *Journal of Philosophy* 72 (1975), 773–778.

Nute, D. E. [1976], letter to James H. Fetzer of 7 September 1976.

Nute, D. E. [1977], 'Simplification and Substitution of Counterfactual Antecedents', *Philosophia* 7 (1977), 317–326.

Nute, D. E. [1978], 'Do Proper Names Always Rigidly Designate?' *Canadian Journal of Philosophy* 8 (1978), 475–484.

Nute, D. E. [1978a], 'Proper Names: How to Become a Causal Theorist While Remaining a Sense Theorist', *Philosophia* 8 (1978), 43–57.

Nute, D. E. [   ], 'Was Aristotle Necessarily a Philosopher?' (unpublished manuscript).

Pap, A. [1963], 'Reduction Sentences and Dispositional Concepts', in *The Philosophy of Rudolf Carnap*, P. Schilpp, ed., La Salle, Illinois: Open Court, 1963, pp. 559–597.

Peirce, C. S. [1955], *Philosophical Writings of Peirce*, J. Buchler, ed., New York: Dover Publications.

Polanyi, M. [1964], *Personal Knowledge*, New York: Harper & Row.

Pollock, J. [1972], 'The Logic of Projectibility', *Philosophy of Science* 39 (1972), 302–314.

Pollock, J. [1974], 'Subjunctive Generalizations', *Synthese* 28 (1974), 199–214.

Pollock, J. [1976], *Subjunctive Reasoning*, Dordrecht, Holland: D. Reidel.

Popper, K. R. [1957], 'The Propensity Interpretation of the Calculus of Probability, and the Quantum Theory', in *Observation and Interpretation in the Philosophy of Physics*, S. Körner, ed., New York: Dover Publications, 1957, pp. 65–70.

Popper, K. R. [1957a], 'Probability Magic or Knowledge Out of Ignorance', *Dialectica* 11 (1957), 354–374.

Popper, K. R. [1959], 'The Propensity Interpretation of Probability', *British Journal for the Philosophy of Science* 10 (1959), 25–42.

Popper, K. R. [1965], *The Logic of Scientific Discovery*, New York: Harper & Row.

Popper, K. R. [1968], *Conjectures and Refutations*, New York: Harper & Row.

Popper, K. R. [1972], *Objective Knowledge*, Oxford: Oxford University Press.

Popper, K. R. [1974], 'Replies to My Critics', in *The Philosophy of Karl Popper*, P. Schilpp, ed., La Salle, Illinois: Open Court, 1974, pp. 961–1197.

Popper, K. R. [1976], *Unended Quest*, La Salle, Illinois: Open Court.

Pratt, J. [1977], ' "Decisions" as Statistical Evidence and Birnbaum's "Confidence Concept" ', *Synthese* 36 (1977), 59–69.

Putnam, H. [1951], *The Meaning of the Concept of Probability in Application to Finite Sequences*, University of California at Los Angeles, unpublished dissertation, 1951.

REFERENCES

Putnam, H. [1963], ' "Degree of Confirmation" and Inductive Logic', in *The Philosophy of Rudolf Carnap*, P. Schilpp, ed., La Salle, Illinois: Open Court, 1963, pp. 761–783.
Putnam, H. [1967], 'Probability and Confirmation', in *The Philosophy of Science Today*, S. Morgenbesser, ed., New York: Basic Books, 1967, pp. 100–114.
Putnam, H. [1973], 'Meaning and Reference', in *Naming, Necessity, and Natural Kinds*, S. Schwartz, ed., Ithaca: Cornell University Press, 1977, pp. 119–132.
Quine, W. V. O. [1951], *Mathematical Logic*, New York: Harper & Row.
Quine, W. V. O. [1953], *From a Logical Point of View*, Cambridge: Harvard University Press.
Quine, W. V. O. [1953a], 'Mr. Strawson on Logical Theory', *Mind* 62 (1953), 433–451.
Quine, W. V. O. [1960], *Word and Object*, Cambridge: The M.I.T. Press.
Quine, W. V. O. [1974], *The Roots of Reference*, La Salle, Illinois: Open Court.
Radnitzky, G. [1973], *Contemporary Schools of Metascience*, Chicago: Henry Regnery Company.
Ramsey, F. P. [1931], *The Foundations of Mathematics*, London: Routledge & Kegan Paul.
Reichenbach, H. [1949], *The Theory of Probability*, Berkeley: University of California.
Reichenbach, H. [1976], *Laws, Modalities, and Counterfactuals*, Berkeley: University of California.
Reinhardt, L. [1979], 'What Reference Can't Be', *Philosophia* 9 (1979), 21–38.
Rogers, B. [1971], 'Material Conditions on Tests of Statistical Hypotheses', in *Boston Studies in the Philosophy of Science*, Vol VIII, R. Buck and R. Cohen, eds., Dordrecht, Holland: D. Reidel, 1971, pp. 403–412.
Rogers, B. [1977], 'The Probabilities of Theories as Frequencies', *Synthese* 34 (1977), 167–183.
Rosenkrantz, R. [1977], *Inference, Method, and Decision*, Dordrecht, Holland: D. Reidel.
Rudner, R. [1966], *The Philosophy of Social Science*, Englewood Cliffs, N.J.: Prentice-Hall.
Russell, B. [1948], *Human Knowledge: Its Scope and Limits*, New York: Simon and Schuster.
Salmon, W. C. [1965], 'The Status of Prior Probabilities in Statistical Explanation', *Philosophy of Science* 32 (1965), 137–146.
Salmon, W. C. [1967], *The Foundations of Scientific Inference*, Pittsburgh: University of Pittsburgh Press.
Salmon, W. C. [1968], 'The Justification of Inductive Rules of Inference', in *The Problem of Inductive Logic*, I. Lakatos, ed., Amsterdam: North-Holland, 1968, pp. 24–43.
Salmon, W. C. [1968a], 'Reply', in *The Problem of Inductive Logic*, I. Lakatos, ed., Amsterdam: North-Holland, 1968, pp. 74–97.
Salmon, W. C. [1970], 'Partial Entailment as a Basis for Inductive Logic', in *Essays in Honor of Carl G. Hempel*, N. Rescher, et al., eds. Dordrecht, Holland: D. Reidel, 1970, pp. 47–82.
Salmon, W. C., ed. [1971], *Statistical Explanation and Statistical Relevance*, Pittsburgh: University of Pittsburgh Press.
Salmon, W. C. [1974], 'Comments on "Hempel's Ambiguity" by J. Alberto Coffa'. *Synthese* 28 (1974), 165–169.

REFERENCES

Salmon, W. C. [1975], 'Discussion: Reply to Lehman', *Philosophy of Science* **40** (1975), 397–402.
Salmon, W. C. [1975a], 'Theoretical Explanation', in *Explanation*, S. Körner, ed., Oxford: Basil Blackwell, 1975, pp. 118–145.
Salmon, W. C. [1976], 'Foreword', in *Laws, Modalities, and Counterfactuals*, by H. Reichenbach, Berkeley: University of California Press, 1976, pp. vii–xliii.
Salmon, W. C. [1977], 'Objectively Homogeneous Reference Classes', *Synthese* **36** (1977), 399–414.
Salmon, W. C. [1977a], 'A Third Dogma of Empiricism', in *Basic Problems in Methodology and Linguistics*, R. Butts and J. Hintikka, eds., Dordrecht, Holland: D. Reidel, 1977, pp. 149–166.
Salmon, W. C. [1978], 'Why Ask, "Why?"?' *Proceedings and Addresses of the American Philosophical Association* **51** (1978), 683–705.
Salmon, W. C. [1978a], 'Unfinished Business: The Problem of Induction', *Philosophical Studies* **33** (1978), 1–19.
Salmon, W. C. [1979], 'Why Ask, "Why?"?' in *Hans Reichenbach: Logical Empiricist*, W. Salmon, ed., Dordrecht, Holland: D. Reidel, 1979, pp. 403–425. This is a slightly revised version of Salmon [1978].
Salmon, W. C. [1979a], 'Propensities: A Discussion Review', *Erkenntnis* **14** (1979), 183–216.
Salmon, W. C. [1980], 'Probabilistic Causality', *Pacific Philosophical Quarterly* **61** (1980), 50–74.
Savage, L. J. [1954], *The Foundations of Statistics*, New York: Dover Publications.
Schilpp, P., ed. [1963], *The Philosophy of Rudolf Carnap*, La Salle, Illinois: Open Court.
Schilpp, P., ed. [1974], *The Philosophy of Karl Popper*, La Salle, Illinois: Open Court.
Seidenfeld, T. [1979], *Philosophical Problems of Statistical Inference*, Dordrecht, Holland: D. Reidel.
Seidenfeld, T. [1979a], 'Why I am Not an Objective Bayesian: Some Reflections Prompted by Rosenkrantz', *Theory and Decision* **11** (1979), 413–440.
Shimony, A. [1955], 'Coherence and the Axioms of Confirmation', *Journal of Symbolic Logic* **20** (1955), 1–28.
Shimony, A. [1976], 'Amplifying Personal Probability', *Philosophy of Science* **34** (1967), 326–332.
Shimony, A. [1970], 'Scientific Inference', in *The Nature and Function of Scientific Theories*, R. Colodny, ed., Pittsburgh: University of Pittsburgh Press, 1970, pp. 70–172.
Sklar, L. [1970], 'Is Probability a Dispositional Property?' *Journal of Philosophy* **67** (1970), 355–366.
Sklar, L. [1973], 'Unfair to Frequencies', *Journal of Philosophy* **70** (1973), 34–48.
Sklar, L. [1974], Review of Mellor's *The Matter of Chance, Journal of Philosophy* **71** (1974), 418–423.
Sklar, L. [1979], 'Probability as a Theoretical Concept', *Synthese* **40** (1979) 409–414.
Skyrms, Brian [   ], 'Higher Order Degrees of Belief', in *Prospects for Pragmatism: Essays in Honor of F. P. Ramsey*, D. H. Mellor, ed., Cambridge: Cambridge University Press, forthcoming.

Smith, C. A. B. [1977], 'The Analogy Between Inference and Decision', *Synthese* **36** (1977), 71–85.
Smokler, H. [1968], 'Conflicting Conceptions of Confirmation', *Journal of Philosophy* **65** (1968), 300–312.
Smokler, H. [1971], Review of M. Swain, ed., *Induction, Acceptance, and Rational Belief*, *Synthese* **23** (1971), 327–334.
Smokler, H. [1979], 'Single-Case Propensities, Modality, and Confirmation', *Synthese* **40** (1979), 497–506.
Sneed, J. [1971], *The Logical Structure of Mathematical Physics*, Dordrecht, Holland: D. Reidel.
Stalnaker, R. [1968], 'A Theory of Conditionals', *American Philosophical Quarterly*, Supplementary Monographs No. 2, Oxford: Basil Blackwell, 1968, pp. 98–112.
Stegmüller, W. [1973], *Personelle und Statistische Wahrscheinlichkeit*, Berlin: Springer-Verlag.
Stegmüller, W. [1973a], *Theorienstrukturen und Theoriendynamik*, Berlin: Springer-Verlag.
Stegmüller, W. [1976], *The Structure and Dynamics of Theories*, New York: Springer-Verlag. This is an English translation of Stegmüller [1973a].
Steinberg, D. [1973], 'Discussion: Explanation and Description-Relativity', *Philosophy of Science* **40** (1973), 403–407.
Stern, C. D. [1981], 'Lewis' Counterfactual Analysis of Causation', *Synthese* **48** (1981), 333–345.
Suppe, F. [1972], 'What's Wrong with the Received View on the Structure of Scientific Theories?' *Philosophy of Science* **39** (1972), 1–19.
Suppe, F., ed. [1977], *The Structure of Scientific Theories*, 2nd ed., Urbana: University of Illinois Press.
Suppes, P. [1966], 'Probabilistic Inference and the Concept of Total Evidence', in *Aspects of Inductive Logic*, J. Hintikka and P. Suppes, eds., Amsterdam: North-Holland, 1966, pp. 49–65.
Suppes, P. [1967], 'What is a Scientific Theory?' in *The Philosophy of Science Today*, S. Morgenbesser, ed., New York: Basic Books, 1967, pp. 55–67.
Suppes, P. [1967a], *Set Theoretical Structures in Science*, Stanford: mimeographed.
Suppes, P. [1970], *A Probabilistic Theory of Causality*, Amsterdam: North-Holland.
Suppes, P. [1973], 'New Foundations of Objective Probability: Axioms for Propensities', in *Logic, Methodology, and Philosophy of Science*, P. Suppes, et al., eds. Amsterdam: North Holland, 1973, pp. 515–529.
Suppes, P. [1974], 'The Measurement of Belief', *The Journal of the Royal Statistical Society*, Series B (Methodological) **36** (1974), 160–191.
Teller, P. [1972], 'Epistemic Possibility', *Philosophia* **2** (1972), 303–320.
Teller, P. [1973], 'Conditionalization and Observation', *Synthese* **26** (1973), 218–258.
Tuomela, R. [1976], 'Causes and Deductive Explanation', in *Boston Studies in the Philosophy of Science*, Vol. XXXII, R. Cohen, C. Hooker, A. Michalos, and J. van Evra, eds., Dordrecht, Holland: D. Reidel, 1976, pp. 325–360.
Tuomela, R. [1976a], 'Morgan on Deductive Explanation: A Rejoinder', *Journal of Philosophical Logic* **5** (1976), 527–543.
Tversky, A. [1974], 'Discussion of the Papers by Professor Tversky and by Professor

REFERENCES

Suppes', *The Journal of the Royal Statistical Society*, Series B (Methodological) 36 (1974), 183–184.
Uchii, S. [1973], 'Inductive Logic with Causal Modalities: A Deterministic Approach', *Synthese* 26 (1973), 264–303.
van Fraassen, B. [1970], 'On the Extension of Beth's Semantics of Physical Theories', *Philosophy of Science* 37 (1970), 325–339.
van Fraassen, B. [1977], 'Relative Frequencies', *Synthese* 34 (1977), 133–166.
van Fraassen, B. [1978], Review of Stegmüller's *Personnelle und Statistische Wahrscheinlichkeit*, *Philosophy of Science* 45 (1978), 158–163.
van Fraassen, B. [1979], 'Relative Frequencies', in *Hans Reichenbach: Logical Empiricist*, W. Salmon, ed., Dordrecht, Holland: D. Reidel, 1979, pp. 129–167. This is an expanded version of van Fraassen [1977].
van Fraassen, B. [1979a], 'Modality', in *Current Research in the Philosophy of Science*, P. Asquith and H. Kyburg, eds., East Lansing: Philosophy of Science Association, 1979, pp. 282–290.
von Mises, R. [1964], *Mathematical Theory of Probability and Statistics*, edited by H. Geiringer, New York: Academic Press.
von Wright, G. H. [1971], *Explanation and Understanding*, Ithaca: Cornell University Press.
Wald, A. [1950], *Statistical Decision Functions*, New York: Hohn Wiley & Sons.
Watkins, J. W. N. [1964], 'Confirmation, the Paradoxes, and Positivism', in *The Critical Approach to Science and Philosophy*, M. Bunge, ed. New York: The Free Press, 1964, pp. 92–115.
Whitehead, A. N. and B. Russell [1910], [1912], [1913], *Principia Mathematica*, Cambridge: Cambridge University Press.
Wolgast, E. [1977], *Paradoxes of Knowledge*, Ithaca: Cornell University Press.

INDEX OF NAMES

Achinstein, P. 147
Ackermann, R. 143, 182, 284
Adler, A. 18
Anderson, A. 137
Aristotle 8, 15, 36
Austin, J. 27
Ayer, A. J. 29, 162, 166

Belnap, N. 137
Bennett, J. 60
Beth, E. W. 178
Birnbaum, A. 236–240, 243–244, 273
Braithwaite, R. B. 29, 106, 203, 221
Buchler, J. 24, 26, 106
Burks, A. 203, 216, 220, 257

Cargile, J. xv
Carnap, R. 15, 23–24, 28, 33–34, 72, 127, 129, 131, 160, 162, 166, 203, 206–208, 214–219, 221, 256, 261, 277, 286, 290
Cartwright, N. 51
Chalmers, A. F. 163, 208
Church, A. 8, 13
Coffa, A. 100–101, 113
Copernicus, N. 8, 15
Copi, I 42–43
Cramér, H. 114–115

de Finetti, B. 203, 207, 219–221, 257, 282
Dias, P. 217
Dretske, F. 161
Dunlop, C. E. M. xv, 193

Eberle, R. 143
Eddington, A. 225
Einstein, A. 8, 228
Euclid 8

Feigl, H. 177
Feller, W. 249–250
Fetzer, J. H. xv, 59, 77, 105, 118, 120
Fisher, R. A. xiii, 203, 233
Fodor, J. 278–279
Freud, S. 18

Galileo, G. 8, 18
Gettier, E. 15
Giere, R. N. xiii, xv, 71, 105, 116, 118–123, 150–151, 165, 203, 221, 238–239, 241–243, 245–246, 262, 284
Gillies, D. A. xiv, 71, 113–114, 184, 203
Gnedenko, B. V. 251
Good, I. J. 203
Goodman, N. xi, 29, 47, 187–192, 196, 211
Graybill, F. A. 250

Haack, S. xv
Hacking, I. xiii, 5, 9, 10, 18, 105–106, 110, 116–117, 122, 203, 205–206, 209, 223, 226, 231, 234, 240, 245, 273
Hájek, J. 251
Harman, G. 268
Heilig, K. 283, 286
Hempel, C. G. xi, xiii, xv, 7, 24, 29, 47, 51, 78, 84, 86, 94–102, 104–106, 124–127, 129, 132–134, 138, 142–145, 162, 164, 166, 179, 181–182, 187, 189, 192–194, 196, 199–200, 211–213, 215, 258, 264, 266, 269–270, 289
Hilpinen, R. 203
Hintikka, J. 203, 206–207, 214, 217, 221, 256, 261
Hodges, J. L. 250

310

INDEX OF NAMES

Hume, D. xi, 56, 117–118, 144, 186–187, 197–200
Humphreys, P. xv, 72, 114, 283

Jabbar, K. A. 31
Jaynes, E. T. 217, 257
Jeffrey, R. 203, 221–222, 272–273, 279, 281–282
Jobe, E. 187
Jung, C. G. 18

Kaplan, D. 143
Kemeny, J. G. 207
Kennedy, J. F. 38
Kepler, J. 8
Kiefer, J. 243
Kim, J. 143
Kolmogorov, A. N. 113–114, 123, 251
Kripke, S. 44, 69
Krüger, L. 134–135
Kuhn, T. S. 166
Kyburg, H. E. xi, xv, 55–58, 72–73, 77, 95, 102, 105, 113, 132, 135, 166–167, 170, 206–207, 219, 221, 258–262, 282–283

Lakatos, I. 165, 184, 213, 215, 218–219, 221–222, 225–226
Lehmann, E. L. 250
Lehmann, R. S. 207
Lehrer, K. 13
Levi, I. xii, 104–105, 111–112, 115, 124, 135–136, 170, 203, 207, 220–221
Lévy, P. 251
Lewis, D. 34–35, 60
Lincoln, A. 41–43
Lindley, D. V. 203, 282

Malinowski, B. 292
Maxwell, G. 280
Mellor, D. H. 105, 115–119, 129, 160, 278–280, 284
Merill, G. H. 168
Montague, R. 143
Mood, A. M. 250

Moore, D. S. 237
Morgan, C. G. 143
Morgenbesser, S. 161
Morris, W. E. xv

Nagel, E. 211, 215–216
Newton, I. 8, 15, 165, 228
Neyman, J. xiii, 203, 233–238, 241–247, 262–268, 272–273
Nickles, T. 143
Niiniluoto, I. xv, 143, 203, 216–217
Nute, D. E. xv, 59–60, 70, 77, 105, 193

Oppenheim, P. 142–143, 213

Pap, A. 33
Pearson, E. xiii, 203, 233–238, 241–247, 262–263, 266–268
Peirce, C. S. xiii, 24–27, 123, 164, 203
Plato 15, 23
Polanyi, M. 17
Pollock, J. 34–35, 60, 195
Popper, K. R. xiii, xv, 22, 34, 45, 105–106, 113, 116–117, 120, 147, 160, 175–176, 180–187, 195–196, 200–201, 202–204, 213, 216–219, 222–229, 233, 248, 254, 273
Pratt, J. 240, 244–245
Ptolemy 15
Putnam, H. 69, 79, 192, 207

Quine, W. V. O. 30, 32, 38, 120, 158, 194, 296

Radnitzky, G. 214
Ramsey, F. P. 203, 219
Reichenbach, H. xiii, 31, 56, 78–79, 81–83, 85–86, 88–89, 102, 111–112, 129–130, 179, 203, 208, 232, 255
Reinhardt, L. 69
Rogers, B. 211, 243
Rosenkrantz, R. 203–204, 206–207, 217, 257
Rudner, R. 33

Russell, B. 23, 31, 102
Ryle, G. 3

Salmon, W. C. xiii, xv, 52, 78, 88–94, 101–103, 113, 118, 124–126, 135, 137–139, 143–147, 152, 177, 179, 203, 206, 208–209, 211, 213, 255, 278, 283–284, 286
Savage, L. 203, 207, 219, 257
Schilpp, P. 183, 196, 215, 219, 290
Seidenfeld, T. xv, 217, 236, 242–243, 245, 274–277
Shimony, A. 203, 207, 217, 220
Sklar, L. 31, 102, 105, 118
Skyrms, B. 121–122, 280
Smith, C. A. B. 236, 244
Smokler, H. 53, 261, 270–271
Sneed, J. 162
Socrates 23
Stalnaker, R. 34–35, 60, 195
Stegmüller, W. 126, 134–135, 162, 165–166
Steinberg, D. 143

Stenner, A. 143
Stern, C. D. 35, 69
Suppe, F. 162
Suppes, P. 51–52, 71, 166, 203, 207, 221

Tarski, A. 23
Teller, P. 7, 9, 207, 209
Tuomela, R. xv, 143, 203, 216
Tversky, A. 280–282, 286

Uchii, S. 216–217, 257

van Fraassen, B. 72–73, 77, 102, 134–135, 166–167, 169–170, 265, 289
von Mises, R. 31, 78, 102, 113–114
von Wright, G. H. 196

Wald, A. xiii, 203, 233–235
Watkins, J. W. N. 196
Whitehead, A. N. 23
Wolgast, E. 6, 9

INDEX OF SUBJECTS

a posteriori evidence *e* 223-224
a posteriori measures of corroboration 226-227
a posteriori probabilities 215, 220
a priori content measure *r* 223-224
a priori degree of confidence 249-254
a priori measures of corroborability 226-227
a priori probabilities 215, 218-220; *see* prior probabilities
abductive inference 270
absolutistic idealism 161, 165
abstract calculus plus empirical interpretation 162-163
acceptability 226-227; *see* acceptance and rejection rules
acceptance and rejection rules 19, 231-263
accepted beliefs 3, 11-14, 16, 20, 23-24, 82, 93-95, 205, 208, 268-272, 275-293
accepting an hypothesis 208, 231-263
accidental generalization 29, 44-46, 211, 152-153, 184-185
actual frequency interpretation of probability 78-103
actual mental states 280-281
actual physical states 150, 280-281
actual world W 54-58, 67-73, 108-109, 194-195, 294-295
ad hoc hypotheses 218
aim of philosophy of science xi
aim of this inquiry xi
Alternate Rule for Inductive Acceptance 259
ambiguous names 50
analytical sentences 290-291
applied scientific rationality 273, 277, 283
arguments 26, 137-139

arrangements 36-37, 150
atomic events 42, 149-150, 294-295
atomic sentence 30, 37-38
attribute class 30, 78-103

background knowledge 224-226, 232, 247
basic statements 180-181
Bayes' theorem 21, 204, 209-210, 215, 219, 283-287
Bayesian conception of rationality 204-208
Bayesian conceptions of confirmation 202-222, 228-230
behavioral conceptions 236-247
Bernoulli's theorem 110, 114, 119
"bleen" 191
bodies and minds 295
broad deductivist conception of scientific inquiry 265, 292-293
broadly analytical sentences 290-291

calculus of causal conditionals of universal strength $\mathscr{U}$ 58-64
calculus of causal conditionals of statistical strength $\mathscr{C}$ 64-67
calculus of causal conditionals of statistical strength $\mathscr{C}^*$ 107-109
calculus of probability 65, 214, 284-285
calculus of subjunctive conditionals $\mathscr{S}$ 58-62
Carnap's logical interpretation of probability 127-131, 214-222
categorical terms, Aristotelian presupposition for 36
causal conditionals 28-29, 47-54, 57-59, 61-67, 70-73, 77-78, 142-146
causal conditionals of probabilistic strength 55-59; *see* propensities

313

INDEX OF SUBJECTS

causal conditionals of universal strength 61-64; *see* causal conditionals
causal connections 107, 197-200, 294-295
causal explanation, Salmon's conception of 143-146
causal explanations 124-146, 169-171, 212-213, 271-272
causal forms 49-50
causal relevance 51-52, 81, 92, 143-146; *see* nomic relevance
causal tendencies 56-58, 62-65, 71-73, 77-78; *see* propensities
causation 25; *see* causal tendencies
cause 197-200
central limit theorem 250
ceteris paribus clauses 51
chance predicates as primitive predicates 135-136
"chance set-up" 116, 281
classes 158-161
classical instrumentalism 161-162
Coffa's reference class density principle 99-101
cognitively insignificant sentences 198-199
cognitively significant sentences 198-199
coherence 205-206, 219-220
coherence condition (CA-1) 274-275
coherent betting quotients 117-118, 219-220, 282-283
competent sign-users 27
complete decidability 187
completely general 29, 44-45, 47; *see* lawlike sentence
complex frequency distributions 249-250
complex probability distributions 249-250
composite alternatives 235
condition of rationality (CR-1) 12, 14-16, 23, 205, 253, 258-259, 270-272, 275-276, 283, 287
condition of rationality (CR-2) 12, 14-16, 23, 205-206, 253, 258-259, 270-272, 275-276, 283, 287
condition of rationality (CR-3) 13-14, 16, 20, 23, 208, 227, 253, 274, 287-288, 293
condition of rationality (CR-4) 13-14, 16, 20, 23, 208, 227, 253, 274, 287-288, 293
condition of rationality (CR-5) 16
conditional probabilities 72, 205, 209-210, 283-286
conditionalization 206-207, 285-286
conditionalization condition (CA-2) 274-275
confidence principle (CA) 244-245
confidence principle (Conf) 237-240
confirmation and corroboration 202-229; *see* corroboration
confirmatory inquiry 241-242, 262-263
conjectures and refutations 22, 227, 261-262, 264, 293
conjunctivitis 261-262
constant conjunctions 56, 149; *see* extensional distributions
content measure for lawlike hypotheses 222-223
context of explanation 86-88, 129, 134, 203, 245-246, 266-272
context of prediction 86-88, 129, 134, 245, 266-272, 287
(continuous) confirmational consistency 207-208, 220, 229
conventions 25, 28
converse consequence condition 269-272
corroborability 218-219
corroboration 221-229, 244-263, 267-273, 287-293
corroboration ratios 226
counterhistorical inference 69
counternomological inference 69
covering laws (CL) 128, 292; *see* explanation
criteria of cognitive significance 182-183
criteria of demarcation 182-183
critical Bayesians 203
crucial experiments 225-226

data domains 182
decision-making 234-247
deduction as truth-preserving 177-179

INDEX OF SUBJECTS

deductive closure (CR-1) 12; *see* condition of rationality (CR-1)
deductive rules of inference 22-23, 28, 288
deductively sound 177
deductively valid 177
definite descriptions 41-44, 67-70, 139-140; *see* event definite descriptions
degree of corroboration 224-225
degree of nomic expectability 127-131, 222-224, 245-246, 269-270
degrees of belief 276-287
demonstrative conventions 27
demonstrative "rules of inference" 18-21
density principle for singular trial descriptions 80, 99-101
descriptive conceptions 3, 9-10, 17-18, 95
descriptive conventions 27
descriptive opacity 53
desideratum of certainty 180-181
determinism 55, 64, 83, 101, 122, 134-135, 275, 278, 285-287, 294-295
dicent signs 26
direct inference 276, 285-286; *see* degree of nomic expectability
direct inference condition (CA-4) 274-275
disjunctive properties 91
"display" 115-119
dispositional language 198-200
dispositional ontology xi, 36, 46, 293-296
dispositional predicates 36-37, 47-49
dispositional properties 36, 68, 295
dispositional properties as single case causal tendencies 56-59, 77-78, 115-123, 150, 280-281, 283-287, 294-295
dispositional term 33
dispositions 36-73, 294-296
drawing-inferences 234-247
Dutch Book theorems 282-283; *see* Ramsey-de Finetti theorem

effect 197-200
empirical adequacy, van Fraassen's sense of 169-170, 265

empirical Bayesians 203, 208-214
empirical predicates 163
epistemic contexts 210
epistemic objectivity 16-17, 198-200, 228-229, 242-244, 274-275, 287-289
epistemic resources, theory of 22, 287-288
epistemically homogeneous reference classes 93-94
equal and independent propensities 112
equivalence condition 192-193
error probabilities 238-247
errors of type I 235
errors of type II 235
essentially dispositional 44, 46; *see* lawlike sentence
eternal sentences 38, 57, 125
event attributions 38
event definite descriptions 41-42
event names 41-42
evidence 7; *see* rules of inference
evidential conceptions 236-247
evidential relevance 86-88
exchangeable events 221
Existential Generalization (EG) 54
existential generalizations 182
Existential Instantiation (EI) 53
exoneration 179-180, 289; *see* problem of induction
experiential findings 19, 22, 288; *see* perceptual rules of inference
explanandum-phenomenon 124; *see* explanation
explanandum-sentence 124-125; *see* explanation
explanation 32, 86-88, 93-94, 166-171, 212-213, 294; *see* symmetry thesis
explanation contexts 225, 245; *see* explanation
explanation requires acceptance 246, 265
explanation-seeking why questions 86-88, 134-135, 212-213
explanatory ambiguity 96-103
explanatory contexts 203-204; *see* explanation
explanatory inquiry 246-247, 253-254

explanatory paradoxes 133-136
explanatory relevance 86-88, 101-103, 110-111, 137-139, 270-271; see nomic relevance
explicit definition of dispositional predicates 33-34, 47-49, 153-154, 289-292
explicit definitions 29, 47-49
exploratory inquiry 241-242, 246-247, 253-254, 262-263
expressive-completeness 29
extensional distributions 149-150, 189-190, 227, 248, 253, 265-266
extensional equivalence 52-53, 109
extensional frequency conception 56; see extensional distributions
extensional generalizations 147-148, 153, 185-187, 193-196
extensional interpretations 104-105
extensional language 29-30, 34
extensional operators across possible worlds 167
extensionality across possible worlds 56, 167-170
"external" requirements 3, 13

F-consequence class for community C 7-8
fair betting quotients 129-131; see rational betting quotients
falsifiability 181-182, 186
falsification 175
finite frequency interpretation of probability 31, 102-103
finite frequency statements 182
finite sequences 78-79
first-order extensional language 142-143, 166-170, 289
Firstness 25
fork operator 47, 54, 58-59; see subjunctive conditionals
freedom from aftereffect 112
frequencies 71-73; see extensional distributions
frequency distribution functions 249-250
frequency interpretation of probability 31, 46, 71-73, 77-103, 110-115

Gödel rule of necessitation (NEC) 59, 61-62, 64
Goodman's degrees of entrenchment 188
Goodman's riddle of induction 187-192
Goodman's theory of projectibility 188
"grue" 191

habits 26
Hacking's law of likelihood 226-227, 231
Hempel's paradox of confirmation 192-196
high probability requirement 52, 96-97, 101-102, 133-135
historical existence condition 43-44; see definite descriptions
historical relations 67-68, 150, 192
historical uniqueness condition 44; see definite descriptions
history of science 14-15
Hume's critique of causation 197-201
Hume's critique of induction 200-201
Humean conceptions 117-118, 144; see extensional distributions
Humphrey's paradox 72, 283-286
hypothetical frequency interpretation of probability 102-103

i-predicate in $\mathcal{K}zt$, Hempel's conception of 94-95
icons 25-26
ideal gas law 144-146
idealism 161-171
identity conditions 41-43, 67-68, 158-161
identity of events 41-42
identity of objects 42-43, 67-68
identity of properties 158-160
identity of scientific theories 159-161
idiolects xii, 8
"imperfect" knowledge 15-16, 293
(imperfect) knowledge context $\mathcal{K}zt$ 16
implicit definitions 29; see explicit definitions
impossibility, historical 55
impossibility, logical 55, 194
impossibility, physical 55, 121-122, 186
impossibility, stochastic 120-122

INDEX OF SUBJECTS

indeterminism 55, 57-59, 64, 102, 134-135, 281, 285-287, 294-295
indices 25-26
individual constants 37, 50, 57
individual objects 41-43
induction as knowledge-expanding 178-179
inductive contexts 203-204; see inductive rules of inference
"inductive" in the broad sense 176
"inductive" in the narrow sense 176
inductive inconsistency 187-188, 211, 255
inductive logic 216-217, 274-275
inductive relevance 86-87; see inductive rules of inference
inductive rules of inference 20, 22-23, 28, 177-201; see confirmation and corroboration
inductively correct 177
inductively proper 177
Inference Rule for Inductive Acceptance 252-254
inference to the best explanation 268-269
infinite sequences 78-79, 113-114
initial conditions 34, 51
insensitivity to ordinal selection 112
instrumentalism xiii, 161-171
intelligence 295
intensional 6, 47-49; see intensionality
intensional dispositional conception 56
intensional generalizations 185-187, 193-196, 227, 248, 253, 265-266
intensional interpretations 104-105
intensional language 34-35
intensional language framework 58-59
intensional operators 58-59
intensional realism 168, 290
intensionality 296; see strongly intensional conceptions and weakly intensional conceptions
internal principles plus bridge principles 162-165
"internal" requirements 3
inverse inference 276, 285-286; see inductive rules of inference

Jaynes' maximum entropy rule 217
kinds of things 40
knowledge context $\mathcal{K}zt$ 3, 9-11, 14, 23-24, 83, 232-233, 260-261, 270-272, 287-288

lambda 215
language framework $\mathfrak{L}zt$ 3, 16, 22, 288; see knowledge context $\mathcal{K}zt$
large numbers condition 208, 247, 253
lawful future worlds 55-56
lawlike generalization 29, 44-46; see lawlike sentence
lawlike sentence 30, 44, 49, 53, 104-105, 154, 184-185
lawlike sentences, basic forms of 49-50
lawlike sentences of "simple" form 187
laws of co-existence or of co-variation 63, 144-146, 283-284
legisigns 25
Leibniz condition 134
Lévy distance 250-254
Lewis' $\mathcal{VC}$ 35, 60
Lewis' $\mathcal{VW}$ 60
liberalized thesis of empiricism 289
likelihood principle (L') 237-240
likelihood principles 240; see corroboration
likelihood ratios 226, 245
likelihoods 204, 223-224, 262
limiting frequencies 56; see extensional distributions
limiting frequency interpretation of probability 31
limiting frequency statements 182
linguistic habits 26-28
Logic of Scientific Discovery 183
logic proper 25
logical Bayesians 202-203, 214-219
logical consistency (CR-2) 12; see condition of rationality (CR-2)
logical empiricism 33
logical empiricist conception of theories 162-165
logical positivism 33

318 INDEX OF SUBJECTS

logical positivist conception of theories 162-163
logical truth 59-62, 217, 257
(logically trivial) scientific explanations 291-292
"long run" propensity interpretation of probability 105-115
lottery paradox 258-261

maximal rationality 14, 287-288
maximally specific reference predicate, Hempel's conception of 51, 95-96
maximizing expected utility 267-269, 281-282
maximum likelihood principles 267-269
meaning postulates 29, 154
meaning postulates (MP) 291-292
measure-theoretical semantics 71-73
mechanistic character of propensities 77-78
meta-language $\mathfrak{M}\text{-}\mathfrak{L}zt$ 21, 222
methodological problem of induction 180
minds and bodies 295
minimal rationality 12, 205-206
minimax loss principle 242
minimax regret principle 242
mixed quantification statements 182
model theory for calculus $\mathscr{C}$ 66-67
model theory for calculus $\mathscr{C}^*$ 107-109
model theory for calculus $\mathscr{S}$ 61-62
model theory for calculus $\mathscr{U}$ 64
modus ponens (MP) 59, 62, 64
modus tollens 195
molecular events 42, 149-150
molecular sentence 30

n-fork operator 48, 54, 58-59; see causal conditionals of probabilistic strength
naive conception of scientific explanation 91-92, 99-100
naming postulates 140
narrow inductivist conception of scientific inquiry 264, 292-293
narrower thesis of empiricism 24, 33, 289, 292

narrowly analytical sentences 290-291
natural (or physical) science 292
natural kinds 39-40
natural object kinds 39
natural property kinds 39
nature's cooperation 293
necessary connections 197-201; see causal tendencies
necessity, causal 216-217
necessity, community 7
necessity, epistemic 16
necessity, historical 55
necessity, impersonal 13
necessity, logical 55, 194
necessity, personal 5
necessity, physical 55, 121-123
necessity, scientific 17
necessity, stochastic 120-122
"negative" causal relevance 51-52
neo-classical instrumentalism 165-171
Neyman-Pearson Lemma 236
Neyman-Pearson procedures 234-247, 262-263, 266-269
nomic expectability 127-131, 224-226; see explanation
nomic irrelevance 52-53, 91-92, 99, 125-126, 131-133; see nomic relevance
nomic relevance 50, 57-58, 86-88, 91-92, 99, 112-113, 141-142, 156-158, 168, 170-171, 254, 294-295
(nomically significant) causal explanations 126-127
(nomically significant) non-causal explanations 141-142
(nomically significant) scientific explanations 170-171, 291-292
(nomically significant) theoretical explanations 156-157
nomological conditionals 49-50, 53-54; see lawlike sentences
nomological irrelevance 52; see nomic irrelevance
nomological relevance 50, 112-113, 254; see nomic relevance
nomological truth 217, 257
non-causal explanations 139-158, 169-171

non-demonstrative "rules of inference" 18-19
non-dispositional properties 36, 150, 160-161, 191-192; see extensional distributions and historical relations
non-extensional operators 58-59
non-observational language 33, 161-168, 180-181, 198-200, 211, 215-216, 255, 287-290
non-rigid definite descriptions 69-70
non-rigid designation 69-70
normal sequence 112, 254
normative conceptions 3, 9-10, 17-18, 95
normative epistemology 229

object hypotheses 40
object kinds 40
object language 222
object language $\mathcal{L}zt$ 21, 28
objective knowledge 288
objective of scientific inquiry 179-180, 183, 211-212, 216, 221, 227, 265
objective predicates 198-199
objectivity 229; see epistemic objectivity
objectivity condition (CA-5) 274-275
observational language 162-164, 198-200
observational term 33; see observational language
occasion sentences 38, 41, 57, 125, 150
ontic contexts 203-204, 210
ontically homogeneous reference classes 83-86
ontology of science 296
ordinary knowledge xii
orthodox statistical hypothesis testing procedures xiii; see Neyman-Pearson procedures

p-worlds 57-59, 107-109
paradox of knowledge 6
paradox of minimal evidence 231-232, 254
paradox of perfect evidence 254
paradox of the explanation of the improbable 133-135
paradoxes of confirmation xi, 187-196

paradoxes of falsification 195-196
partial decidability 187
partition of a reference class 88-89
pattern's ultimate configuration 77-78
perceptual rules of inference 19-20, 23-28, 287-289
"perfect" knowledge 15-16
(perfect) knowledge context $\mathcal{K}zt$ 16
permanent properties 38-39, 68-73, 147-149, 151-152, 296
permissible predicates 83-86, 100-101, 160-161, 199; see projectible predicates
personal implication 5
personal language 5; see idiolects
photons 41
physical law 29, 47; see lawlike sentence
physical object 295; see identity of objects
Pollock's $\mathcal{SS}$ 35, 60
Popperian procedures of corroboration xiii, 228-230; see corroboration
"positive" causal relevance 51-52
possibility, community 7
possibility, epistemic 16
possibility, historical 55
possibility, impersonal 12
possibility, logical 4, 55, 194, 201, 293
possibility, personal 5, 12
possibility, physical 4, 54-55, 201, 294
possibility, scientific 17
possible world's instrumentalism 168
possible worlds 54-55, 158-161
power of a test 235-236
practically homogeneous reference classes 93-94
pragmatics 27
predicate constant 37
prediction 86-88, 93-94, 166-171, 212-213, 294; see symmetry thesis
prediction contexts 245; see prediction
preferability 226-227; see acceptability
presentness 25
primitive terms 24
principle of conditionality 237-238
Principle of Confidence 251-254
principle of confidence (Conf) 262
principle of conjunction 261-262

principle of empiricism 186-187, 200, 221
principle of hypothetical syllogism (H.S.) 193-194
Principle of Maximizing Empirical Content 290-291
Principle of Relativistic Realism 161
principle of strengthening of subjunctive antecedents (SSA) 193-194
principle of sufficiency 237-238
principle of unbiasedness (U) 237-240
principles of decision 268-269; see Neyman-Pearson procedures
principles of inference 268-269; see Neyman-Pearson procedures
prior probabilities 204-206, 208, 211, 214, 228-229
probabilistic acceptance and rejection rules 215, 258-262, 277, 287
probabilistic beliefs 277-287
probabilistic causal calculus $\mathscr{C}$ 59-67, 123, 142, 167
probabilistic causal calculus $\mathscr{C}$* 106-109
probabilistic causation 284-285
probabilistic conceptions of confirmation xiii, 256-258; see Bayesian conceptions of confirmation
probabilistic framework, thesis of 22
probabilistic intuitions 218-219, 227-228
probabilities given facts 205, 228, 278
probabilities of events 21, 204, 275-276
probabilities of hypotheses 21, 204-205, 275-276
probability, community 7
probability, epistemic 16, 276-277
probability, impersonal 13, 21
probability, personal 5, 274-287
probability, scientific 17
probability distribution functions 249-250
probability learning 280
probable knowledge 202
problem of acceptance 231; see acceptability
problem of induction xi, 175-201, 288-289

problem of preference 231; see preferability
program 177-180
projectible predicates 188-192
propensities 71-73, 105-123, 281; see single-case causal tendencies
propensities, Gier's conception of 119-123
propensities, Mellor's conception of 115-119, 130, 279-280
proper names 43-44, 139-140
properties 158-161
property hypotheses 40
property identification 154
property specification 154
propositional attitudes 278-279
propositions 3
psychological propensities 277-287; see indeterminism
psychological properties 278-279
pure dispositions as universals 190-192
pure grammar 25
pure rhetoric 25
pure scientific rationality 273, 277
purely dispositional predicates 160-161, 191-192

qualified-instance confirmation 218-222
qualisigns 25
quantum theory 40-41, 83, 122; see indeterminism
quasi Bayesians 203
quasi-accidental generalizations 153-154, 156
quasi-lawlike generalizations 153-154, 156

radical instrumentalism 163-164
Ramsey-de Finetti theorem 219
random sample 115, 248-249
random sequence 112, 254
rational betting quotients 218-222, 277-278
rational responsibility 20
real causal connections between events 107; see necessary connections
realism 161-171

INDEX OF SUBJECTS

reason-seeking why questions 86–88, 134–135, 212–213
reduction sentences 29, 33
reference class 30, 78–103
"reference class under a description" 160–161
referential transparency 53
regular association 197
Reichenbach's logical interpretation of probability 130–131, 210
rejection region 233–234
relative frequencies 31, 46, 56, 71–103, 107, 109–115; see extensional distributions
requirement of maximal specificity 50, 60, 62, 125–126
requirement of strict maximal specificity (RSMS) 125–126, 137–139, 270
requirement of total evidence 129, 214
resemblance 197
retrodiction 32–33, 169
revised requirement of maximal specificity (RMS*) 96, 125, 213
rhemes 26
rigid definite descriptions 69–70
rigid designation 43–44, 68–70
"rules of inference" 13, 16–18; see deductive rules of inference, inductive rules of inference, and perceptual rules of inference

satisfaction criterion of confirmation 189-190
scientific community Z 17, 288, 293
scientific conditionals 49–50, 184–185, 195; see nomological conditionals
scientific explanations 77–171, 294–295; see explanation
scientific inquiries xi, 266–273, 287–293
scientific knowledge xi–xii, 11, 14–24, 264–273, 287–293
scientific practice 17–18, 163–164, 228
scientific rationality 265–273; see conditions of rationality (CR-1)–(CR-5)
scientific theories 157–171, 199–200, 211, 215–216, 221, 255, 258, 265–273

second-order extensional language 166–170, 289–290
Secondness 25
selection functions f, g, s 67–73
semantics 27
semiotic 24–25
sentential functions 37–38, 278
set of beliefs $\mathscr{B}zt$ 3, 260–261; see knowledge context $\mathscr{K}zt$
severe tests 51, 224–226, 229, 267–268
Short Run Principle 114–115, 123, 252
significance level 234–236, 241–244
simple alternatives 235
simple forms 49–50
simple frequency distributions 249–250
simple probability distributions 249–250
simultaneity 150
sincerity 224, 248, 256
Single Case Principle 114–115, 123
"single case" propensity interpretation of probability 105–123; see propensities
single-case causal tendencies 77–78; see probabilistic causal calculus $\mathscr{C}$
singular events 41–42, 80–81, 84–86, 91–92, 110–111, 124–131, 140–142, 276–277, 294–295
sinisigns 25
size of a test 234–236
Skyrms' paradox 121–122
social (or behavioral) science 292
special consequence condition 269–272
Stalnaker's $\mathscr{C}$ 2 35, 60
"standard conception" of definition 33
statistical ambiguity 96–102
statistical disposition 46, 151; see propensities
statistical explanations, Hempel's conception of 94–96
statistical explanations, Salmon's conception of 90–91
statistical generalization 30; see lawlike sentence
statistical lawlike statements 182; see lawlike sentences
statistical laws 46; see lawlike sentence
statistical relevance 92, 112–113, 143, 254; see explanatory relevance and nomic relevance

statistical relevance, Hempel's criterion of 95-96
statistical relevance, Reichenbach's criterion of 81-82
statistical relevance, Salmon's criterion of 88-89
statistical strength dispositional properties 37; *see* propensities
statistical-probabilistic explanations 128-129, 132-133
Stegmüller's paradox 126, 133-135
stochastic system 120-122
strict coherence 207-208
strongly intensional conceptions 72-73, 77-78, 148, 166-171, 193-195, 265-266, 289-290, 295
"strongly intensional" language frameworks 167-170, 290
subjective Bayesians 203, 219-222
subjective epistemology 229
subjective predicates 198-199
subjunctive conditionals 28, 48, 58-62, 193-196; *see* calculus of subjunctive conditionals $\mathscr{S}$
subjunctive generalizations 41-54, 183-187, 193; *see* lawlike sentence
supplementary definition of epistemic homogeneity 93-94, 98-99
symbols 25-26
symmetry thesis of explanation and prediction 86-90, 93, 101-102, 129, 161-162, 212-213, 245-248, 267-273, 276-277, 294-296
syntax 28
system's initial conditions 77-78
systematic power of lawlike hypothesis 222-223

technical language 11
technological innovations 293
teleological character of frequencies 77-78
Tentative Rule for Inductive Acceptance 258-259
testability 218-219, 223
theoretical explanations 153-157
theoretical language 33, 162-164, 198-200; *see* scientific theories

theoretical realism 161-171
theoretical structures plus empirical claims 162-165
theoretical term 33, 162-163; *see* theoretical language
theory of signs 24
thesis of conditionalization 202
thesis of the probabilistic framework 202
things of kinds 40
Thirdness 25
total evidence condition (CA-3) 274-275
traditional conception (P) 237-240
traditional principles of induction xiii, 208-214, 227-228, 255-256
traditional problem of induction 180
transformation rules 28
transient properties 39, 68-70, 147-149, 151-152, 292
transposition 193-196
truth 6, 14-15, 27, 194-195
truth-conditions for nomological conditionals 49-59; *see* intensionality
truth-functional language xiii, 30, 47-49; *see* scientific conditionals
two-alternative decision problem 237
two-hypothesis inference problem 236-237

u-fork operator 58-59; *see* causal conditionals of universal strength
undefined terms 24
universal disposition 46, 151; *see* dispositions
Universal Generalization (UG) 53, 148-149
Universal Generalization (UG*) 148-149
universal generalizations 31; *see* lawlike sentence
Universal Instantiation (UI) 54
universal lawlike statements 182; *see* lawlike sentence
universal material generalizations 185-186
universal strength dispositional properties 37; *see* dispositions
universal-deductive explanations 127-128, 131-132, 291-292

INDEX OF SUBJECTS

universals as pure dispositions 190–192
unprojectible predicates 191–192

validation 176–177; *see* problem of induction
verifiability 181–182, 199
verification 175
vindication 176–177, 209, 289; *see* problem of induction

weakly intensional conceptions 72–73, 77–78, 148, 166–171, 265–266, 289–290
"weakly intensional" language frameworks 167–170
weight, Reichenbach's conception of 79–86, 110–111
wide variety condition 208, 247, 253

BOSTON STUDIES IN THE PHILOSOPHY OF SCIENCE

Editors:
ROBERT S. COHEN and MARX W. WARTOFSKY
(Boston University)

1. Marx W. Wartofsky (ed.), *Proceedings of the Boston Colloquium for the Philosophy of Science 1961-1962.* 1963.
2. Robert S. Cohen and Marx W. Wartofsky (eds.), *In Honor of Philipp Frank.* 1965.
3. Robert S. Cohen and Marx W. Wartofsky (eds.), *Proceedings of the Boston Colloquium for the Philosophy of Science 1964-1966. In Memory of Norwood Russell Hanson.* 1967.
4. Robert S. Cohen and Marx W. Wartofsky (eds.), *Proceedings of the Boston Colloquium for the Philosophy of Science 1966-1968.* 1969.
5. Robert S. Cohen and Marx W. Wartofsky (eds.), *Proceedings of the Boston Colloquium for the Philosophy of Science 1966-1968.* 1969.
6. Robert S. Cohen and Raymond J. Seeger (eds.), *Ernst Mach: Physicist and Philosopher.* 1970.
7. Milic Capek, *Bergson and Modern Physics.* 1971.
8. Roger C. Buck and Robert S. Cohen (eds.), *PSA 1970. In Memory of Rudolf Carnap.* 1971.
9. A. A. Zinov'ev, *Foundations of the Logical Theory of Scientific Knowledge (Complex Logic).* (Revised and enlarged English edition with an appendix by G. A. Smirnov, E. A. Sidorenka, A. M. Fedina, and L. A. Bobrova.) 1973.
10. Ladislav Tondl, *Scientific Procedures.* 1973.
11. R. J. Seeger and Robert S. Cohen (eds.), *Philosophical Foundations of Science.* 1974.
12. Adolf Grünbaum, *Philosophical Problems of Space and Time.* (Second, enlarged edition.) 1973.
13. Robert S. Cohen and Marx W. Wartofsky (eds.), *Logical and Epistemological Studies in Contemporary Physics.* 1973.
14. Robert S. Cohen and Marx W. Wartofsky (eds.), *Methodological and Historical Essays in the Natural and Social Sciences. Proceedings of the Boston Colloquium for the Philosophy of Science 1969-1972.* 1974.
15. Robert S. Cohen, J. J. Stachel and Marx W. Wartofsky (eds.), *For Dirk Struik. Scientific, Historical and Political Essays in Honor of Dirk Struik.* 1974.
16. Norman Geschwind, *Selected Papers on Language and the Brain.* 1974.
18. Peter Mittelstaedt, *Philosophical Problems of Modern Physics.* 1976.
19. Henry Mehlberg, *Time, Causality, and the Quantum Theory* (2 vols.). 1980.
20. Kenneth F. Schaffner and Robert S. Cohen (eds.), *Proceedings of the 1972 Biennial Meeting, Philosophy of Science Association.* 1974.
21. R. S. Cohen and J. J. Stachel (eds.), *Selected Papers of Léon Rosenfeld.* 1978.
22. Milic Capek (ed.), *The Concepts of Space and Time. Their Structure and Their Development.* 1976.
23. Marjorie Grene, *The Understanding of Nature. Essays in the Philosophy of Biology.* 1974.

24. Don Ihde, *Technics and Praxis. A Philosophy of Technology.* 1978.
25. Jaakko Hintikka and Unto Remes, *The Method of Analysis. Its Geometrical Origin and Its General Significance.* 1974.
26. John Emery Murdoch and Edith Dudley Sylla, *The Cultural Context of Medieval Learning.* 1975.
27. Marjorie Grene and Everett Mendelsohn (eds.), *Topics in the Philosophy of Biology.* 1976.
28. Joseph Agassi, *Science in Flux.* 1975.
29. Jerzy J. Wiatr (ed.), *Polish Essays in the Methodology of the Social Sciences.* 1979.
32. R. S. Cohen, C. A. Hooker, A. C. Michalos, and J. W. van Evra (eds.), *PSA 1974: Proceedings of the 1974 Biennial Meeting of the Philosophy of Science Association.* 1976.
33. Gerald Holton and William Blanpied (eds.), *Science and Its Public: The Changing Relationship.* 1976.
34. Mirko D. Grmek (ed.), *On Scientific Discovery.* 1980.
35. Stefan Amsterdamski, *Between Experience and Metaphysics. Philosophical Problems of the Evolution of Science.* 1975.
36. Mihailo Marković and Gajo Petrović (eds.), *Praxis. Yugoslav Essays in the Philosophy and Methodology of the Social Sciences.* 1979.
37. Hermann von Helmholtz: *Epistemological Writings. The Paul Hertz/Moritz Schlick Centenary Edition of 1921 with Notes and Commentary by the Editors.* (Newly translated by Malcolm F. Lowe. Edited, with an Introduction and Bibliography, by Robert S. Cohen and Yehuda Elkana.) 1977.
38. R. M. Martin, *Pragmatics, Truth, and Language.* 1979.
39. R. S. Cohen, P. K. Feyerabend, and M. W. Wartofsky (eds.), *Essays in Memory of Imre Lakatos.* 1976.
42. Humberto R. Maturana and Francisco J. Varela, *Autopoiesis and Cognition. The Realization of the Living.* 1980.
43. A. Kasher (ed.), *Language in Focus: Foundations, Methods and Systems. Essays Dedicated to Yehoshua Bar-Hillel.* 1976.
46. Peter L. Kapitza, *Experiment, Theory, Practice.* 1980.
47. Maria L. Dalla Chiara (ed.), *Italian Studies in the Philosophy of Science.* 1980.
48. Marx W. Wartofsky, *Models: Representation and the Scientific Understanding.* 1979.
50. Yehuda Fried and Joseph Agassi, *Paranoia: A Study in Diagnosis.* 1976.
51. Kurt H. Wolff, *Surrender and Catch: Experience and Inquiry Today.* 1976.
52. Karel Kosík, *Dialectics of the Concrete.* 1976.
53. Nelson Goodman, *The Structure of Appearance.* (Third edition.) 1977.
54. Herbert A. Simon, *Models of Discovery and Other Topics in the Methods of Science.* 1977.
55. Morris Lazerowitz, *The Language of Philosophy. Freud and Wittgenstein.* 1977.
56. Thomas Nickles (ed.), *Scientific Discovery, Logic, and Rationality.* 1980.
57. Joseph Margolis, *Persons and Minds. The Prospects of Nonreductive Materialism.* 1977.
58. Gerard Radnitzky and Gunnar Andersson (eds.), *Progress and Rationality in Science.* 1978.

59. Gerard Radnitzky and Gunnar Andersson (eds.), *The Structure and Development of Science.* 1979.
60. Thomas Nickles (ed.), *Scientific Discovery: Case Studies.* 1980.
61. Maurice A. Finocchiaro, *Galileo and the Art of Reasoning.* 1980.
62. William A. Wallace, *Prelude to Galileo.* 1981.
63. Friedrich Rapp, *Analytical Philosophy of Technology.* 1981.
64. Robert S. Cohen and Marx W. Wartofsky (eds.), *Hegel and the Sciences.* (Forthcoming).
65. Joseph Agassi, *Science and Society.* 1981.
66. Ladislav Tondl, *Problems of Semantics.* 1981.
67. Joseph Agassi and Robert S. Cohen (eds.), *Scientific Philosophy Today.* (Forthcoming).
68. Władysław Krajewski (ed.), *Polish Essays in the Philosophy of the Natural Sciences.* (Forthcoming).
69. James H. Fetzer, *Scientific Knowledge.* 1981.

RAYMOND H. FOGLER LIBRARY
DATE DUE

**BOOKS ARE SUBJECT TO
RECALL AFTER TWO WEEKS**